16 —
Aviation

U.S.
CIVIL
AIRCRAFT
VOL. 2

This work is dedicated to the preservation and perpetuation of a fond memory for the men and the planes that made a future for our air industry. And, to help kindle a knowledge and awareness within us of our debt of gratitude we owe to the past. . . . .

*Library of Congress Catalog Card Number
62-15967*

Printed and Published in the United States by Aero Publishers, Inc.

U.S.
CIVIL
AIRCRAFT

VOL. 2

(ATC 101–ATC 200)

By

JOSEPH P. JUPTNER

AERO PUBLISHERS, INC.

ACKNOWLEDGEMENT

Any historian soon learns that in the process of digging for obscure facts and information, he must oftentimes rely on the help of numerous people; unselfish and generous people, many who were close to or actually participated in various incidents or events that make up this segment of history recorded here, and have been willing to give of their time and knowledge in behalf of this work. To these wonderful people I am greatly indebted, and I feel a heart-warming gratitude; it is only fitting then, that I proclaim their identity in appreciation.

My thanks to Chas. W. Meyers; George A. Page Jr.; Earl C. Reed; and Wm. "Billy" Parker; to F. J. Delear of Sikorsky Aircraft Div.; Pan American World Airways; Lillian Keck of American Airlines; Harvey H. Lippincott of Pratt & Whitney Aircraft Div.; Philip S. Hopkins and staff at Smithsonian Institution, National Air Museum; Ken M. Molson of National Aviation Museum of Canada; Fairchild Aircraft Div.; Beech Aircraft Corp.; Phillips Petroleum Co.; Eastern Air Lines; Lockheed Aircraft Corp.; United Air Lines; Northwest Orient Airlines; Alexander Film Co.; Michigan Department of Aeronautics; Boeing Airplane Co.; American Aviation Historical Society; Eby Photo Service of Santa Ana, Calif.; and the following group of dedicated aviation historians; Stephen J. Hudek; Roy Oberg; Wm. T. Larkins; Marion Havelaar; John W. Underwood; Truman C. Weaver; Dave Jameson; Gordon S. Williams; Peter M. Bowers; and Gerald H. Balzer.

FOREWORD

We know that aviation was literally dropped with a big flop right after World War I and would have been left to slowly die had it not been for a hand-full of adventurous young-bloods who had aviation in their heart and soul; stout-hearted men who could not stand by and see it come to such an ignoble end.

Like ambassadors of good-will toward aviation, they chose to roam the country-side with their craft, acquainting many thousands of people with their first-hand look of an airplane and spreading the gospel of aviation as ardently as any dedicated evangelist. By the early "twenties" a good deal of progress had already been made towards the preservation of aviation in itself, but there was slowly rising a keen yearning among these men to develop more useful aircraft, more practical craft to replace the tired and worn war-surplus aircraft that had been the backbone of the so-called "commercial aviation" they were trying to sell. By far the easiest and simplest approach to this problem was the modification or modernization of these existing aircraft into somewhat more useful vehicles, but several men of wider and keener vision felt this was surely not the answer; undertaking to design and develop aircraft on entirely new lines and principles. At first, some of these new designs seemed to be only little removed from those they were to replace, but several break-throughs resulted in the latter half of the "twenties" and a whole new concept of design was beginning to take shape. The immediate success of several of the new designs gave added impetus to the new movement and by the time 1927 rolled around, there was a clamoring evidence that commercial aviation, though yet an infant, would soon be growing into a strapping youngster of capable proportion.

The year 1927, though blessed with fervent bally-hoo for public attention, was quite like a dawning of purpose in commercial aviation, a beginning that was soon to blossom in the next few years from a hobby-like backyard operation into one of the up-and-coming industries of our nation; a happy young giant that had found use for it's services and was attracting world-wide notice. Opportunities, especially for aircraft manufacture during the next two years were never better, and new airplanes to fit all the varied services were pouring out of factories both large and small, by the hundreds and the thousands. Along with this phenomenon of aircraft manufacture that mushroomed as the months went by, came a surging wave of sporting people who wanted to fly and to own their own airplanes; so it is recognizable that every facet of commercial avia-tion, be it airplane manufacturer or dealer, air-line or air-service operator, flying-school operator, or be it Joe Propwash who was hopping passengers on week-ends for $ 3.00 a ride, all were in clover and were having their hey-dey.

Best remembered by me as sort of a symbol of these times, was the happy young pilot that always used to taxi up to the line with a grand flourish, raise up his goggles and flash that big boyish smile that reflected an unfettered joy; people seemed to be so happy in their new-found joy of flying or just any form of participation in aviation for that matter, and to me that flash of boyish smile remains as a symbol that portrayed the general feeling of these times. Aviation was again proving useful, everyone felt a pride in being a part of it, and above all, flying was great fun and each flight like a new adventure.

Joseph P. Juptner

TERMS

To make for better understanding of the various information contained herein, we should clarify a few points that might be in question. At the heading of each new chapter, the bracketed numerals under the ATC number, denote the first date of certification; any ammendments made are noted in the text. Unless otherwise noted, the title photo of each chapter is an example of the model that bears that particular certificate number; any variants from this particular model, such as prototypes and special modifications, are identified. Normally accepted abbreviations and symbols are used in the listing of specifications and performance data. Unless otherwise noted, all maximum speed, cruising speed, and landing speed figures are based on sea level tests; this method of performance testing was largely the custom during this early period. Rate of climb figures are for first minute at sea level, and the altitude ceiling given is the so-called service ceiling. Cruising range in miles or hours duration is based on the engine's cruising r.p.m., but even at that, the range given must be considered as an average because of pilot's various throttle habits.

At present, the specifications and performance data are given in a running text, but numerous requests indicate that there is a preference for this information in a tabular form; we plan to comply with this request in Volume 3. At the ending of each chapter, we show a listing of registered aircraft of a similiar type; most of the listings show the complete production run of a particular type and this information we feel will be valuable to historians, photographers, and collectors in making correct identification of a certain aircraft by it's registration number.

In each volume there are separate discussions on 100 certificated airplanes and we refer to these discussions as chapters, though they are not labeled as such; at the end of each chapter there is reference made to see the chapter for an ATC number pertaining to the next development of a certain type. As each volume contains discussions on 100 aircraft, it should be rather easy to pin-point the volume number for a chapter of discussion that would be numbered as A.T.C. # 93, or perhaps A.T.C. # 176, as an example. The use of such terms as "prop", "prop spinner", and "type", are normally used among aviation people and should present no difficulty in interpreting the meaning.

TABLE OF CONTENTS

BRUNNER-WINKLE "BIRD", MODEL A

Fig. 1. The Brunner-Winkle "Bird" with Curtiss OX-5 engine had well-planned proportions; note visibility.

The Brunner-Winkle "Bird" was a 3 place open cockpit biplane that was powered with an 8 cylinder Curtiss OX-5 engine; seemingly just an added entry in an already over-crowded field, but actually the "Bird" was not "just another OX-5 powered airplane", it was one among the few that came out with a little something extra to offer. Had the "Bird" biplane come out about a year earlier, with a little promotion, it certainly would have been far up the ladder in sales and popularity by this time. Carefully planned with definate purpose in mind, it's configuration was aerodynamically efficient and structurally sound; bordering on the "sesqui-plane" arrangement, the "Bird" had nearly 70% of it's lifting area in the thick-sectioned upper wing. A sensible arrangement that offered both aerodynamic and structural advantages. The fuselage structure was faired out neatly and was fairly typical for a ship of this type, except for the use of a tunnel-type engine coolant radiator; this was a feature reminiscent of "Army Pursuit" design of a few years back. In all things, the "Bird" Model A poses as a clean design; a graceful and

well-proportioned airplane of somewhat better than average performance.

First introduced along about September of 1928, the balance of that year was spent in extensive testing to pass requirements for certification. Local interest in the new prototype (X-7878, serial # 1000) was running high and pilots who flew it had nothing but praise for it's good handling characteristics and performance. Being well aware of the fact that a certain amount of bally-hoo was necessary to bring a new "type" to public attention, Brunner-Winkle had launched a series of flights with their new craft; hoping to recieve some national acclaim. Elinor Smith, a youthful and daring aviatrix, flew the "Bird" biplane to an altitude record for OX-5 powered airplanes; also established a non-refueled solo endurance record of over 13 hours for light planes in this class. (A record that was formerly held by the winsome "Bobbie" Trout in a "Golden Eagle", and then was recaptured by her again about 2 weeks after the flight of Elinor Smith in the "Bird"; such was the tempo of aircraft development and advancement in this period). These flights

Fig. 2. Prototype of the Brunner-Winkle "Bird" A series.

helped considerably, but even at that the Brunner-Winkle "Bird" was still pretty much unknown country-wide until the genial Wm. E. Winkle made it's formal introduction to the flying public at the 1929 National Air Races held at Cleveland, Ohio. The "Bird" created a good deal of interest at this showing and won the hearts of many with it's pleasant character and exceptional performance. Though nimble and fast for a plane of this type, the "Bird" had take-off (100 feet in no wind), climb-out, and slow-speed (full control at 40 m.p.h.) characteristics that were really quite remarkable.

With ambition and enthusiasm to spare, the Ford-Leigh entry (a Brunner-Winkle "Bird", somewhat modified) was groomed for participation in the Guggenheim Safe Airplane Contest of 1929; unofficially, the "Bird" took highest ratings for conventional type aircraft among the entries from the U.S.A.

The first nine "Birds" of the Model A type (serial # 1000 thru # 1008) were covered by a Group 2 approval numbered 2-33 that was issued in January of 1929; all subsequent "Model A" type were manufactured under A.T.C. # 101 which was also issued in January of 1929. The "Bird A" series were built on through 1931, but the stockpile of Curtiss OX-5 engines was already gone, so the customer had to furnish his own engine and propeller for installation at the factory. The "Bird" biplane was manufactured by the Brunner-Winkle Aircraft Corp. at Glendale, Brooklyn, New York. Just how Mr. Brunner fit into this organization is hard to say; the only person from the organization ever identified publicly with the "Bird" airplane was

William E. Winkle. Michael Gregor has often been credited with the design of the "Bird" biplane series, but it has been difficult to uncover whether he was their chief engineer, or even if he was actively connected with the firm at any time. We believe so, but cannot find accurate record to prove it.

Listed below are specifications and performance data for the Brunner-Winkle "Bird" Model A as powered with the Curtiss OX-5 engine; length overall 22'3"; height 8'8"; span upper 34'; span lower 25'; chord upper 69"; chord lower 48"; airfoil "U.S.A. 40B Mod."; area upper 184 sq.ft.; area lower 82 sq.ft. total wing area 266 sq.ft.; wt. empty 1315; useful load 835; payload 360; gross wt. 2150 lbs.; max. speed 105; cruising speed 88; land 37; climb 520 ft. first min. at sea level; service ceiling 12,500 ft.; gas cap. 45 gal.; oil cap. 4 gal.; cruising range 450 miles; price at the factory with new OX-5 engine (when available) was $3150, later boosted to $3295; price without engine or propeller was $2300, reduced to $2195 in 1931. In it's earlier prototype stage the "Bird" was also offered with the 80 h.p. Anzani (French) engine for $3500, but no examples of this combination were ever reported built.

The fuselage framework was built up of welded chrome-moly steel tubing in a Warren truss form, faired to shape with wood fairing strips and fabric covered. The entire top section of the fuselage was covered with aluminum panels; the metal paneled turtleback was quickly removable for inspection and maintenance of control operating mechanism, etc. The wing panels were built up of heavy spruce spars of solid section, with spruce and plywood truss-type ribs; the leading edge was covered to the front spar with dural sheet to preserve airfoil form and the completed framework was then fabric covered. The small lower wings and the large amount of stagger between panels enabled the pilot to see his wheels at all times; a feature which was quite helpful in taxiing, during take-off and in landing. The ailerons in the upper wing, and the elevators were operated by metal push-pull tubes; the rudder was actuated by stranded cable. The split-axle landing gear was of the cross-axle type and employed an "oil-strut" encased with rubber compression-rings for absorbing the shocks; the tail-skid was of the simple, trouble-free, spring-leaf type with a removable hardened "shoe". A fair-sized baggage compartment was provided in the upper fuselage section just behind the pilot's station; the front control "stick" and the rudder pedals were quickly detachable when carrying joy-riding passengers. The main fuel tank was in the forward section of the fuselage, with provision for a 5 gallon "reserve fuel" tank. The fabric covered tail-group was also built up of welded chrome-moly steel tubing; the vertical fin was ground adjustable to compensate for "torque" and the horizontal stabilizer was adjustable in flight. It might be well to note that the "Bird" biplane was of a rugged structure; minimum average safety factor was reported to be 8 plus.

We can surely understand Brunner-Winkle's enthusiasm and great pride in their "Bird" biplane, but their initially published performance figures were wholly dominated by pride and enthusiasm; more enthusiasm than fact! For example: top speed 120 m.p.h.; cruising speed 100; landing speed 35; climb, a rousing 1000 feet per minute; cruising range at least 9 hours; this may sound beautiful, but it just wasn't so. Compare actual performance figures in the paragraph above. However, we must temper this outburst somewhat by saying that the Brunner-Winkle "Bird" was a remarkable airplane in many instances, with a performance potential that was somewhat better than the average.

With the supply of Curtiss OX-5 engines becoming rather skimpy and uncertain in the latter part of 1929, a new "Bird" with the 5 cylinder Kinner K5 engine of 100 h.p. was announced to be the forthcoming standard production model in October of 1929. This interesting development will be discussed in the chapter for A.T.C. # 239. The "Bird A" was continued in production, of course, and at least 85 examples of this popular model were built; of these, several have been restored and are probably flying yet!

Fig. 3. The Ford-Leigh entry in the Guggenheim safe airplane contest scored high in unofficial tests.

Fig. 4. The cockpit arrangement and baggage compartment of the "Bird" was neat and roomy.

A.T.C. #102
(1-29)
LOCKHEED "AIR EXPRESS"

Fig. 5. The romance of the parasol wing monoplane was personified in the Lockheed "Air Express", which stimulated a revival of this popular configuration.

The unique Lockheed "Air Express" was a glamor-girl from the very out-set. It's name alone implied capability of efficient transport at high speed, and it's "parasol" configuration in cantilever form was yet novel enough to create notice and romantic comment; a configuration that seemed to suggest swash-buckling dash and romance, a colorful quality that took everyone's fancy. Designed by the incomparable John K. Northrop, the "Air Express" was a departure from the familiar "Vega" form; studies had proven that they could use the basic fuselage with only minor modification, and could actually put the wing in any one of three positions. The proof of this fact is reflected in such various and differing models as the hi-wing "Vega", the parasol-type "Air Express", and the low-wing "Sirius"; all using the same basic fuselage and the same basic wing. This indeed was ingenuity at it's best.

The Lockheed "Air Express" was a parasol-type cabin monoplane with the familiar all-wood one piece full cantilever wing; a wing that was suspended or perched high above the fuselage by heavy "N type" center-section struts. The fuselage, as was mentioned, was of the basic Lockheed full-monococque

type; it was cigar-shaped and of all-wood construction, the only difference from the "Vega" was the wing attach and the cockpit cut-out. There was seating for four in the cabin section up forward, with provisions to fold or remove two of the seats for carrying mail or extra baggage; oddly enough, the pilot was placed in an open cockpit just aft of the cabin section. The standard "Air Express", in most cases, was powered with the Pratt & Whitney Wasp engine of 425—450 h.p.; the later models in this series were using the full N.A.C.A. type "low drag" engine cowling and some were also equipped with "wheel pants". This airplane's performance was exceptional and bordered on the fantastic, as was proven time and again by it's many record-breaking flights across our country in 1929 and 1930. Speed dashes about the country kept the "Air Express" in front-page news for some time, and many noted pilots basked in it's reflected glory, but, like the bright star that it was, it eventually faded from the scene and others came by to take it's place. Before it left the lime-light completely tho', the "Air Express" left many indelible memories and probably fostered a popular revival in the parasol-wing configuration, as evident in the coming of such models as the Davis, Inland, Doyle, and many others.

The "Air Express" was primarily a design suggested by Western Air Express to carry mail, cargo, and passengers, but fate soon had the "Air Express" romping around and across the country-side setting new records; the airplane "type" was actually used very little for it's intended purpose. The swift cigar-shaped monoplane first came to national fame when Frank Hawks flew "Texaco # 5" across the country and broke the trans-continental record held by the "Yankee Doodle", a "Wasp" powered "Vega"; Hawks set a time of just over 18 hours from coast to coast. Hawks set this record in February of 1929, then broke his own record later in June of that year. Henry J. Brown, in a hairy-chested "Air Express" that was power-ed with a "Hornet" engine, run off with honors in the Los Angeles to Cleveland Air Derby, with two Wasp-Vegas closely behind. Some time later, the genial Roscoe Turner began making mad-dashes across the country in quest of records and so it went; these "gentlemen in a hurry" kept the "Air Express" basking in the lime-light of fame for some time. Yet, we must not fail to mention that not all of the "Air Express" type were

out making mad-dashes across the country; one (the prototype) had flown for Western Air Express (now Western Air Lines) car-rying mail and passengers in regular scheduled service. Western Air Express contracted for the first of this type, shown here, in October of 1927 and it was delivered to them in March of 1928; W.A.E. sold this first model back to Lockheed later on in 1929.

In July of 1928, the price of the "Air Express" with P & W "Wasp" engine was an-nounced as $19,500, the "Hornet" powered version was to be available for $22,500; in May of 1929, these prices were lowered by $1,000. Early "Air Express" models did not mount the N.A.C.A. "low drag" engine cow-ling, nor the "wheel pants"; performance without these extras was slightly inferior. The N.A.C.A. engine cowling boosted the maxi-mum speed some 18 m.p.h., and the "pants" added another 6 m.p.h. The type certificate number for the Lockheed "Air Express" was issued in January of 1929 and the model was built by the Lockheed Aircraft Co. at Los Angeles, California. F. E. Keeler was presi-dent; Alan H. Loughead was V.P. in charge of sales; Ben S. Hunter was secretary-trea-surer; Eddie Bellande was chief test pilot for a time, and Gerrard "Gerry" Vultee was

Fig. 6. Frank Hawks and the famous "Texaco #5", a record-breaking team of 1929; wheel pants and N.A.C.A. cowling added speed.

Fig. 7. The prototype "Air Express" was built to specifications suggested by Western Air Express for hauling mail and passengers.

chief engineer. John K. Northrop had left Lockheed by this time, but before he left he had laid out the design for the "Air Express" and a basic low-wing design for a number of models, which revolved around the "Sirius" type.

Listed below are specifications and performance data for the "Wasp" powered Lockheed "Air Express"; length overall 27′6″; hite 9′6″; span 41′(43′); chord at root 102″; chord at tip 63″; total wing area 275(288) sq.ft.; airfoil at root Clark Y-18; airfoil at tip Clark Y-9.5; empty wt. 2533; useful load 1842; payload 982; gross wt. 4375 lbs.; max. speed 185 (167); cruising speed 150 (135); landing speed 55; climb 1460 feet first min. at sea level; service ceiling 22,000 ft.; gas cap. 100 gal.; oil cap. 10 gal.; range 725 miles; price at the factory was $19,500 with "Wasp" engine, and $22,500 with the 525 h.p. "Hornet" engine; in May of 1929 these prices were lowered by $1,000. The bracketed figures of span, indicate a big-wing model and the bracketed figure of area indicates 13 sq.ft. of extra area; the bracketed figures of top and cruise speed indicate model without N.A.C.A. cowl and wheel-pants. The following figures are for the early prototype version, less "low drag" cowling and wheel-pants; empty wt. 2050; useful load 1625; payload 765; gross wt. 3675 lbs.; max. speed 167; cruising speed

135; land 52; climb 1600 ft. first min., ceiling 25,000 ft.; range 675 miles; all dimensions and other data were typical.

The fuselage framework was built up of two 5/32″ laminated spruce plywood "shells" that were formed to shape in a mold that was called a "concrete tub", and then assembled over circular shaped laminated spruce formers that were held in line by several wood stringers; after the shells were joined, cut-outs were formed for cabin windows, entry door and pilot's cockpit. The cantilever wing framework was built up of spruce and plywood box-type spars and wing ribs were built up of laminated plywood webs with spruce capstrips; the completed framework was then covered with 3/32″ laminated spruce plywood veneer. The tail-group was all-wood and of similiar construction; the horizontal stabilizer was adjustable in flight. Early models of the "Air Express", without the N.A.C.A. type engine cowling, employed an exhaust collector-ring, but the later "cowled" versions usually used individual stacks. The two fuel tanks of 50 gallon capacity each, were of 16 gauge aluminum and were mounted in the wing; (mounted thus for gravity feed); one each side of center-line. The split-axle landing gear of 96 inch tread had 32 x 5 wheels and individual wheel brakes were standard equipment; a metal propeller and inertia-type engine star-

Fig. 8. This "Air Express" was delivered in Texas, flying 1200 miles at an average of 151.6 m.p.h. and using 17 gallons of fuel per hour.

ter were also furnished as standard equipment. For the next Lockheed development, see the chapter for A.T.C. # 140 in this volume, which will discuss the "Vega" as powered with the 9 cylinder Wright J6 engine of 300 h.p.

Listed below are "Air Express" entries that were gleaned from the Aeronautical Chamber of Commerce aircraft register; as far as is known, these were the only examples built:

X-4897; Air Express (# 1) Wasp, this was prototype used by W.A.E.

NR-7955; Air Express (# 2) Wasp, "Texaco" # 5", flown by Frank Hawks.

X-7430; Air Express, serial # 19, Hornet C, Wm. Thaw II, also as Vega 2.

NC-514E; Air Express, serial # 65, Wasp, N.Y.R.B.A. Line.

NR-3057; Air Express, serial # 75, Hornet A, Gilmore Oil & General Tire.

NC-306H; Air Express, serial # 76, Wasp, Texas Air Transport & S.A.T.

NC-307H; Air Express, serial # 77, Wasp, N.Y.R.B.A. Line.

Fig. 9. The Curtiss "Falcon Mailplane" with 12 cylinder "Liberty" engine.

Stemming from a very illustrious breed of airplane, the Curtiss "Falcon Mailplane" had all the "romance of airmail" built into it's every line. Designed to transport heavy airmail and air-cargo loads at a high cruising speed, the "Falcon" mailplane series were basically a development from the well-known U.S. Army Air Corps "Falcon" that was used as an "observation" airplane in the service, since back in 1925. The "Mailplane" was quite typical in most respects, including the pronounced sweep-back in the upper wing panel which gave the "Falcon" distinction and a jaunty air. The mailplane version had a single open cockpit located far aft in the fuselage, with two large hatch-covered strap-cinched mail and cargo compartments up forward that displaced a large capacity, even for bulky loads. Because of the sizeable stock of engines still left on hand, this "Mailplane" version was powered with the time-honored "Liberty 12" of 400-435 h.p., and could carry a sizeable and profitable payload at a maximum speed of over 140 m.p.h. These "Falcon" mailplanes, put into regular service in the early part of 1929, became a familiar sight

on the National Air Transport (N.A.T.) route between New York and Chicago and return, carrying the night mail. Sometime later, they were also used regularly on the N.A.T. route from Chicago to Dallas; in all, the N.A.T. line had at least eleven of these "Falcon Mailplanes" in regular scheduled service. These eleven aircraft were just about the bulk of the "Liberty-Falcon" that were built.

The "Falcon" was an ideal mailplane, it had sure-footed performance and easily predictable handling characteristics which made it a favorite among the pilots who plied the mail routes at night.

The "Liberty" powered Curtiss "Falcon Mailplane" (also "Commercial Falcon") was first awarded an approval on a Group 2 certificate numbered 2-31 but this was immediately superseded by A.T.C. # 103 which was issued in January of 1929; the type certificate was amended and re-issued in July of 1929 to cover installation of the "Liberty 12A" of 435 h.p., night-flying equipment, and a higher permissable gross load. Curtiss also had built a special "Falcon Cargoplane" with a Curtiss D-12 engine on a Group 2 certificate numbered 2-37, and another "Falcon Mailplane" that was powered with the 12 cylinder "Geared Conqueror" engine of 600 h.p. on A.T.C. # 213. The "Falcon" mailplane series were manufactured by the Curtiss Aeroplane & Motor Co. of Garden City, Long Island, New York. The venerable Charles S. "Casey" Jones was chief pilot and manager of sales, and T. P. Wright, one of the most outstanding men in the design field, was chief engineer in charge of the design and development section.

Listed below are specifications and performance data for the "Liberty" powered Curtiss "Falcon Mailplane"; length overall 27'7"; hite overall 10'3"; span upper 38'; span lower 35'; chord upper 70"; chord lower 54"; total wing area 352.7 sq.ft.; airfoil "Clark Y"; wt. empty 3179-3341; useful load 1931-1924; payload 841-834; gross wt. 5110-5265 lbs.; max. speed 146; cruising speed 124; landing speed 66; climb 940 ft. first min. at sea level; service ceiling 14,000-13,500 ft.; gas cap. 135 gal.; oil cap. 13 gal.; range 600 miles. The fuselage framework was built up of riveted duralumin tubing, with main fittings at various high-stress points of welded chrome-moly steel sheet-stock and tubing; framework was faired to shape with fairing strips and fabric covered. The upper section of the fuselage was covered with dural panels clear back to the tail-post. The wing panel framework was built up of heavy-sectioned spruce box-type spars, with spruce and plywood girder-type wing ribs, also fabric covered; ailerons of the Freise "balanced-hinge" type were attached to both upper and lower panels and were connected with a push-pull strut. The fabric covered tail-group was built up of riveted duralumin tubing, with the rudder being of the "balanced-horn" type; the horizontal stabilizer was adjustable in flight. The bulk of the fuel load was carried in a large "belly tank" that was built into the contours of the under-side of the fuselage; a gravity-feed fuel

tank was built into the center-section of the upper wing. The "belly tank" had a large dump-valve which was controllable by the pilot; this was a safety precaution that would allow the pilot to jettison the biggest portion of his fuel load in case of an emergency forced landing. It is a known fact that many fearless pilots were deathly afraid of fire during a forced landing, so the dump-valve would be used to prevent or at least lessen this dire circumstance, especially when flying at night. Individual wheel brakes, metal propeller, and inertia-type engine starter were standard equipment; the pilot's cockpit was supplied with exhaust-manifold heat which helped to some degree, but it was a token gesture and certainly wouldn't permit anything approaching shirt-sleeve flying in cold winter weather! As a consequence, mail-flying in winter weather called for good hardy men and plenty of warm clothing; a little flask of "spirits" was known to help a good deal. For the next Curtiss development see chapter for A.T.C. # 159 in this volume, which discusses the Curtiss "Challenger" powered "Thrush" monoplane.

Listed below are "Liberty-Falcon" entries that were gleaned from the Aeronautical Chamber of Commerce aircraft register; as far as is known, these were the only examples built:

C-112E; Liberty-Falcon (# 7) N.A.T.
C-208E; Liberty-Falcon (# 8) N.A.T.
C-209E; Liberty-Falcon (# 9) N.A.T.
C-210E; Liberty-Falcon (# 10) N.A.T.
C-211E; Liberty-Falcon (# 11) N.A.T.
C-212E; Liberty-Falcon (# 12) N.A.T.
C-213E; Liberty-Falcon (# 13) N.A.T.
C-214E; Liberty-Falcon (# 14) Aviation Exploration
NC-255H; Liberty-Falcon (# 17) N.A.T.
NC-256H; Liberty-Falcon (# 18) N.A.T.
NC-257H; Liberty-Falcon (# 19) N.A.T.
NC-258H; Liberty-Falcon (# 20) N.A.T.

Fig. 10. This view shows the cargo hatches and "belly tank"; note sweep-back in upper wing.

Fig. 11. The "Falcon" carried an 800 pound payload of airmail and cargo.

A.T.C. #104
(1-29)
MAHONEY-RYAN "BROUGHAM", B-3

Fig. 12. The Mahoney-Ryan B-3 was indeed a rare sight; originally powered by the Wright J5 engine, it is shown here with a Wright J6-9-300.

The Mahoney-Ryan "Brougham" model B-3 was a trim and lady-like 5 to 6 place high wing cabin monoplane of typical "Ryan" construction and of a familiar configuration, except for only a few slight modifications over the previous model B-1 design (see chapter for A.T.C. # 25 in U.S. CIVIL AIRCRAFT, Vol. 1); these modifications included such items as larger tail-surfaces for better directional control, an improved air-oil strut landing gear, a swiveling tail-wheel, rounded off nose-section cowling without the familiar "prop spinner", and a wider and larger cabin interior that had ample provisions to seat six. The Wright "Whirlwind" J5 powered model B-3 was a rare type indeed, and almost an "orphan" in the Ryan family of airplanes; it was designed and developed during a

period of change-over from company ownership and factory location, and this had a tendency to stifle it's development and sales promotion. The genial B. F. "Frank" Mahoney, due to illness and a desire to relax from constant business pressures, decided to sell out to a St. Louis group of financiers. They of course, had elected to move the plant machinery and other operations to their home-grounds. A few of the first B-3 type did come out with the Wright J5 engine, but these aircraft were very soon modified by the installation of the new 9 cylinder Wright "Whirlwind" J6 series engine of 300 h.p. and eventually, this modification led to the direct development of the popular "Ryan B-5". The model B-3 was then more or less a transition of the earlier model B-1 into the new model

B-5.

The type certificate number for the Mahoney-Ryan "Brougham" model B-3 as powered with the 9 cylinder Wright "Whirlwind" J5 engine of 220 h.p., was issued in January of 1929. It was manufactured by the Mahoney-Ryan Aircraft Corp. at San Diego, California, and later at Lambert-St. Louis Airport, Anglum (St. Louis County), Missouri. The name "Mahoney" was soon dropped from the company mast-head and it became the Ryan Aircraft Corp., a division of the Detroit Aircraft Corp. John C. Nulsen was V.P. and general manager; W. A. "Bill" Mankey was chief engineer; and J. J. "Red" Harrigan was chief pilot.

The last of the "Brougham" model B-1 was serial # 178; the first of the Wright J5 powered model B-3 was serial # 179. Three of the model B-3, as modified with the installation of the Wright J6-9-300 engine, were approved under a Group 2 approval numbered 2-50 which was issued in March of 1929; these were aircraft serial # 184-185-186, approved as 5 place airplanes. Later, this approval was to include serial # 183. A model B-3A as a 6 place airplane with the Wright J5 engine (serial # 210) was under a Group 2 approval numbered 2-105 which was issued in July of 1929.

Listed below are specifications and performance data for the Mahoney-Ryan "Brougham" model B-3 as powered with the Wright J5 engine; length overall 28'3"; hite overall 8'10"; span 42'4"; chord 84"; total wing area 280 sq.ft.; airfoil "Clark Y"; wt. empty 2114; useful load 1590; payload 895; gross wt. 3704 lbs.; max. speed 125; cruising speed 105; landing speed 50; climb 750 ft. first min. at sea level; service ceiling 15,500 ft.; gas cap. 80 gal.; oil cap. 5 gal.; range approx. 700 miles; price at the factory field was $12,250. The fuselage framework was built up of welded chrome-moly (molybdenum) steel tubing, faired to shape and fabric covered. Cabin walls were sound-proofed and insulated against noise, vibration, and temperature changes; the interior was quite neatly equipped with comfortable seats of the bucket and bench type. The semi-cantilever framework of the wing, consisted of two heavy-sectioned solid spruce spars that were routed to an "I beam" section, and wing ribs were built up of spruce members and plywood gussets in truss-type form; the completed framework was then fabric covered. Two fuel tanks, flanking the fuselage, were built into the root ends of each wing. The fabric covered tailgroup was also built up of welded chrome-moly steel tubing; the rudder had aerodynamic balance and the horizontal stabilizer was adjustable in flight. The robust landing gear of 120 inch tread, employed "Gruss" air-oil struts for absorbing the shocks; a swiveling tail-wheel was provided, wheels were 30x5 (Bendix) and individual wheel brakes (Bendix) were standard equipment. A Hamilton wood propeller was provided, but a metal propeller was optional. The next "Brougham" development was the Ryan B-5; see chapter for A.T.C. # 142 in this volume.

Listed below are the Mahoney-Ryan model B-3 entries that were gleaned from the Aeronautical Chamber of Commerce aircraft register; as far as is known, these were the only examples built:

C-114E; Ryan B- 3(# 179) Wright J5, later modified with J6-9-300

NC-7732; Ryan B-3 (# 180) Wright J5

NC-7733; Ryan B-3 (# 181) Wright J5

NC-7734; Ryan B-3 (# 182) Wright J5, not completely verified

NC-7735; Ryan B-3 (# 183) Wright J5, later modified with J6-9-300

NC-7736; Ryan B-3 (# 184) Wright J6, donated to Kelly & Robbins

NC-7737; Ryan B-3 (# 185) J6-9-300, Group 2-50

NC-7738; Ryan B-3 (# 186) Wright J6, not completely verified

NC-311K; Ryan B-3A (# 210) Wright J5, as 6 pl., owned by S. L. Lambert

X-8321; serial # 187 was prototype of Ryan B-5.

Fig. 13. The Swallow TP with Curtiss OX-5 engine; a sensible approach to efficient student instruction.

The first real step towards the realization of practical "commercial aviation" in this country was the development of the three place biplane, and it is noteworthy to dwell upon the fact that the first practical airplane of this type, bore the name of "Swallow". The greatest testimonial to this first "Swallow's" supremecy (1924) was the almost immediate formation of other aircraft manufacturing companies for the express purpose of building a similiar airplane. Now, some years after "Swallow's" first great contribution towards the development of practical commercial activities in aviation, this same organization, with the same integrity of purpose, has again offered a new type airplane. An airplane that has been designed to fulfill a great need — an airplane that is again destined to create somewhat of an upheaval in aircraft manufacturing circles. This airplane is the "Swallow TP", and again the Swallow organization anticipates the subtle flattery of other manufacturers following suit. The foregoing statements are almost a direct quote, but actually they bear much truth.

The new "Swallow TP" (Training Plane) was indeed an interesting and new approach in a civilian trainer-type airplane; taking precedent from the basic "Air Corps" trainer-type design, the TP was made simple and rugged with no superfluous features or niceties that wouldn't be needed in this type of airplane. Bare of frills and fancy finery, the TP was designed for efficient and economical pilot-training, while well suited too as a low-cost sportplane that offered the minimum in maintenance and a structure rugged enough to forgive an occasional pilot error. Concieved while Amos O. Payne was with "Swallow" as chief engineer, the TP was designed with all this in mind, and proved itself well.

As pictured here, the TP trainer was an open cockpit biplane that seated two in tandem; cockpits were roomy and coamings were well padded to prevent injury from occasional nip-ups that occur in pilot-training. Instruments were actually of very little interest to a fledgling learning to fly, so only the necessary basic instruments were installed, where they could be seen from either cockpit. The fuselage and wing construction was rugged and beefed-up throughout, even down to the tough and simple straight-axle landing gear; yet this rugged structure did not hamper the "Swallow's" easy-flying characteristics, nor it's easy manuverability. The TP performed deftly at the 1929 National Air Races held at Cleveland, Ohio; it won and hi-scored in dead-stick landing and balloon-bursting contests consistently, showing proof of sure-footed manuverability and sharp, concise control.

The "Swallow TP" made it's formal appearance before the public in December of

1928 at the Chicago Air Show and caused quite a stir in aviation circles; it was an immediate success and very soon became a familiar sight at many flying-schools across the country. Almost 200 of the OX-5 powered TP were built and actually it would have fared much better if the supply of OX-5 engines hadn't run out; the TP was also offered with the 7 cyl. Siemens-Halske engine, and later with the 5 cyl. Kinner and 7 cyl. Warner engines, but these models proved a bit expensive for the average small-operator at this time and not a great number of these were built. The type certificate number for the OX-5 powered "Swallow TP" was issued in January of 1929 and it was manufactured by the Swallow Airplane Company at Wichita, Kansas. W. M. Moore was president; C. A. Noll was V.P.; Geo. R. Bassett was secretary and general manager; M. D. Kirkpatrick was factory superintendant; "Jay" Sudowsky was their chief pilot, and Dan Lake was now the chief engineer.

Listed below are specifications and performance data for the OX-5 powered "Swallow TP"; length overall 23'7"; hite overall 8'10"; span upper 30'11"; span lower 30'3"; chord both 60"; total wing area 296 sq.ft.; airfoil "U.S.A.27 Mod.") wt. empty 1283; useful load 542; payload 178; gross wt. 1825 lbs.; max. speed 90; cruising speed 75; landing speed 35; climb 650 ft. first min. at sea level; service ceiling 15,500 ft.; take-off run 150 ft.; landing roll 150 ft.; gas cap. 28 gal.; oil cap. 4 gal.; range 300+ miles; price at the factory in January of 1929 was $1795 less engine and propeller; in May of 1929 the price was upped to $2045 less engine and propeller. The fuselage framework was built up of welded steel tubing of various grades to suit the purpose; framework was lightly faired to shape and fabric covered. The wing panels were built up of solid spruce spar beams and spruce and plywood built-up ribs, also fabric covered. Wing brace and center-section struts were of chrome-moly steel tubing in streamlined section, but the interplane bracing wires

were of plain stranded aircraft cable. The fabric covered tail-group was also built up of welded steel tubing; the horizontal stabilizer was adjustable in flight. The rugged and simple landing gear was built up of welded chrome-moly steel tubing, braced with stranded aircraft cable; the wheels were connected to a rather out-moded straight-axle and two spools of rubber shock-cord were used to snub the hard bumps. The engine coolant radiator which was faired into the nose-section, has adjustable shutters for temperature control; engine cowling had large cut-outs around the upper ends of the engine cylinders for easy inspection and maintenance of the assorted plumbing and the valve gear. Twenty-eight gallons of fuel was carried in a fuselage tank that was mounted high up forward of the front cockpit; a quick-release was provided on the student's control stick and rudder pedals to forestall dire catastrophe in case of student freeze-up on the controls during a moment of panic. All controls were cable operated, with numerous cut-outs provided for easy inspection. Summing up the discussion, we can see that the TP was built and arranged for ease of maintenance and inspection right on the flying-line, to keep the hangar-time at a bare minimum. For the next "Swallow" development, see chapter for A.T.C. # 125 in this volume which will discuss the 3 place Swallow F28-AX as powered with the 7 cyl. Axelson engine; for the next TP development, see chapter for A.T.C. # 186, also in this volume.

Fig. 14. *Bare of frills and fancy finery, the Swallow TP offered low-cost sport flying.*

Fig. 15. The Boeing 95 with "Hornet" A engine of 525 h.p.; the "95" introduced several new concepts in mail transport design.

Towards the close of year 1928, air-mail and air-cargo loads on certain routes across the country were gradually on the increase and showing promising signs of leveling off at high tonnage; existing airplanes that were flying the airways at this time were proving somewhat unsuitable to handle this type of traffic. What was needed was an airfreighter that would handle large loads economically and at a fair cruising speed to maintain fast schedules. Well versed in the needs of air-traffic, through experience gained in millions of miles of flying with "Boeing" types on the vast Boeing Air Transport System, it was a logical and almost imperative step for Boeing to lay plans for an airplane suitable to meet these oncoming needs. Boeing engineers developed the "Model 95" mailplane to serve in this capacity; it is self-evident from the history left behind by this airplane that Boeing engineers had analyzed the problem accurately and came up with the correct answer.

Of necessity, the new "Ninety Five" was quite a good-sized airplane; more or less the same size as the earlier "40 series", the parentage of which still showed up here and there in the new design. Though somewhat similiar, the "95" was a much cleaner and simpler design than any of the "40 type" and was built specifically for the hauling of heavy loads of air-mail and air-cargo. The wing panels on this new craft were in normal biplane arrangement of the single-bay type, with the upper wing now being the much larger; wing tips were well rounded off, and in contrast to the previous "40 type", the lower wings on the "95" were comparitively small and had a large amount of interplane stagger. The pilot was still seated in the traditional open cockpit that was located well aft in the fuselage, and enough cargo space (89 cu.ft.) was up forward in hatch-covered compartments to carry a 1600 pound payload, if need be. The "Boeing 95" was powered with a 9 cyl. Pratt & Whitney "Hornet" engine of 525 h.p., a compatible combination that offered surprisingly good performance for an aircraft of this size; other built-in attributes offered economical transport on the pounds-per-mile basis, rapid servicing and loading features, with facilities provided for almost care-free night flying.

The Boeing System used several of these craft on their San Francisco to Chicago route, and the "Ninety Five" was also in use by National Air Transport, Western Air Express, and Robertson Aircraft. The type certificate number for the Model 95 was issued in January of 1929, and re-issued in August of 1929 for an extra fuel capacity of 25 gallons; a total capacity of 155 gallons. A total production of 25 of the Model 95 type were reported built by the Boeing Airplane Co. of Seattle, Washington. W. E. "Bill" Boeing was now chairman of the board; Philip Gustav Johnson was Pres.; L. R. Tower was chief pilot; the affable Erik Nelson of 1924 round-the-world fame was sales manager, and C. N. Monteith continued in his engineering chores.

Listed below are specifications and performance data for the "Hornet" powered Boeing "Model 95"; length overall 31'11"; hite overall 12'1"; span upper 44'3"; span lower 39'6"; chord upper 86"; chord lower 66"; total wing area 490 sq.ft.; airfoil "Boeing"; dihedral 2 deg. in both wings; wheel tread 88"; wt. empty 3196-3222; useful load 2644-2618; payload 1610-1434; gross wt. 5840 lbs.; max. speed 142; cruising speed 120; landing speed 56; climb 950 ft. first min. at sea level; climb in 10 min. 7350 ft.; climb time to 10,000 ft. is 15.4 min.; service ceiling 16,000 ft.; take-off run 540 ft. (11.6 sec.); fuel capacity 130-155 gal.; oil cap. 12.5 gal.; cruising range 520 miles; price at the factory field was $24,500. The fuselage framework was built up of welded chrome-moly steel plate and tubing in the forward section and at all main stress points and fittings; the aft section was built up of bolted duralumin tubing, with steel tie-rod bracing. The forward portion of the fuselage was largely covered with aluminum panels and hatch covers; most of the rear portion was faired to shape and covered with fabric. The wing panels were built up of heavy spruce spar beams, with spruce and plywood girder-type ribs; completed framework was fabric covered. The ailerons were metal-framed and covered with corrugated "Alclad' sheet; they were of the offset-hinge type for aerodynamic balance. The fuel tank was built into the center-section panel of the upper wing. The tail-group, including all fixed and movable surfaces was metal framed and covered with corrugated Alclad sheet; the horizontal stabilizer was adjustable in flight. The robust landing gear was of the cross-axle type with "oleo" shock-absorbers; individual wheel brakes were standard equipment. The Model 95 was bonded and shielded throughout for the installation of radio equipment, and complete night-flying equipment was available; metal propeller, inertia-type engine starter, and an exhaust-manifold type cockpit heater were also provided as standard equipment.

The "Boeing 95" was introduced along in December of 1928 and showed a good deal of design influence from the sporty "83" and "89" models which were forerunners to the fabulous "Boeing 100", the P-12 and the F4B series; the model "203", a sport-trainer version, was practically a "95" in miniature. For reference to prevoius Boeing models, see

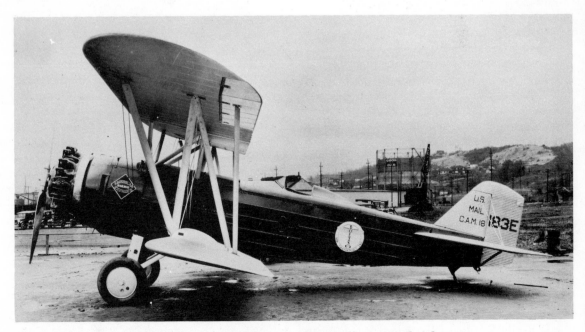

Fig. 16. The Boeing 95 carried 1600 pounds of airmail and cargo.

Fig. 17. The "95" played an important role in pioneering speedy transport of airmail and cargo.

Fig. 18. The "95" hauled mail at night from New York to Chicago.

chapters for A.T.C. # 2 - 23 - 27 - 54 - 64, all in U. S. CIVIL AIRCRAFT, Vol. 1; for the next Boeing development, after the "95" discussed here, see chapter for A.T.C. # 133 in this volume which discusses the "Boeing 100".

Listed below are the "Boeing 95" entries that were gleaned from the Aeronautical Chamber of Commerce aircraft register; as far as is known, these were the only examples built:

C-183E; Boeing 95 (# 1046) Boeing A.T.
NC-184E; Boeing 95 (# 1047) Boeing A.T.
NC-185E; Boeing 95 (# 1048) Boeing A.T.
NC-186E; Boeing 95 (# 1049) not verified
NC-187E; Boeing 95 (# 1050) Boeing A.T.
NC-188E; Boeing 95 (# 1051) Boeing A.T.
NC-189E; Boeing 95 (# 1052) Boeing A.T.

NC-190E; Boeing 95 (# 1053) Boeing A.T.
NC-191E; Boeing 95 (# 1054) N.A.T.
NC-192E; Boeing 95 (#   ) probable
C-412E; Boeing 95 (# 1058) N.A.T.
NC-413E; Boeing 95 (# 1059) Boeing A.T.
C-414E; Boeing 95 (# 1060) N.A.T.
NC-415E; Boeing 95 (# 1061) Boeing A.T.
NC-417E; Boeing 95 (# 1056) Boeing A.T.
NC-418E; Boeing 95 (# 1062) N.A.T.
C-419E; Boeing 95 (# 1063) W.A.E.
C-420E; Boeing 95 (# 1064) W.A.E.
C-421E; Boeing 95 (# 1065) W.A.E.
C-422E; Boeing 95 (# 1066) W.A.E.
C-423E; Boeing 95 (# 1067) N.A.T.
C-424E; Boeing 95 (# 1068) N.A.T.
C-425E; Boeing 95 (# 1069) N.A.T.
C-426E; Boeing 95 (# 1070) N.A.T.

Fig. 19. The New Standard D-24 carried five with power of "Hisso" engine; ship shown was modified with Wright J5.

The well-known "Standard" biplane of some ten years previous (1917-18), was graceful, pleasant, and obedient; the "New Standard" biplane was like a re-incarnation of this favorite, just done up in more modern dress. The "New Standard" model D-24, the first of the series, was an unusual type of airplane for these times; largely reminiscent of such older types as the 5 place Pitcairn "Fleetwing", the "Sikorsky S-32", and a few others that were built back in the 1924-25-26 era. Planes that carried four passengers and a pilot, all in open cockpits. Outmoded as this may now have appeared, it actually was a very sensible arrangement that worked out quite well for joy-riding paying passengers on short hops around the airfield, and enabled an operator to make a "fast buck" on a busy Sunday afternoon.

The "New Standard" model D-24 was a large 5 place open cockpit biplane seating four passengers, all in one open cockpit up forward, and the pilot was seated in an open cockpit situated aft. First, as an economy measure to hold down the first-cost, the model D-24 was powered with the watercooled 8 cylinder vee-type "Hisso" (Hispano-Suiza) model E engine of 180 h.p. The engine was cowled in neatly and the engine coolant ra

diator was of the tunnel-type that blended nicely into the front-end contours of this airplane. Largely concerned with a substantial payload and good short-field performance, the D-24 was endowed with an overly-generous amount of wing area; the upper wing was extra large and had a graceful sweep, the lower wing was a good bit smaller and approached the "sesquiwing" arrangement. Though large, and somewhat bulky looking, the "New Standard" series were graceful and amiable airplanes that were very efficient and offered a surprisingly good performance. Of durable and rugged construction, the wing panels were of the normal wood spar and wood rib make-up, but the novel fuselage framework was built up of duralumin angle and open-section members that were riveted and bolted together. Quite a new concept at this time, this was a simple construction that was easily repaired without a large lay-out of special tools and special equipment; easily repaired with tools one might find in any tool box.

The distinctive "New Standard" series were concieved and designed by Charles Healy Day, who will be remembered as one of our leading aeronautical engineers. He had designed and developed a good number of airplanes, but his most famous design was probably the old and beloved "Standard" swept-wing biplane which was a popular training plane of World War 1 vintage; a plane that was used by barnstormers, flying-schools, and early air-lines, for nearly ten years afterwards. The "New Standard" series, were first reared, early in 1928, by Charles Healy Day as the GD-24, as built by the Gates-Day Aircraft Co.; the "Gates" half of the company being Ivan Gates, who was long famous for his "Gates Flying Circus" troupe of the early "twenties". Registration number C-2220 was awarded to a vintage "Hisso-Standard" that flew with the "Gates Flying Circus", when it cracked-up and was "washed-out", it seems that the registration number was then awarded to the GD-24 prototype. The GD-24 (D-24) was Charlie Day's 24th successful aircraft design, up to now an impressive array of outstanding contributions in the annals of aeronautical development. From the short-lived "Gates-Day" venture had blossomed the New Standard Aircraft Corp. of Paterson, New Jersey which now manufactured the various models in the "New Standard" series. The type certificate number for the model D-24, as powered with the "Hisso" E of 180 h.p., was issued in February of 1929. Three aircraft of this model (serial # 101-102-104) were previously built and were certificated under a Group 2 approval numbered 2-38 which was issued in January of 1929; altogether some six of this "New Standard" model were built, and some of these were later modified to the D-25 type as powered with the Wright J5 engine.

Listed below are specifications and performance data for the "New Standard" model D-24 as powered with the Hispano-Suiza model E engine of 180 h.p.; length overall 26'6"; hite overall 10'2"; span upper 45'; span lower 32'6"; chord upper 70"; chord lower 50"; area upper 240 sq.ft.; area lower 110 sq.ft.; total wing area 350 sq.ft.; airfoil "Goettingen 533"; wt. empty 2066; useful load 1334; payload 784; gross wt. 3400 lbs.; max. speed 110; cruising speed 95; landing speed 37; climb 640 ft. first min. at sea level; service ceiling 14,000 ft.; gas cap. 64 gal.; oil cap. 5-6 gal.; range 550 miles; price at the factory was $3500 less engine and propeller; price was $4600 with "Hisso E" engine (usually the Wright-Martin version manufactured in the U.S.A.); the earlier GD-24 was also offered with the Hispano-Suiza model A engine of 150 h.p. for $4250, and with 210 h.p. Mercedes (German) engine for $4800.

The novel fuselage framework was built up of open-sectioned "dural" members in angle, channel, and tee shapes that were bolted and riveted together into a simple structure that was easy to maintain and repair; the completed framework was lightly faired to shape and fabric covered. On the prototype GD-24, there was one great big, elongated, cockpit that seated 5 persons all together; (best described as 5 men in a tub) the only separation between the pilot and his passengers was an instrument panel and a small windshield. On the newer D-24, the pilot had a bit more privacy. The wing panels were built up of heavy-sectioned solid spruce spars that were routed to an "I beam" section, with bass-wood and plywood girder-type ribs; completed framework was fabric covered, a chore that required almost 100 yards of cotton fabric cloth! The gravity-feed fuel tank was built into the center-section panel of the upper wing. The landing gear was of the split-axle long-leg type; the earlier GD-24 used rubber rings in compression for shock absorbers, but the later D-24 had an "oleo strut" landing gear with Bendix .wheels and brakes. Wheel tread extended to 96 inches, and wheels were 30x5. The fabric covered tail-group was built up of "dural" members in a construction quite similiar to that used on the fuselage; horizontal stabilizer was adjustable in flight. A wooden propeller was provided as standard equipment, but a metal propellor, and hand-crank inertia-type engine starter were optional. The next development in the "New Standard" series was the model D-25, see chapter for A.T.C. # 108 in this volume.

Listed below are model D-24 entries that were gleaned from the Aeronautical Chamber of Commerce aircraft register, as far as is known, these were the only examples built:

X-7286; Model GD-24 (# 101) Hisso E

C-193E; Model D-24 (# 102) Hisso E

NC-442; Model D-24 (# 104) Hisso E, converted to 3 pl. D-26 with J5.

NC-9756; Model D-24 (# 105) Wright-Hisso E, later had J5 as D-25

NC-9794; Model D-24 (# 107) Wright Hisso E-4

C-9102 Model D-24 (# 112) Hisso E

Fig. 20. The graceful D-25 was designed to revive the barnstroming era.

The water-cooled Hispano-Suiza ("Hisso") engine used in the "New Standard" model D-24 was becoming somewhat outmoded, so the next logical development in the "New Standard" series was the model D-25; a craft that was powered with the popular 9 cylinder air-cooled Wright "Whirlwind" J5 engine of 220 h.p. Typical in most all respects, the D-25 with it's seating for five, was actually designed in anticipation of the revival of the "barnstorming era" of a few years back. "Joy-riding" was still quite popular with the folks, and the D-25, on a busy Sunday afternoon, could carry up to 40 paying "joy-riders" in an hour; this wasn't bad compensation, even at the cut-rate prices of "a buck a ride"!

The model D-25 was certainly a big airplane, as no one could doubt, but it looked and handled quite gracefully despite it's apparent bulk and it's above-average size. The Wright J5 engine offered a little more reliability and a bit more horsepower, so consequently the model "D-25" was a bit more sprightly and had a somewhat better performance. Some 40 or so of this popular model were built and at least two of these were in

service up north with the Alaskan Airways. The type certificate number for the J5 powered model D-25 was issued in February of 1929 and it was manufactured by the New Standard Aircraft Corp. at Paterson, New Jersey. Charles Healy Day was the V.P. and chief engineer, and Louis Randall was the general manager.

Listed below are specifications and performance data for the "New Standard" model D-25 as powered with the Wright J5 engine; length overall 26'6"; hite overall 10'2"; span upper 45'; span lower 32'6"; chord upper 70"; chord lower 50"; area upper 240 sq.ft.; area lower 110 sq.ft.; total wing area 350 sq.ft.; airfoil "Goettingen 533"; wheel tread 96"; wheels 30x5; wt. empty 2010; useful load 1390; payload 810; gross wt. 3400 lbs.; max. speed 110; cruising speed 98; landing speed 37; climb 750 ft. first min. at sea level; service ceiling 18,000 ft.; gas cap. 64 gal.; oil cap. 5 gal.; cruising range 490 miles. Price at the factory was $9750 in 1929; the earlier Gates-Day version was offered with the Wright J5 engine for $7500, but none were built in this version. The model D-25 on twin "floats" was designated the model D-30;

Fig. 21. The seating capacity of the D-25 was ideal for "joy-riding" passengers.

Fig. 22. The sesquiplane arrangement of the New Standard placed the bulk of the area in upper wing.

naturally, there were some weight and performance differences.

The novel fuselage framework of the "New Standard" series was built up of "dural" (duraluminum) members of open-sectioned angles, channels, and tees, that were bolted and riveted together into a simple yet sturdy structure that offered easy maintenance and repair with a minimum of know-how and a few hand tools; the framework was lightly faired to shape and fabric covered. A compartment for tools and other odds and ends was provided in the turle-back just in back of the pilot's seat. The wing framework was built up of heavy-sectioned spruce spars that were routed to an "I beam" section; wing ribs were built up of plywood gusseted bass-wood into a truss-type form and the completed framework was then fabric covered. The plan-form of the wings was quite graceful, with semi-elliptical wing tips; the large ailerons which were in the upper wing, were operated by torque-tubes and bellcranks through a push-pull strut similiar to that used on many other aircraft at this time. The wide-tread landing gear was built up of welded chrome-moly steel tubing, using "oil-draulic" shock-absorber struts to snub the taxii and landing shocks; 30x5 Bendix wheels

were equipped with Bendix brakes as standard equipment. The fuel supply of 64 gallons was held in a tank that was built into the center-section panel of the upper wing; wings were wired for navigation lights. For entry into the front cockpit, there was a door on either side, with a step about half-way up to lessen the awkward stretch that would be necessary to get in; once seated, the passengers were comfortable but in rather chummy quarters.

The simple and rugged structure of the "New Standard" was certainly a boon to the fixed-base operator, and it was also very conducive to longevity; it is not surprising then that many of these lasted for a good number of years. After ten years there were still about twenty-two of the model D-25 in active service, and a few of these, somewhat modified by now, are still active and profitable as "crop sprayers" in the south-west section of our country. The "New Standard" biplane was offered in six different models which all had the same basic configuration except for the number of seating, and the installation of float-gear. The next development in this series was the three place model D-26; see chapter for A.T.C. # 109 in this volume.

Fig. 23. The New Standard D-26 carried three and 470 pounds of cargo.

The rare "New Standard" model D-26 was a further development of the basic D-25 design, that is, the model D-26 was typical except that it was a three place open cockpit airplane with provisions for extra baggage and a sizeable load of cargo, in addition to the passengers carried. Two passengers were carried in the front cockpit and the pilot was seated in the rear open cockpit; a hatch-covered compartment in the forward portion of the fuselage had the capacity to carry either a large amount of baggage for extensive traveling, or a cargo payload of some 470 pounds. A smaller baggage and tool compartment was located high up in the turtle-back section of the fuselage, just in back of the pilot's station. Also powered with the Wright "Whirlwind" J5 engine of 220 h.p., performance of the three-place version was fairly typical except for a slightly higher maximum and cruise speed, which was largely due to the lesser amount of turbulence created by

the smaller front cockpit. The three-place D-26 version was no doubt designed and developed for shuttle-type service on the smaller air-lines that fed into the main transcontinental systems, or perhaps for extensive traveling in the line of business, but the need for this type of craft never did seem to materialize. Consequently, only one known example of this model was built; it was NC-442 which was serial # 104, as listed in the aircraft register of the Aeronautical Chamber of Commerce.

The type certificate number for the Wright J5 powered model D-26 was issued in February of 1929 which superseded a Group 2 approval numbered 2-28; an approval that was slated to be issued for the model D-26 but was cancelled by the issuance of A.T.C. # 109. From the aircraft listing on Group 2 approvals, and the listing of aircraft by registration numbers, it would appear that the only known example of the D-26 (serial #

Fig. 24. The J5-powered D-26 on floats was known as the Model D-28.

104) actually started out in life as a 5 place "Hisso" powered model D-24! The model D-26 was also manufactured by the New Standard Aircraft Corp. at Paterson, New Jersey and the roster of company officers was the same at this time.

Listed below are specifications and performance data for the three place "New Standard" model D-26 as powered with the Wright J5 engine; length overall 26'6"; hite overall 10'2"; span upper 45'; span lower 32'6"; chord upper 70"; chord lower 50"; area upper wing 240 sq.ft. area lower wing 110 sq.ft.; total wing area 350 sq.ft.; airfoil "Goettingen" 533; wheel tread 96"; wheels 30x5; wt. empty 2010; useful load 1390; payload of 2 passengers & 470 lbs. of cargo, or a total payload of 810 lbs.; gross wt. 3400 lbs.; max. speed 114; cruising speed 100; landing speed 37; climb 750 ft. first min. at sea level; service ceiling 18,000 ft. at gross load; gas cap. 64 gal.; oil cap. 5-6 gal.; range at cruise r.p.m. was 500 miles; price at the factory with standard equipment was $9750.

Except for cockpit arrangement and baggage-cargo compartment, the construction details of the model D-26 were typical as for all other models in the "New Standard" series discussed here; for further details refer to previous chapter discussing the model D-25. The interesting arrangement of the "New Standard" biplane with it's large upper wing and small lower wing, was approaching the "sesqui-plane" configuration; due to the small area of the lower wing and it's great amount of stagger in relation to the upper wing, the pilot was able to see his wheels easily from his position in the cockpit. This was helpful during take-offs and landing. The model D-26 when equipped with twin "floats" for over-water flying, was designated the model D-28. The next development in the basic "New Standard" series was the D-27 mailplane, see chapter for A.T.C. # 110 in this volume.

Fig. 25. The J5-powered D-27 was an economical mail-carrier, ideal for feeder routes.

Still another version of the basic "New Standard" biplane series was the model D-27, which was typical to the model D-25 and D-26 in most all respects except that it was a single-place airplane with an open cockpit for the pilot and large hatch-covered cargo compartments up in the forward section for carrying air-mail and air-express cargo. The large cargo hold was spacious enough for even bulky loads, and could carry payload up to 810 pounds. Performance of the model D-27 was slightly better in the top speed and cruising speed range because of the absence of large, gaping, open cockpits, as on the D-25, which created turbulence and drag of considerable proportions; due to the better streamlining of the forward fuselage section, the model D-27 cruised at least 5 m.p.h. faster. Other inherent characteristics such as quick take-off and slow landing speed, which added up to good short-field performance, were typical. Clifford Ball, well-known air-mail operator, had four of the D-27 type which were slated to be used on his airmail

route; whether they were actually used or not, we cannot determine. Incidently, these four aircraft were the only known examples of this model to be built. The type certificate number for the "New Standard" model D-27 as powered with the Wright "Whirlwind" J5 engine of 220 h.p. was issued in February of 1929, and it was also manufactured by the New Standard Aircraft Corp. at Paterson, New Jersey.

Listed below are specifications and performance data for the Wright J5 powered "New Standard" model D-27; length overall 26'6"; hite overall 10'2"; span upper 45'; span lower 32'6"; chord upper 70"; chord lower 50"; area upper wing 240 sq.ft.; area lower wing 110 sq.ft.; total wing area 350 sq.ft.; airfoil "Goettingen" 533; wheel tread 96"; wheels 30x5; wt. empty 2010; useful load 1390; payload 810; gross wt. 3400 lbs.; these wts. would vary somewhat with the addition of full night-flying equipment; max. speed 118; cruising speed 105; landing speed 37; climb 750 ft. first min. at sea level; service

Fig. 26. The D-27 carried heavy loads with quick take-off and low landing speeds, ideal for night flying.

ceiling 18,000 ft. at gross load; gas cap. 64 gal.; oil cap. 5-6 gal.; range at cruising r.p.m. was 525 miles; basic price at the factory was approx. $9750, depending on extra or special equipment that was installed.

Except for the cockpit arrangements, construction details were typical for all models in the "New Standard" series discussed here; see previous chapter discussing the model D-25 for details. The landing gear (sometime called the undercarriage) on the model D-27 was quite typical except for the streamlining of the vee-struts and metal streamlined "cuffs" over the long "oleo struts"; the D-27 was also completely wired for lights, with two large landing-lights built into the leading edge of the upper wings for night-flying operations. Other features that are typical of all models in this series are, fuel tank in the center-section panel of the upper wing; small baggage and tool compartment in turtleback of fuselage just behind the pilot; and a large exhaust collector-ring that expelled fumes well below the cockpit level. The large and graceful wings on the "New Standard" were semi-

elliptical in plan-form and had an upward sweep at the tips, which acted as extra dihedral angle. Dihedral on the upper wing was 2 degrees: dihedral on the lower wing was 4 degrees; gap between panels was 72 inches, and the stagger between panels was 34 inches; these geometrics added up to a stable, easy-to-fly, and well-behaved airplane. The next development in the "New Standard" series was the 2 place "Cirrus" powered sport-trainer model D-29, see chapter for A.T.C. # 198 in this volume.

Listed below are entries of the "New Standard" model D-27 that were gleaned from the Aeronautical Chamber of Commerce aircraft register; as far as is known, these were the only examples built:

NC-9121; New Standard D-27 (# 114) Wright J5, Clifford Ball
NC-9122; New Standard D-27 (# 116) Wright J5, Clifford Ball
NC-9123; New Standard D-27 (# 117) Wright J5, Clifford Ball
NC-9124; New Standard D-27 (# 118) Wright J5, Clifford Ball

Fig. 27. The Travel Air SC-2000 was a rare type; performance was excellent.

In the pursuit of data, and during much discussion about this particular model, there was very little evidence uncovered to give rhyme or reason for an ATC approval on this extremely rare version, except perhaps to embrace the aircraft of this type that already existed. The Travel Air model SC-2000 was indeed an extremely rare type and it was, of course, powered with an engine of some questionable choice for this late in the game. One known example of this type was NC-8110 (serial # 882) that was operated by International Airways in Washington, D. C. during 1930; another example was NC-7574 as shown here, that was apparently modified around an old basic airframe that carried serial # 137 of 1926. Other Travel Air versions that were of somewhat similiar type, were X-6299 (serial # 746) which was a model 2000 powered with a 6 cylinder water-cooled Aeromarine B engine; this craft was owned by Doug Davis (later of air-race fame) and flown in the 1928 Air Derby from New York to Los Angeles, but failed to finish. Another typical version was X-6416 (serial # 757) that was originally a model 4000 modified with the installation of an Aeromarine 6 water-cooled engine. These four similiar aircraft were the only ones to appear in the

registration records.

The model SC-2000 seemed to be a standard version of the 3 place open cockpit Travel Air biplane of the type currently produced, except for minor modifications necessary to accomodate the 6 cyl. in-line engine. The powerplant for the model SC-2000 was the nearly out-moded 6 cylinder in-line water-cooled Curtiss C-6 engine of 160 h.p. Though the illustrious C-6 had been an outstanding engine in it's time, and had a very colorful background and history, it would certainly not be very appealing to the average buyer at this time, especially since the "air-cooled radials" were now beginning to dominate this power range. The Curtiss C-6 engine with it's 4 valves per cylinder and it's overhead camshaft, was a "tinkerer's" delight and was hard to beat when "in perfect tune", but was more than the average fly-guy would care to put up with and therefore, this would almost label the SC-2000 as special-category. The type certificate number for the Travel Air model SC-2000 as powered with the 160 h.p. Curtiss C-6 engine, was issued in February of 1929 and at least two examples of this model were known to exist. Approval for this craft was issued to the Travel Air Company at Wichita, Kansas. Walter Beech was president; Herb

Fig. 28. The SC-2000 was powered by a Curtiss C-6 engine of 160 h.p.

Rawdon was the chief engineer; and Clarence E. Clark was chief pilot in charge of test and development.

The first Travel Air biplane to be powered with the Curtiss C-6 engine was a special version that was built back in 1925, and was flown by hard-flying Walter Beech in the Ford Air Tour of that year; both ship and pilot doing very nicely by running up a near-perfect score in this rugged "tour" about the country. Kind words and admiration are really owing, because the Curtiss C-6 was really quite an engine in it's time, and it's most avid and persistent booster was one Chas. "Casey" Jones of the Curtiss Flying Service; a man who was "the scourge of the air" during air-race time of the early "twenties" in his special clipped-wing "Oriole" There was a story going at that time, saying that if "Casey" thought the boys were gaining on him a little bit, he'd just trim off a little more wing from his already abbreviated "Oriole", just to gain a few extra miles per hour and remain the top-dog!

Listed below are specifications and performance data for the Travel Air model SC-2000 as powered with the 160 h.p. Curtiss C-6 engine; length overall 24'2"; hite overall 9'2";

wing span upper 34'8"; wing span lower 28'8"; wing chord upper 66"; wing chord lower 56"; wing area upper 178 sq.ft.; wing area lower 118 sq.ft.; total wing area 296 sq.ft.; airfoil "Travel Air # 1"; wt. empty 1659; useful load 941; payload with 42-60 gal. fuel was 480-370 lbs.; gross wt. 2600 lbs.; max. speed 120; cruising speed 102; landing speed 48; climb 850 ft. first min. at sea level; climb in 10 min. was 7000 ft.; service ceiling 15,000 ft.; gas cap. 42-60 gal.; oil cap. 5 gal.; cruising range at 10 gal. per hour was 400-575 miles; price at the factory field would be approx. $4000.

The fuselage framework was built up of welded chrome-moly steel tubing, lightly faired to shape with wood fairing strips and fabric covered. The wing framework was built up of solid spruce spar beams that were routed-out for lightness, with spruce and plywood built-up wing ribs; the completed framework was fabric covered. The main fuel tank was mounted high in the fuselage just ahead of the front cockpit, and extra fuel was carried in a tank that was mounted in the center-section panel of the upper wing. The split-axle. landing gear on one known example was of the outrigger type, but this seems to be at variance with the standard set-up. The fabric covered tail-group was built up of welded chrome-moly steel tubing and sheet steel former ribs; the fin was ground adjustable and the horizontal stabilizer was adjustable in flight. Visible modifications such as metal fuselage panels, tail-wheel, and sectional windshield on the rear cockpit, were probably added at a later date. The next development in the Travel Air biplane was the Warner "Scarab" powered model W-4000, see the chapter for A.T.C. # 112 in this volume.

Fig. 29.   An early version of the Travel Air with C-6 engine; Walter Beech in right foreground.

Fig. 30. The Travel Air W-4000 with 110 h.p. "Scarab" engine; a good flying combination.

The three-place open cockpit "Travel Air" biplane had already been offered in six different versions, all basically the same airplane with only such modifications as were necessary for a certain engine combination; the model W-4000 was the 7th to be offered so far, and there were at least seven more yet to follow. The model W-4000 was a three place open cockpit biplane of the type that was currently built by "Travel Air" during this period; it was typical except for the engine installation which in this case was the increasingly popular 7 cylinder Warner "Scarab" engine of 110 h.p. Combining the lovable "Travel Air" biplane with the delightful "Scarab" engine made a very pleasant airplane that proved popular for personal airplane-travel or just plain sportflying, as a training ship, or a sort of do-all for the small operator. Dr. Lee DeForrest, often called the father of the radio tube, had a W-4000 he used for travel about the country, or for sport-flying around the home base; his praises for the airplane were very high. "Billy" Parker of Phillips Petroleum Co. had one of the first few that were built. A number of flying schools about the country were very much impressed with it's money-making efficiency and it's versatility as an all-around training airplane. One of the first airplanes of this type (NC-6401, serial # 747, Warner #

15) was delivered to the San Diego Air Service in September of 1928; they used it as a trainer and due to it's suitability for this type of work, ordered a good many more.

The prototype version of the model W-4000 (X-6269, serial #708) was introduced towards the latter part of 1928, and was flown to 3rd place by Wm. H. Emery, Jr. in the 1928 Transcontinental Air Derby (Class A) from New York to Los Angeles. This airplane in particular, was specially trimmed for this race, and was probably the most beautiful airplane in this whole series. The earliest models of the standard W-4000 had the familiar "elephant ear" ailerons, and so did a few more that were built after, but the later models finally discarded the "ears" (balance-horns) and used the rounded wing tips which were more or less called the speedwing type. Several of the "ears type" W-4000 were built prior to the type certification; these were issued a Group 2 approval numbered 2-35 which was issued in February of 1929 to cover aircraft built prior to 2-16-29. All successive airplanes were manufactured under A.T.C. # 112 which was also issued in February of 1929, and many of these were of the rounded-tip speed-wing type. The comely W-4000 surely had features that gave it a likeable feminine quality; especially in it's "party dress" of pure white and gold! Built

Fig. 31. The prototype of the "Scarab"-powered Travel Air biplane specially faired for the 1928 Air Derby.

into 1931, at least 30 of the model W-4000 were built, and maybe more.

The first Warner "Scarab" (serial #1), a 7 cylinder air-cooled "radial" engine of 110 h.p. was flight-tested for 150 hours late in 1927 by Walter J. Carr in his own "Travel Air" biplane (X-3642); though this was not a Travel Air factory conversion, it might very well be considered as the actual prototype for the W-4000 series. "Travel Air" model designation which was started out in numbers of one-thousand, had already reached "9000" and with the prospect of many different models yet to come, they had to resort to a different model designation because the "thousand series" were rather impractical beyond 9000. To remedy this situation, they reserved the number "2000" for the OX-5 powered model, "3000" for the "Hisso" (Hispano-Suiza) powered models, "4000" for the standard Wright J5 powered model, and then used 4000 with a prefix letter for all of the other models that were powered with American "radial" engines. For example: there was the W-4000 that was powered with the Warner engine; the A-4000 that was powered with the Axelson engine; the C-4000 that was powered with the Curtiss "Challenger" engine; and so on up the line of various developments. The model W-4000 was manufactured by the Travel Air Company at Wichita, Kansas. Walter Beech was president; Herb Rawdon was chief engineer, and Clarence E. Clark was chief test-pilot for a time.

Listed below are specifications and performance data for the "Scarab" powered "Travel Air" model W-4000; length overall 24'8"; hite overall 8'11"; span upper 34'8" (33'); span lower 28'8"; chord upper 66"; chord lower 56"; total wing area 296 (289) sq.ft.; airfoil T.A.#1; wt. empty 1370; useful load 906; payload 446-356; gross wt. 2276 lbs.; max. speed 108; cruising speed 93; landing speed 42-44; climb 560 ft. first min. at sea level; service ceiling 11,000 ft.; gas cap. 42-57 gal.; oil cap. 4-6 gal.; range approx. 400-600 miles; price at the factory field was $5575. In 1930, the W-4000 was also listed with the following table of weights; with 30 gal. fuel wt. empty 1337, useful load 778; gross wt. 2115; with 42 gal. fuel, wt. empty 1337, useful load 850, gross wt. 2187; with 57 gal. fuel, wt. empty 1337, useful load 940, gross wt. 2277 lbs.; the various gross weights would of course affect certain performance figures to a degree. The fuselage framework was built up of welded chrome-moly steel tubing, faired to shape with wood fairing strips and fabric covered. A baggage compartment, with access door on the side, was located behind the pilot. The wing panel framework was built up of laminated spruce spars and spruce and plywood built-up ribs; the completed framework was fabric covered. Ailerons were in the upper wing panel; early models had the "elephant ear" (balanced-horn) type, and the later models with the rounded wing tips discarded the "horns" and had the off-set hinge type of balance, which actually was a better type of aerodynamic balance because it compensated and balanced pressures around the control surface hinge-line. The fabric covered tail-group was also built up of welded chrome-moly steel tubing, with sheet steel former ribs; the fin was ground adjustable and the horizontal stabilizer was adjustable in flight. Wheel brakes, metal propeller, and inertia-type engine starter were optional equipment. The next "Travel Air" development was the "Wasp" powered A-6000-A monoplane, see chapter for A.T.C. # 116 in this volume.

Fig. 32. The Velie-powered Monocoupe 113, the most popular light monoplane of this period.

The merry little "Monocoupe" was a lovable craft that was making it's way into the hearts of thousands of good flying-folk all over the country-side, and wherever it went it was making new friends; new friends that added up to more sales. Well over a hundred of the "Monocoupe" had already been built by this time and people were still clamoring for more. To keep pace with the general trend for improvement, Mono Aircraft had introduced the model "113" as the improved offering for 1929; the configuration was basically the same as earlier models but to it were added some improvements that brought about a bit more utility and slightly better performance. The most easily recognizable change was the long-leg landing gear which was a definate improvement over the axle and spreader-bar landing gear of the earlier models; another change of some significance was the increased area of a 32 foot wing, which added up to better stability and somewhat better performance. The structure of the model "113" was built robust and rugged, and was well able to withstand the usual and unusual stresses of "stunt-flying" as normally practiced by the average pilot, who was quite often goaded into this type of thing by the eager handling characteristics of the "Monocoupe"; this was certainly quite commendable

for a ship of this type.

The model "113", as shown here in various views, was a chummy two place cabin monoplane with side-by-side seating, and it was powered with the 5 cylinder air-cooled Velie M-5 which was rated up to 65 h.p. The type certificate number for this model was issued in February of 1929 and it was manufactured by the Mono Aircraft Corp. of Moline, Illinois; a subsidiary of Allied Aviation Industries. Former car-manufacturer W. L. Velie was president; Don A. Luscombe was V.P. and sales manager; versatile Vern Roberts was chief pilot in charge of test and development, and Clayton Folkerts was in charge of design and engineering. Clayton Folkerts, who had been pretty much in the background before this, was quite a remarkable young fellow who was fast developing into an outstanding design-engineer; his creations in the next few years attested to his flair for a design style that bordered the work of genius.

The remarkable Phoebe Fairgrove Omlie, a gracious lady and a great pilot, was still acting as "ambassadoress" of good will, for the promotion of "Monocoupe" airplanes, and as usual, she was doing a good job of convincing the people of it's many merits. Inherently endowed with exceptional perfor-

Fig. 33. A 32 ft. wing and long leg landing gear
improved Monocoupe performance.

mance and good, sharp manuverability, the
"113" was naturally attracted to air-races
and other events where it could show off;
many "Monocoupes" were head-liners during
the 1929 air-race season, and some of the
"Monocoupe Specials" were outstanding. The
"113", which was designated for it's A.T.C.
number, was also well adapted to the requirements of pilot training, and schools all over
the country were using the little ship in their
primary phases of flight instruction; later in
the year, the new "Monoprep", developed
from the basic "113" design, was introduced
as a training plane to handle this chore. The
"Monoprep", being developed from the
"Monocoupe", was basically similiar but was
a semi-cabin airplane arranged into a parasol-
type monoplane with only a large windshield
in front; the sides and the rear of the cockpit
were open. The interesting "Monoprep" will
be discussed later on Group 2 approval numbered 2-90 and also on A.T.C. # 218. Manufacture of the "113" series continued un-
checked into late 1929; by then the improved

"Monocoupe 90" was being groomed to take
over. In 1930, Mono Aircraft was hit by the
same sag of orders that had befallen nearly
all other aircraft manufacturers due to the
"depression" of that period; in an attempt to
raise money for the development of new models, which were hoped to help stimulate
buying, the last 25 or so of the model "113"
still left on hand were offered in a clearance
sale at $1895!!

Listed below are specifications and performance data for the "Monocoupe 113" as powered with the Velie M-5 of 55-65 h.p.; length
overall 19'9"; hite overall 6'3"; wing span
32'; wing chord 60"; total wing area 143
sq.ft.; airfoil "Clark Y"; wt. empty 848; useful load 502; payload 172; gross wt. 1350 lbs.;
max. speed 98; cruising speed 85; landing
speed 37; climb 580 ft. first min. at sea level;
service ceiling 11,000 ft.; gas cap. 25 gal.; oil
cap. 2 gal.; range up to 500 miles, or 20
m.p.g.; price at the factory field was $2675,
and had gone up to $2835 later in the year;
the airplane and the engine were guaranteed
as a complete unit. The fuselage framework
was built up of welded steel tubing, faired to
shape with formers and fairing strips, and
then fabric covered. The cabin interior was
done up in a tasteful and comfortable manner, and there was a large luggage shelf in
back of the seat; a large sky-light was provided in the cabin roof for vision overhead.
The wing framework was built up of solid
spruce spars and spruce and bass-wood built-

Fig. 34. One of the last examples of the Monocoupe 113; the remaining few sold for $1895.

Fig. 35.   A pair of sweethearts; the Monocoupe 113 and its lovely pilot.

up ribs, also fabric covered; fuel was carried in two tanks, one in each wing root flanking the fuselage. The wing brace struts were heavy chrome-moly steel tubes that were faired to a streamlined shape with a welded steel tube framework that was fabric covered. The split-axle landing gear was made up with two long "oleo-spring" legs that were connected to the upper fuselage; the legs, streamlined by long metal "cuffs", were of fairly wide tread and long travel, this made even a so-so landing feel like a good one. The fabric covered tail-group was built up of welded steel tubing, and the horizontal stabilizer was adjustable in flight. The standard color scheme was the familiar shiny black of the fuselage and the empennage, with bright orange-yellow wings.

Reasonably stable and quite easy to fly, the "Monocoupe 113" racked up a good safety record; this naturally promoted longevity and it was not unusual to see scores of "Monocoupes" all over the country-side, even in the late "thirties" — some are flying even yet! The next development in the Mono Aircraft line was the Wright J5 powered "Monocoach", a fourplace cabin monoplane that will be discussed in the chapter for A.T.C. # 201.

Fig. 36. Bach "Air Yacht" 3-CT-6 with "Hornet" engine in the nose and "Comet" engines in the wings.

The Bach "tri-motor" which was designed as a medium capacity air transport that would be capable of high performance, was an unusual aircraft in many ways. Probably the most noticeable departure from the average was the mixed power combinations that were used; most of the tri-motors up to now had used three matched engines of identical horsepower, but Bach in their several different models, always had one big powerful brute in the nose in combination with "two little helpers" in the wing nacelles. This arrangement did not seem to create any outstanding problems, but the choice of such a set-up could be questioned, especially in case of failure to any one engine. However, with all three engines going, the Bach "Air Yacht" was a craft with plenty of muscle and hard to beat in all-round performance. Built at a time when most everyone else was making the big switch to metal-framed fuselage structures, the Bach was built of wood to deaden the noise and absorb normal operating vibrations; though not up-to-date in comparison to the general trend, the Bach wood structure had considerable merit in it's favor and worked out very well in practice and in principle.

The first air-line to use Bach tri-motored equipment was the West Coast Air Transport, a division of the Union Air Lines system, which operated a line from Seattle to San Francisco through Portland, Oregon. Pickwick Airways, an affiliate of the Pickwick Stage bus system, studied the market carefully for a craft suitable to their particular type of service and picked the high performance of

the Bach "Air Yacht" for their Los Angeles to San Diego route; three round trips were made daily, with first flight at 6:30 a.m. Pickwick put one 3-CT-6 into immediate service and put in an order for several more which were delivered during the year; records seem to indicate that Pickwick Airways had a total of five of the 3-CT-6, which constituted the total production of this particular Bach version. The W.C.A.T. operated with earlier versions of the Bach that were powered with one Wright J5 and two Ryan-Siemens 9 engines, or one Pratt & Whitney "Wasp" and two Ryan-Siemens 9 engines; these were the models 3-CT-2 and 3-CT-4. As of early 1929, Bach had 3 of their tri-motors out in regular service and 7 more were going through the plant in various stages of construction and various stages of modification; like most manufacturers at this time, Bach had several models in development and also had a proposed 35 passenger "air liner" already on the drawing boards.

Morton Bach, designer and developer of the "Air Yacht" series, operated an aircraft repair shop in early 1927 on the old Clover Field in Santa Monica, Calif.; the shop handled normal aircraft repairs, modifications of older type aircraft, and manufactured aircraft for several designers to their specifications. Several special craft were manufactured and Bach also built a few of his own designs, among these were the CS-1 which was a 3 place cabin biplane of wood construction with a 120 h.p. Bristol "Lucifer" engine, and another craft designated the C-4 with a "Hisso" engine; development of the "Air Yacht" series began later in the year. Incorporated towards the end of 1927, Bach operations continued on at Clover Field and moved over to the Los Angeles Metropolitan Airport in Van Nuys, Calif. early in 1929 where a good-sized plant was erected for "Air Yacht" manufacture. Waldo Waterman, well-known "Early Bird" pilot in California flying circles, was airport manager at the time and also came on as Bach's test-pilot, flying most of the certification tests for the various "Air Yacht" versions.

Early versions of the tri-motored 3-CT had Wright J5 and Siemens engine combinations, P & W "Wasp" and Siemens engine combinations, and the 3-CT-6 was powered with one 525 h.p. Pratt & Whitney "Hornet" engine in the nose and two 130 h.p. "Comet" engines in the wing nacelles. The Bach model 3-CT-6 was a high wing monoplane transport with seating for 10 in good comfort and practical elegance, with plenty of power reserve for a rather high performance. The combined power for this version was a combination that totaled 785 h.p. and this translated into short and quick take-offs, excellent climb-out, and a top speed that was well above the average for a craft of this type during these times. Though sparse in actual number built, the Bach "Air Yacht" had a personality that carved itself a memorable niche in aviation history of this period. The type certificate number for the Bach "Air Yacht" model 3-CT-6 as powered with one "Hornet" engine and two "Comet" engines, was issued in February of 1929 and some 5 or more examples of this model were manufactured by the Bach Aircraft Co., Inc. at Van Nuys, Calif. H. J. Heffron was the president; L. Morton Bach was V.P. in charge of engineering; C. W. Faucett was secretary-treasurer; C. W. White was

Fig. 37.   An early version of the Bach with a "Wasp" in the nose and Siemens engines in the wings.

sales manager; and the well-known Waldo Waterman was the chief pilot in charge of test and development.

Listed below are specifications and performance data for the Bach "Air Yacht" model 3-CT-6 as powered with one "Hornet" engine and 2 "Comet" engines; length overall 36'10"; hite overall 9'9"; wing span 58'5"; wing chord at root 132"; wing chord at tip 96"; total wing area 512 sq.ft.; airfoil "Clark Y"; wing tapered in planform only; wt. empty 4739; useful load 3261; payload with 200 gal. fuel was 1750 lbs.; gross wt. 8000 lbs.; max. speed 154; cruising speed 126; landing speed 60; climb 950 ft. first min. at sea level; service ceiling 18,000 ft.; gas cap. 200 gal.; oil cap. 18 gal.; cruising range at 40 gal. per hour was 600 miles; price at the factory field was $39,500.

The fuselage framework was built up of 6 wooden longerons that were bolted together with steel plates and fittings; the completed framework was covered with plywood veneer and an extra outer covering of fabric for smoothness and added strength. The cabin was fitted with shatter-proof glass and arranged with 8 passenger seats; the pilot and co-pilot were seated in a separate section up front and the baggage compartment of 50 cu. ft. capacity was to the rear of the cabin and was accessible from the inside or the outside. The wing framework was built up of spruce

Fig. 38.   The 3-CT-6 offered rapid and comfortable service on the West Coast.

*Fig. 39. The cabin interior of the Bach "Air Yacht";
wood paneling was in vogue.*

and plywood box-type spar beams, with plywood wing ribs that were reinforced with
spruce diagonals and cap-strips; the leading
edge was covered with plywood and the completed framework was fabric covered. The
wing was braced with heavy chrome-moly
steel tubes that were faired to an Eiffel 376
section for stability and added lift; the engine
nacelles were built into a truss framework
with the wing brace struts. The fuel tanks
were mounted in the root ends of each wing
half, and the fuel flowed through a selector
valve in the cockpit. The plywood covered
tail-group was built up in a wood framework;
the fin was built integral to the fuselage and
the horizontal stabilizer was adjustable in
flight. The split-axle landing gear of 18 foot
tread was fastened from the fuselage to the
engine nacelles and used rubber-hydraulic
shock absorbers; wheels were 36x8 and Bendix brakes were standard equipment. The
next development in the Bach "Air Yacht"
series was the 3-CT-8; for a discussion of
this version see the chapter for A.T.C. # 172
in this volume.

Listed below are Bach 3-CT-6 entries that
were gleaned from various records; this listing appears to be the total production of
this model.

C-388 ; 3-CT-6 (# 5) Hornet & 2 Comet.
C-302E; 3-CT-6 (# 6) Hornet & 2 Comet.
C-539E; 3-CT-6 (# 7) Hornet & 2 Comet.
C-850E; 3-CT-6 (# 9) Hornet & 2 Comet.
C-219H; 3-CT-6 (# 10) Hornet & 2 Comet.

Fig. 40. The Arrow "Sport" with LeBlond 60 engine; note lack of wire bracing.

By the very nature of it's make-up and disposition, the Arrow "Sport" was a real "chum", an airplane that was easy to "buddy-up" with and a pleasure to own and to fly; it's owners, scattered over the countryside, were known as a legion of loyal fans and enthusiastic boosters. Friendly and eager, the little "Sport" became a favorite with "Sunday pilots" and was even used with great success in pilot training. The "Sport" was an open cockpit biplane seating two side-by-side, and therein lie it's great appeal. Many people favored the companionship of the side-by-side seating, especially in the freedom of an open cockpit; for those so inclined, the Arrow "Sport" was the answer. Arrow Air-craft proudly announced a prediction during the 1929 Air Show season, that the "Sport" would surely go straight to the heart of America; from the history left behind by this airplane, we could say that it certainly did, as predicted.

As it is shown here in various views, we can see that this little craft was blessed with a handsome configuration that made good sense; the tapered wing panels of cantilever construction were still quite novel and gave it that look and carriage of a hot-performing "pursuit ship". The full-cantilever principle, as used in the wings, actually did away with the usual "landing wires" and "flying wires", but

"N type" wing struts were still used to tie the wing panels together at their outer portion. Hard to go unnoticed was the unusually wide tread landing gear that was of the long-leg split-axle type; the jolts from taxiing or landing, were ironed out by two spools of rubber shock-cord (later called bungee-cord). Built robust in all respects, these little ships were quite rugged and could absorb a good amount of rough handling, but there was one irrateable feature present. This was the lower wing and ailerons, which were very low to the ground, and had the habit of getting fouled up in the tall grass, or hay, that most always prevailed on pasture-airports of this early period. The Arrow "Sport" A2-60, as pictured here, was powered with the 5 cylinder LeBlond engine of 60 h.p.; a combination that flew and handled quite well, with a rather sprightly performance.

The prototype version of the Arrow "Sport" was built in 1926; based on designs by Swen S. Swanson, it was an open cockpit biplane that seated two side-by-side, and also had the tapered cantilever wing panels. Oddly enough, it had no interplane struts to tie the wing panels together; the upper wing was perched atop a pair of center-section struts and the lower wing panel was bolted directly to the bottom of the fuselage frame. Naturally, both wing panels were built in one con-

Fig. 41. The "Sport" was easy to "buddy-up" with and a real pleasure to fly.

tinuous piece. A. H. G. "Tony" Fokker tried this unbraced wing cellule on one of his early "D-7" type of World War 1, and although this method was thoroughly practical and quite safe, it was frowned upon to say the least! Likewise, this unstrutted feature on the early "Sport" was greeted with some skepticism; needless to say, they added the interplane struts to forestall uneasiness in prospective buyers. Yet, the interplane struts were offered as optional equipment, even in the latter part of 1928! The diminutive prototype of 1926 was powered with a 3 cylinder Anzani (French) engine of 35 doubtful horsepower, and though quite lively with only one person aboard, it had a tendency to bog down a bit with a full gross load. Swen S. Swanson who had designed the "Sport", and other early Arrow models, was an extremely modest man of very great talent; his next effort was the design and development of the Kari-Keen "Coupe" which reflects a designing style borrowed from the Arrow "Sport" and also carried into such later designs as the graceful Swanson "Coupe", the Fahlin "Coupe", and the "Plymacoupe".

The early Arrow "Sport" of 1928 was first offered with the 6 cylinder Anzani engine of some 60 h.p., and also with the Detroit "Air Cat" engine, but the 5 cylinder LeBlond engine of 60 h.p. soon became the standard powerplant installation. One example of the "Sport" had been powered with the 5 cylinder Velie engine for test, and later in 1929-1930, the "pursuit" version of the "Sport" was powered with the 5 cylinder Kinner K5 engine of 90-100 h.p. The Arrow

"Sport" A2-60, which was powered with the 5 cyl. LeBlond engine, received it's approval for a type certificate number in February of 1929, and it was manufactured by the Arrow Airplane & Motors Corp. at Havelock, Nebraska. Mark M. Woods was president; J. B. "Johnnie" Moore was V. P., chief engineer, and also in charge of production; F. Pace Woods was sales manager, and the veteran Joe Lowrey was chief pilot.

Listed below are specifications and performance data for the LeBlond powered Arrow "Sport" model A2-60; length overall 19'3"; hite overall 7'5"; span upper 25'10"; span lower 25'4"; chord both 62" at root and 38" at tip (M.A.C. 48"); airfoil "Eiffel 385"; wing area upper 96 sq.ft.; wing area lower 87 sq.ft.; total wing area 183 sq.ft.; wt. empty 811; useful load 459; payload 187; gross wt. 1270 lbs.; max. speed 98; cruising speed 82; landing speed 35; climb 680 ft. first min. at sea level; service ceiling 14,000 ft.; gas cap. 16 gal.; oil cap. 2-3 gal.; cruising range at 4½ gal. per hour was 280 miles plus; price at the factory field was $3131 and later raised to $3485. As was usually the case in all new designs, the prototype LeBlond powered "Sport" was somewhat smaller and a good bit lighter; it was quite a bit more sprightly in the consequence. Some of the later versions of the A2-60 were a good bit heavier than th figures shown, and performance suffered for it in some respects.

The robust fuselage framework was built up of welded steel and chrome-moly steel tubing in a Warren "truss" form; the framework was faired to shape with steel tube

Fig. 42. An early example of the Arrow "Sport" with LeBlond engine.

fairing strips and then fabric covered. A baggage compartment, with an access door on the side, was located in back of the cockpit. The cantilever wing framework was built up of heavy spruce box-type spars and spruce and plywood truss-type ribs, also fabric covered. The wings, which were built into one piece, were tapered in plan-form and section. The "N type" wing struts of streamlined steel tubing, were standard equipment on all "Sports" from 1929 and on, but even in late 1928, these struts were offered as optional equipment! The unusually wide split-axle landing gear had a wheel tread of 78 inches; the wheels were 26x4 and the shock-absorbers were two spools of rubber shock-cord. The fabric covered tail-group was also built up of welded steel tubing; the horizontal stabilizer was adjustable in flight. The fuel tank of 16 gallon capacity was mounted in the upper portion of the fuselage, just ahead of the cockpit. A total of 100 or so of the LeBlond powered A2-60 "Sport" were built, and at least 40 of these were still in active operation in 1939, some 10 years later; even a few of these have weathered the ravages of time and have been rebuilt to fly again! The next development in the Arrow "Sport" series was the Kinner powered "Sport Pursuit", this model will be discussed in the chapter for Group 2 approval numbered 2-110 which was issued in August of 1929.

Fig. 43. A restored Arrow Sport, rebuilt in 1961 and still flying.

Fig. 44. The Travel Air Model A-6000-A offered high performance for business or pleasure.

There is actually very little to get excited about in a large cabin monoplane; that is, to the average flyer at least, who would lean more towards the "sport-type" biplane, but here we have an exception because the "Wasp" powered "Travel Air" monoplane, model A-6000-A, really was an exciting airplane. A good-looking airplane with plush appointments to suit the most impeccable; with comfort enough to please even the most fussiest, and an exhilerating performance that would even warm the heart of a "pursuit-pilot". Basically the same as the 6 place model 6000 that was powered with a Wright J5 engine, the "Wasp" powered A-6000-A was developed largely in answer to the urgent demands circulated from the various air-lines for a single-engined cabin monoplane with extra performance enough to help hold a competitive advantage with higher cruising speeds and a faster climb-out. An on-coming need for a performance that would enable these lines to advertise and to offer faster and more frequent service. A high-performance cabin monoplane such as this, would also attract private-ownership amongst a certain class of people that could afford and would enjoy a luxurious airplane that also had plenty of pep and spice to offer in the way of good handling characteristics and above average performance; a pride of owner-ship that appealed to such as Wallace Beery, noted movie-actor who was one of the film-colony's most avid aviation enthusiasts. Beery was a fine pilot

and loved a good airplane; his comments on the A-6000-A were complimentary to say the least. Central Air Lines, Overland Airways, Northwest Airways, and others, were using the A-6000-A on scheduled routes with great success. Progressive business firms were also attracted to this airplane and it was soon used by such as, the Phillips Petroleum Co, the Ruckstell Corp., and the Hal Roach film studios in Hollywood.

An interesting story heard long ago, tells of a customer who was quite pleased with the demonstration he received, and immediately put in his order for a shiny new A-6000-A. Though this customer was not a suspicious soul by nature, he did like to see the inside details of airplanes as they were being built. Consequently, he spent a week in Wichita as an interested by-stander while his new purchase, A "Travel Air" monoplane with "Wasp" engine, progressed by the usual stages through the Travel Air factory. Upon it's completion and test, he flew it home, happy in the knowledge that he had seen it built from the start to the finish.

The model A-6000-A, slightly larger but basically typical to the earlier model 6000, was a strut-braced high wing cabin monoplane of rather large proportions, with a handsome interior that had ample room and comfort for seating six. The big "Wasp" engine was muffled well, and the cabin walls were sound-proofed and insulated to keep extreme temperatures, noise, and vibrations, at a fairly low

Fig. 45. The Model A-6000-A was powered by a 450 h.p. "Wasp" engine.

level. The performance of this big beauty was certainly very adequate and it's agility in the air was quite surprising for a craft of this size and type. Altogether, some 25 examples of this model were built; some 9 of these were 7 place airplanes on duty with various air-lines. Travel Air production at the beginning of 1929 was averaging about 8 planes a week, in the various types; shortly after, factory space was more than doubled and the production potential, with 3 shifts operating around the clock, went up to 30 complete airplanes per week. Introduced in the latter part of 1928, the type certificate number for the "Wasp" powered model A-6000-A was issued in February of 1929 and it was manufactured by the Travel Air Company at Wichita, Kansas; an air-minded metropolis that was surely becoming the air capitol of America.

Listed below are specifications and performance data for the "Wasp" powered "Travel Air" monoplane model A-6000-A; length overall 31'2"; hite overall 9'3"; wing span 54'5"; wing chord 84"; total wing area 340 sq.ft.; airfoil "Clark Y-15"; wt. empty 3225; useful load 2025; payload 995; gross wt. 5250 lbs.; empty weight does not include landing lights, flares, battery or generator, when these are added the payload suffers accordingly; max. speed 140; cruising speed 120; landing speed 60; climb 1,000 ft. first min. at sea level; service ceiling 18,000 ft.; gas cap. 130 gal.; oil cap. 10 gal.; range 680 miles; price at the factory field was $18,000. The fuselage framework was built up of welded chrome-

moly steel tubing, faired to shape with formers and fairing strips and then fabric covered. Cabin walls were sound-proofed and insulated with an upholstered interior; seats were easily removable to provide clear floor space for hauling bulky cargo. On the 7 place version for air-line work, there was an extra seat in the extreme rear of the cabin section. The semi-cantilever wing framework was built up of spruce box-type spars and spruce and plywood truss-type ribs, also fabric covered. The fuel supply was carried in two gravity-feed wing tanks of 65 gallon capacity each; the ailerons were of the Freise "balanced-hinge" type, an aerodynamical "balance" that was far superior to the old "projecting horn" type. The fabric covered tail-group was also built up of welded chrome-moly steel tubing and sheet; the horizontal stabilizer was adjustable in flight. The wide tread landing gear was of the outrigger type, and used "Aerol" shock absorber struts (air-oil); tail-wheel was full-caster. A large baggage compartment was provided and on the later models, there were lavatory facilities. Standard color scheme was a shiny black fuselage with bright orange-yellow wings, but other color combinations were of course optional. A metal propeller, inertia-type engine starter, and wheel brakes were standard equipment; night-flying equipment, which included storage battery and engine-driven generator, were optional. The "Wasp" powered model A-6000-A was also offered on twin "floats" as the model SA-6000-A, see chapter for A.T.C. # 175 in this volume. The next

Fig. 46. The Travel Air A-6000-A operated on frequent schedules with Northwest Airways.

"Travel Air" development was the Wright J6 powered B-6000 (S-6000-B) which is discussed here in this volume, see chapter for A.T.C. # 130.

Fig. 47. The General "Aristocrat" Model 102-A was neat and efficient; it carried three in comfort.

The General "Aristocrat" model 102-A was a neat and trim looking light cabin monoplane that incorporated many excellent features of construction and design into a well balanced and efficient configuration; a pleasant configuration that was somewhat reminiscent of the form that was popularized by Bellanca. Seating three with ample room and comfort, the 102-A had plenty of muscle which showed up in good performance and manuverability, with a utility of operation that was primarily leveled at the private-owner or the small business man; especially those who had need for a personal air-taxi that would be ready to go on call. Powered with the 7 cylinder Warner "Scarab" engine of 110 h.p., the "Aristocrat" asked no favors from landing-strip or pilot, and operated happily under any and all conditions. One of the earliest examples in the model 102-A was donated to Com. Richard E. Byrd for use on his Antarctic expedition, and a fleet of 8 "Aristocrats" was bought by the General Tire & Rubber Co. that were used on a 50,000 mile tour of the nation during the latter part of 1929; a nation-wide tour that was to help demonstrate the reliability of air-travel to all points, and under all manner of conditions. A good demonstration to the feasability of travel by air to all parts of the nation was certainly accomplished, but it is un-derstandable that the prime object of this tour was to test various new rubber products that were developed for use in aircraft. Exhibited at the Cleveland Air Show in late 1929, the General "Aristocrat" attracted much attention from the average flying-folk and several were sold for personal use, others were soon being used by flying-schools to teach cabin monoplane flying techniques. A good number of the model 102-A were built in a rather brief period of manufacture, but they were scattered pretty thin across the country and remained quite rare.

The General Airplanes Corp. was formed in Buffalo, New York in about June of 1928; among the several experimental aircraft that were first developed, was a folding-wing light cabin monoplane which was known as the model "102". Another craft was the model "101", a large twin-engined high wing cabin monoplane proposed for photographic use and aptly called the "Surveyor"; further development of this design was to be modified into a twin-engined or tri-engined passenger-carrying version called the "Pullman". Several other passenger-carrying transports were proposed, among these was an early "Aristocrat" version that was finally developed intó the model 102-A as described here. There was rumored to have been a 3 place cabin biplane that was fashioned along the "Aristocrat" lines, and accounts of this illusive craft definately say "biplane", but no information or data could be found to substantiate the existance of such an airplane. The type certificate number for the General "Aristocrat" model 102-A as powered with the Warner engine, was issued in March of 1929, and some 26 or more of this model were manufactured by the General Airplanes Corp. at Buffalo, New York. C. S. Rieman was the president; L. A. Listug was V. P. and treasurer; A. Francis Arcier was V. P. in charge of engineering; and George A. Townsend was the secretary and sales manager. The engineering staff was headed by A. Francis Arcier who had been an engineer for Handley-Page in Britain during the first World War, and also an engineer for A. H. G. "Tony" Fokker (Atlantic Aircraft) here in the U.S.A.

Listed below are specifications and performance data for the General "Aristocrat" model 102-A as powered with the 110 h.p. Warner "Scarab" engine; lenght overall 25'4"; hite overall 7'7"; wing span 36'4"; wing chord 72"; total wing area 195.4 sq.ft.; airfoil "GAC-500"; wt. empty 1327; useful load 783; payload with 40 gal. fuel was 350 lbs.; gross wt. 2110 lbs.; max. speed 109; cruising speed 90; landing speed 45; climb 647 ft. first min. at sea level; service ceiling 14,370 ft.; gas cap. 40 gal.; oil cap. 4 gal.; cruising range 490 miles; price at the factory field was an average of $6,000. The fuselage framework was built up of welded chrome-moly steel tubing, faired to shape with formers and fairing strips and then fabric covered. The pilot sat in front on a bucket-type seat, and the two passengers sat on a bench-type seat in back; windows were of shatter-proof glass and could be rolled up and down. There was a wide door on each side for cabin entry, and a baggage compartment behind the rear seat; the rear seat was removable for hauling cargo. The semi-cantilever wing framework was built up of heavy-sectioned solid spruce spars with spruce and plywood built-up wing ribs; the leading edges were covered to the front spar with "dural" sheet to preserve the airfoil form and the completed framework was fabric covered. Two gravity-feed fuel tanks were mounted in the root ends of each wing, and ailerons were of the aerodynamically balanced "offset hinge" type. The fabric covered tail-group was built up of welded chrome-moly steel tubing and both rudder and the elevators were aerodynamically balanced; the fin was ground adjustable and the horizontal stabilizer was adjustable in flight. The split-axle landing gear of 84inch tread was built up of two forged aluminum alloy cantilever legs that were faired to a streamlined shape by sheet metal "cuffs"; the shock absorbers were rubber "donut rings" in compression, individual wheel brakes and a swiveling tail-wheel provided excellent ground manuvering. A novel engine starting crank was in the forward cabin within easy access to the pilot, and the engine was well muffled by a large volume collector-ring that permitted normal conversation in the cabin at all times. Flight controls were of the joy-stick type and a dual set was available. Wheel brakes, engine starter, navigation lights, and a metal propeller were standard equipment. The next develop-

Fig. 48. The "Aristocrat" 102-A in Canada; note shape of fuselage, which was designed to provide lift.

Fig. 49.   The twin-engined General "Surveyor", an early experiment.

ment in the "Aristocrat" monoplane was the model 102-E as powered with the 5 cylinder Wright J6 engine of 165 h.p.; discussion of this version will be in the chapter for A.T.C. # 210.

Listed below are "Aristocrat" model 102-A entries that were gleaned from various records; this list is not quite complete, but it does show the bulk of this model that were built.

X- 6788; 102 (# 1) 3 pl. fold-wing mono.

X-7942; 102 (# 3) Warner, 3 pl. mono.

X-216E; 102-A(# 1) Warner 110

C-8650; 102-A(# 2) Warner 110

C-8651; 102-A(# 3) Warner 110

C-8652; 102-A (# 4) Warner 110

C-8653; 102-A (# 5) Warner 110

C-8654; 102-A(# 6) Warner 110

C-8655; 102-A(# 7) Warner 110

NC-8656; 102-A(# 8) Warner 110

C-8657; 102-A)# 9) Warner 110

C-8658; 102-A (# 10) Warner 110

C-8659; 102-A(# 11) Warner 110

C-9445; 102-A(# 12) Warner 110

C-9446; 102-A(# 13) Warner 110

C-9447; 102-A(# 14) Warner 110

C-9448; 102-A(# 15) Warner 110

C-9449; 102-A (# 16) Warner 110

NC-275H; 102-A(# 17) Warner 110

NC-276H; 102-A(# 18) Warner 110

NC-277H; 102-A(# 19) Warner 110

NC-278H; 102-A(# 20) Warner 110

NC-279H; 102-A(# 21) Warner 110

NC-454K; 102-A(# 35) Warner 110

NC-455K; 102-A(# 36) Warner 110

NC-456K; 102-A(# 37) Warner 110

NC-457K; 102-A (# 38) Warner 110

NC-458K; 102-A(# 39) Warner 110

Serial # 4-6-10-11-12-13-14-15 were "General Tire Fleet".

Serial # 23-24-25-26-40-41-42 were 102-E on A.T.C. # 210.

Serial # 22-27-28-29-30 were 102-F with Continental A-70.

Serial # 31-32-33-34 are unknown as to model or type.

Fig. 50.   The Command-Aire 3C3-A had a 110 h.p. Warner engine; a popular combination.

The familiar OX-5 powered "Command-Aire" biplane had been selling quite well up to now, and at least 100 had been built by this time, but the supply of Curtiss OX-5 engines was uncertain and beginning to run short so Command-Aire, Inc. like many another manufacturer began developing new models that were powered with the medium horse-power air-cooled "radial" engines; among the first of these to appear was the model 3C3-A which was powered with the 7 cylinder Warner "Scarab" engine of 110 h.p. Only slightly modified from the basic 3C3 design, the 3C3-A was also a three place open cockpit biplane of the familiar "Command-Aire" configuration. Designed to perform the varied needs of the average small-operator, the 3C3-A was a pleasant craft with an efficient performance that was delivered in good economy. Altogether, some 20 of the model 3C3-A were built and being spread rather thin about the country, they remained a rather rare type. The type certificate number for the Warner powered 3C3-A was issued in March of 1929 and it was manufactured by the Command-Aire, Inc. at Little Rock, Arkansas. Company officers were changed and shuffled around a bit since late 1927, and as of January 1929 were as follows: R. B. Snowden

Jr. was president; Chas. M. Taylor was V.P.; and Major J. Carroll Cone was in charge of sales. Albert Voellmecke had been with "Command-Aire" since it's inception in 1927 and was still the chief engineer; the devil-may-care Wright "Ike" Vermilya was chief pilot in charge of test and development.

Listed below are specifications and performance data for the 3 place model 3C3-A as powered with the 110 h.p. Warner engine; length overall 24'10"; hite overall 8'4"; wing span upper and lower 31'6"; chord both 60"; area upper 169 sq.ft.; area lower 134 sq.ft.; total wing area 303 sq.ft.; airfoil "Aeromarine 2A"; wt. empty 1282 (1357); useful load 865 (815); payload 365 (360); gross wt. 2147 (2172) lbs.; max. speed 105; cruising speed 92; landing speed 36; climb 525 ft. first min. at sea level; service ceiling 13,350 ft.; gas cap. 43-48 (43) gal.; oil cap. 5 gal.; range at 7 gal. per hour was approx. 520 miles; price at the factory field ranged from $5600 to $4675; figures above in brackets were as listed for latest models as of 1930. The fuselage framework was built up of welded chrome-moly steel tubing, lightly faired to shape with wood fairing strips and fabric covered. The entire top section of the fuselage was covered with metal panels; quite novel was the metal turtle-

Fig. 51. The Warner-powered 3C3-A was a pleasant craft with good performance.

Fig. 52. The 3C3-A was ideally suited for general purpose service.

back that was quickly removable for inspection and maintenance to various components in the rear section of the fuselage. With the use of metal engine-cowling, metal cockpit cowling, and the metal turtle-back panel, actually less than half of the fuselage frame was fabric covered. The wing framework was built up of solid spruce spars and spruce and plywood girder-type ribs, completed framework was also fabric covered. The slotted-hinge ailerons were of a welded chrome-moly steel tube framework, also fabric covered. Actuation of all controls was by push-pull tubes and bellcranks; no cables nor pulleys were used. The fabric covered tail-group was built up of welded chrome-moly steel tubing; the horizontal stabilizer was adjustable in flight. The extra wide center-section cabane was built up of the usual "N type" struts, with an extra strut attached at the top wing and fastened to a fitting on the lower fuselage longeron; this method of bracing eliminated cross-wires in the center-section bay and offered clear and uncluttered vision to the front. The split-axle landing gear used two spools of rubber shock-cord to absorb the bumps; wheel tread was 80 inches. Both cockpits were quite roomy, comfortable, and well upholstered; the baggage compart had a capacity for four cubic feet. The fuel tank was high up in the fuselage, ahead of the front cockpit. A fuel-level gauge, which was fastened to the tank and protruded through the top-section of the cowling, was visible from either cockpit. (Dash-board fuel gauges had not yet come to popular use). The sales and distribution for the various models in the "Command-Aire" series was now handled by the Curtiss Flying Service through most of 1929. A model 3C3-A on Edo "floats" (serial # W-79) was

issued a Group 2 approval numbered 2-137 with an allowable gross load of 2305 pounds. The Warner powered 3C3 as a 2 place trainer was the model 3C3-AT, see chapter for A.T.C. # 151 in this volume. The next development in the "Command-Aire" biplane series, after the 3C3-A, was the Siemens-Halske powered model 3C3-B; see chapter for A.T.C. # 120 in this volume.

Listed below are model 3C3-A entries that were gleaned from the Aeronautical Chamber of Commerce aircraft register; as far as is known, these were the only examples built. With 19 "Command-Aire" aircraft unaccounted during this production period, it is quite possible that there may have been a few more 3C3-A than are listed here.

NC-514; 3C3-A (# W-53) Warner 110, later as 2 pl. model 3C3-AT

C-543; 3C3-A (# W-57) Warner 110

NC-545; 3C3-A (# W-59) Warner 110

NC-608E; 3C3-A (# W-66) Warner 110

NC-609E; 3C3-A (# W-68) Warner 110

C-910E; 3C3-A (# W-73) Warner 110

C-911E; 3C3-A (# W-75) Warner 110

C-915E; 3C3-A (# W-77) Warner 110

C-916E; 3C3-A (# W-79) Warner 110

C-918E; 3C3-A (# W-80) Warner 110

C-923E; 3C3-A (# W-83) Warner 110

NC-924E; 3C3-A (# W-84) Warner 110

C-927E; 3C3-A (# W-85) Warner 110

C-929E; 3C3-A (# W-87) Warner 110

C-930E; 3C3-A (# W-89) Warner 110

NC-931E; 3C3-A (# W-90) Warner 110

NC-979E; 3C3-A (# W-117) Warner 110

C-980E; 3C3-A (# W-118) Warner 110

NC-981E; 3C3-A (# W-119) Warner 110

NC-982E; 3C3-A (# W-120) Warner 110

all airplanes listed here were 3 pl.

NC-330V; 3C3-A (# W-139) Warner 110

Fig. 53. The Wallace "Touroplane" B had a 100 h.p. Kinner engine; it was built by American Eagle as the Model 330.

The Wallace "Touroplane" was a comely little cabin monoplane with seating for a pilot and two passengers, another example of a new trend that was slowly forming to provide more of the indoor comforts for the private-owner while engaged in flying. It seems the industry, following a pattern that was set long ago, was now slowly beginning to realize that not everyone enjoyed the drafty thrill of the open cockpit airplane. Designed primarily for the private-owner or the small fixed-base operator, who usually had need for a three place airplane, the "Touroplane" was a compact strut-braced high wing cabin monoplane that seated three in good comfort, and delivered a good performance in the best economy. Arranged so that the pilot and one passenger sat up forward with good visibility, and the second passenger sat in back, sideways, on a folding seat. Dual controls were available to handle student training, and the rear seat could be folded or quickly removed to provide 30 cubic feet for hauling light cargo to the amount of 390 lbs. The configuration and the construction of the "Touroplane" was quite normal to the general pattern, but an added feature was the folding wings, which could be quickly folded back by one man to a width of just over 12 feet; an area of 13x25 feet was ample space for tie-down or storage.

Designed by Stanley Wallace, who already

had 13 years of aircraft design and engineering experience up to this time, the "Touroplane" had been in development since early in 1928, and was being well recieved in flying circles about the mid-west as the type of airplane that would inevitably replace the 3 place open cockpit biplane. The prototype airplane, as shown here, was powered with the 6 cylinder Anzani (French) engine of 80 h.p., and was also offered with the 7 cyl. Ryan-Siemens engine, or the popular Curtiss OX-5, but subsequent airplanes were being powered with the 5 cyl. Kinner engine and it soon became the standard power installation. The second "Touroplane" was powered with the Kinner engine and was delivered to a dealer on the western coast; the balance of the year was spent in tests and further development of the type. The type certificate number for the Wallace "Touroplane" B, as powered with the Kinner K5 engine of 90-100 h.p., was issued in March of 1929 and it was manufactured by the Wallace Aircraft Co. in Chicago, Illinois. During the early part of 1929, negotiations were under way to absorb the Wallace firm into the American Eagle Aircraft Corp. of Kansas City, Kansas; Stanley Wallace was of course retained in the engineering capacity to design and develop new models, but manufacturing facilities were removed from Chicago to Kansas. Hardriding E. E.

Fig. 54. The "Touroplane" was designed to replace the open cockpit biplane in general service.

Fig. 55. The Wallace "Touroplane" prototype had an 80 h.p. Anzani engine.

Porterfield Jr. was the president; D. L. Chick was V.P. and treasurer; Stanley Wallace was chief engineer for the Wallace division; Daniel Noonan, formerly of Alexander Aircraft, was production manager; and Larry Ruch was chief pilot in charge of test and development.

Listed below are specifications and performance data for the Wallace "Touroplane" B (American Eagle 330) as powered with the 5 cyl. Kinner K5 engine of 90-100 h.p.; length overall 23'10"; hite overall 7'6"; wing span 37'; wing chord 70"; total wing area 212 sq.ft.; width wings folded 12'4"; wt. empty 1320; useful load 780; payload 390; gross wt. 2100 lbs.; max. speed 105; cruising speed 90; landing speed 43; climb 630 ft. first min. at sea level; time to 10,000 ft. is 35 min.; service ceiling 14,000 ft.; gas cap. 32 gal.; oil cap. 4 gal.; cruising range at 5.5 gal. per hour was 500 miles; price at the factory field was $5795. The production version of the "Tour-

oplane" B (American Eagle 330) outweighed the prototype by some 600 lbs. which was a natural phenomenon in itself, but we must mention that the prototype only carried 1½ passengers! (Or, 3 passengers at 143 lbs. each). The fuselage framework was built up of welded chrome-moly steel tubing, faired to shape and fabric covered. The cabin section was insulated and upholstered neatly, there were safety glass windows, and a baggage compartment was provided in back of the rear seat. The rear seat could be folded or quickly removed to provide 30 cubic feet of clear cargo space amounting to some 390 lbs. A fairly large cabin door and a convenient step was provided on each side for easy entrance and exit. The wing framework was built up of solid spruce spar beams and spruce and plywood truss-type ribs, and then was fabric covered. Hinged panels allowed folding of the wings to a width of just over 12 feet,

Fig. 56. The "Touroplane" offered comfortable flying for three.

Fig. 57. Wings of the "Touroplane" folded for space-saving storage.

an area of 13x25 feet would be sufficient for storage. The fuel tank was built into the center-section panel of the wing, which also formed the roof of the cabin. The fabric covered tail-group was built up of welded chrome-moly steel tubing, and the horizontal stabilizer was adjustable in flight. The split-axle landing gear was of the normal 3-member inverted tripod type, using air-oil shock absorber struts; the tail-skid was of the leaf-spring type, with a removable hardened shoe. An exhaust gas collector-ring, which muffled exhaust noise to a rather low level, was faired into the front of the engine. Wheel brakes, metal propeller, and engine starter, were offered as optional equipment. In an effort to capture more of the business-flying market, which seemed more lucrative, American Eagle enlarged the "Touroplane" to seat 4 places and installed more power; thus, the next development in this series, bar one or two experimentals, was the American Eagle D-430 which was powered with the Wright J6-5-165, see A.T.C. # 301. The next development in the American Eagle biplane was the Kinner powered A-129; see chapter for A.T.C.

124 in this volume.

Listed below are "Touroplane" B entries that were gleaned from the Aeronautical Chamber of Commerce aircraft register:

X-4253; Touroplane B (# 7) Anzani 80, prototype.

-6842; Touroplane (# 8) Kinner, & Mac Clatchie "Panther".

NC-7742; Touroplane (# 9) Kinner

C-7740; Touroplane (# 10) OX-5, also Kinner K5.

-7987; Touroplane (# 11) OX-5

NC-276K; Touroplane (# 12) Kinner K5

NC-744K; Touroplane (# 14) Kinner K5

NC-571H; Touroplane (# 15) Kinner K5

NC-584H; Touroplane # 16) Kinner K5

NC-209N; Touroplane (# 17) Kinner K5

NC-211N; Touroplane (# 18) Kinner K5

NC-566H; Touroplane (# 19) Kinner K5

NC-580H; Touroplane (# 20) Kinner K5

NC-590H; Touroplane B (# 21) Kinner K5

Numbers 1 through 6 were other Wallace developments before the introduction of the "Touroplane" series; at least one of these was a 3 place open cockpit biplane with the Curtiss OX-5.

Fig. 58.   The Command-Aire 3C3-B had a 7 cylinder Siemens Halske engine; it is shown here at the factory.

Manufacturers of the popular three-place open cockpit biplane type had one decided advantage in their favor, and that was the apparent ease of bringing out new models by using the design of a basic "type", almost intact, and just modifying the configuration or structure slightly to suit the installation of several different powerplants. Many times a modification of this sort would only require a new engine mount and new engine cowling to suit. Some manufacturers, by this time, were already able to stretch the basic design of a "type" into at least a half-dozen different models. Likewise; a good number of the manufacturers of open cockpit biplanes, during this period, managed to have at least one model that was powered by the "Siemens" engine, and "Command-Aire" was certainly no exception. They had two.

The model 3C3-B, as discussed here, was an all-purpose 3 place open cockpit biplane of typical "Command-Aire" configuration and construction; it was typical to current production models in most all respects except for the engine installation which in this case was the 7 cylinder Siemens-Halske (SH-14) engine of 105-113 h.p. The "Siemens-Halske" was an air-cooled "radial" engine of German manufacture that was now distributed here in the U.S.A. by K. G. Frank as the "Yankee-Sie-

mens" in 5 cyl., 7 cyl., and 9 cyl., models. The performance in this Siemens powered combination (3C3-B) was really quite good and all of the good characteristics of the "Command-Aire" biplane were retained and perhaps somewhat emphasized; but, the model did not sell in any great number, and remained a very rare type. The engine installation actually could have been this airplane's undoing because of the often uncertain supply of parts and accessories from abroad, and then too, American built air-cooled "radials" in this general power range were coming along nicely by now and these would naturally be preferred. Command-Aire also had another rare type in their early line-up; this was a 3 place open cockpit biplane that was powered with the 9 cylinder "Walter" (Czechoslovakian) engine of 120-135 h.p. This was the model 4C3, as shown here. One of the 4C3 type with "Walter" engine was delivered to a distributor in the state of Washington, and this was the only known example (X-70E, serial # W-51).

The type certificate number for the model 3C3-B, as powered with the 7 cylinder Siemens-Halske engine, was issued in March of 1929 and only about 3 of this type were built. The models 3C3-B and the 4C3 were both manufactured by Command-Aire, Inc. at

Fig. 59. A rare Command-Aire version, the 4C3, powered by a Walter engine.

Little Rock, Arkansas. R. B. Snowden Jr. a devout air-enthusiast for a number of years, was president; Albert Voellmecke, well experienced in aircraft design both here and abroad, was the chief engineer; Wright "Ike" Vermilya, the fearless one, who used to ride the top of the fuselage while a "Command-Aire" was flying by itself to demonstrate it's stability, was the chief pilot in charge of test and development; Major John Carroll Cone was sales manager.

Listed below are specifications and performance data for the model 3C3-B as powered with the Siemens-Halske engine of 105-113 h.p.; length overall 24'7"; hite overall 8'6"; span upper and lower 31'6"; chord both 60"; wing area upper 169 sq.ft.; wing area lower 134 sq.ft.; total wing area 303 sq.ft.; airfoil "Aeromarine 2A"; stagger between upper & lower wings 6"; dihedral lower wing 2½ deg.; wt. empty 1325; useful load 808; payload 350; gross wt. 2133 lbs.; max. speed 102-106; cruising speed 90-92; landing speed 36; climb 525-550 ft. first min. at sea level; service ceiling 13,350-14,000 ft.; gas cap. 43 gal.; oil cap. 4 gal.; approx. range 500+ miles; price at the

factory field was about $5500.

The fuselage framework was built up of welded chrome-moly steel (molybdenum) tubing in a truss-type form, no tie-rods or wire bracing was used; the framework was lightly faired to shape and fabric covered. The wing framework was built up of heavy-sectioned solid spruce spars and a combination of spruce and plywood girder-type wing ribs; completed framework was fabric covered. The two upper wing panels were joined together at the center-line; there was no separate center-section panel, as on most biplanes. The fuel tank of 43 gallon capacity was mounted high up in the fuselage, just ahead of the front cockpit. All other construction details were identical to the model 3C3-A; refer to chapter for A.T.C. # 118 in this volume. The model 3C3-B was wired for navigation lights; a Hamilton wood propeller was standard equipment, but a metal propeller and inertia-type engine starter were optional. The landing gear tread was 78", shock-absorber spools were streamlined with metal "cuffs", and wheels were 28x4. For the next development in the "Command-Aire" biplane series, see chapter for A.T.C. # 150 in this volume, which discusses the OX-5 powered 2 place "Trainer".

Listed below are 3C3-B entries that were gleaned from the Aeronautical Chamber of Commerce' aircraft register; as far as is known, these were the only examples built:
C-516 ; 3C3-B (# W- 55) Siemens-Halske
C-610E; 3C3-B (# W- 69) Siemens-Halske
C-958E; 3C3-B (# W-104) Siemens Halske, K. G. Frank.

Fig. 60. This 3C3-B gave many years of service; rugged structure withstood hard knocks.

A.T.C. #121
(3-29)
AEROMARINE-KLEMM AKL-25A

Fig. 61. An Aeromarine-Klemm AKL-25A with 9 cylinder 40 h.p. Salmson engine.

The AKL-25A was an unusual and very interesting light airplane that was an American adaption of the popular Klemm-Daimler; a very successful German design that dates back to the early part of 1921. Actually born of a motored-glider, the Aeromarine-Klemm monoplane had one aim and purpose for it's existance; to offer care-free flying in it's strictest economy with a bare minimum of repair and maintenance. By far it's most remarkable feature was the responsive performance it could coax out of "40 horses". With flight instruction costs still pretty much out of reach for the average man in the street, the AKL-25A posed as an ideal training plane that could offer day by day operation that ran a good 30% below the average; an economy of operation that could be reflected into cheaper instruction time and cheaper solo-flying time. Built rugged enough to absorb rough or inexperienced handling, and with aerodynamic characteristics built in to enable the little craft to practically fly by itself, it was also ideal for the "week-end pilot" who just flew around to pile up time and to enjoy himself.

The Aeromarine-Klemm model AKL-25A, as pictured here, was a two place open cockpit low-wing monoplane of surprisingly large dimensions; though a light-plane in weight, it was hardly a typical light-plane in size. Motive power for this combination was the 9 cylinder Salmson AD-9 (French) "radial" en-

gine of 40 h.p. The air-cooled Salmson AD-9, as many will recall, was just about the cutest and smoothest running engine that one would ever see; a powerplant that was used very often on experimentals and home-builts of all sorts for the next 20 years. The simple and uncluttered Aeromarine-Klemm derived much of it's rugged character from a vibration-absorbing and shock-deadening all-wood construction; the wing was of full cantilever design and could be quickly dismounted for transport or space-saving storage. In Germany, it was not unusual to see a Klemm with it's wings lashed to the sides of the fuselage, and being towed to a flying site that was probably a good distance from where the airplane was stored. This type of thing was not practiced, nor even necessary here in the U.S., but the extra utility was there in the design, if ever needed.

Introducd here in the U.S.A. sometime in 1928, the "AKL" was flown about the country on demonstration to test reaction. Making an appearance at the National Air Races for 1928, that were held in Los Angeles, Calif., the new craft was entered in the dead-stick (power off) landing contests and consistently came out on top. The best effort by the Aeromarine-Klemm was 4 feet and 5 inches from the mark, and the next best landing by another airplane was 14 feet from the mark; the glider parentage of the "AKL" certainly proved quite suitable for a contest of this

Fig. 62. The Klemm motor-glider was the basis for the design of the Aeromarine-Klemm.

type. The type certificate number for the AKL-25A (sometimes listed as L-25-A) as powered with the Salmson AD-9 of 40 h.p., was issued in March of 1929; A Group 2 approval numbered 2-47 was also issued in March of 1929 for serial # 1-2-4. A Group 2 approval numbered 2-87 was issued in July of 1929 for the AKL-25A on twin floats for over-water flying (serial # 2 and up). Altogether about 40 of the Salmson powered AKL-25A were built, and they were manufactured by the Aeromarine-Klemm Corp. at Keyport, New Jersey. "Aeromarine" was a great name in early aviation, and many will recall the Aeromarine 39-B and the Aeromarine flying boats; the company was one of our oldest aircraft manufacturing concerns here in the U.S.A. The honorable Inglis M. Uppercu was the president.

Listed below are specifications and performance data for the AKL-25A as powered with the Salmson AD-9 of 40 h.p.; length overall 24'6"; hite overall 6'6"; wing span 40'2"; wing chord 79" at root, tapered to tip; total wing area 210 sq.ft.; airfoil "Goettingen 387" modified; wt. empty 815; useful load 510; payload 230; gross wt. 1325 lbs.; max. speed 85; cruising speed 75; landing speed 35; climb 375 ft. first min. at sea level; service ceiling 9,000 ft.; gas cap. 15 gal.; oil cap. 2 gal.; approx. range 325 miles; price at the factory field was $3500, later reduced to $3350 in 1930; the Salmson AD-9 engine cost $1400. The angular fuselage framework was built up of wooden longeron members, plywood formers and bulkheads, and was covered with laminated wood veneer. The full cantilever wing framework was built up of spruce and plywood box-type spars and the wing ribs were of laminated plywood with lightening

Fig. 63. The AKL-25A on Edo floats; performance with 40 h.p. was remarkable.

Fig. 64. The AKL-25A was an ideal trainer and good for low-cost sport flying.

holes in the form of cut-outs; completed framework was covered with laminated wood veneer. The wing was in two panels that were attached to a center-section panel that was attached to the underside of the fuselage. The split-axle landing gear which was made up of 3 members on each side, was attached to this center-section stub wing. The veneer covered tail-group was also of cantilever, internally braced construction, and was made up of wood members and plywood formers; the horizontal stabilizer was adjustable in flight. A fuel tank of 15 gallon capacity was high up in the fuselage just ahead of the front cockpit; a small luggage compartment was in the turtle-back section of the fuselage, just in back of the rear cockpit. The Aeromarine-Klemm was built into 1931 with at least 6 or 7 different versions; probably a total of 75 or more airplanes were built in all. For the next development in the "AKL" series, refer to A.T.C. # 203 and A.T.C. # 204 which cover the LeBlond 60 powered AKL-26 and the AKL-26A.

Listed below are AKL-25A entries that were gleaned from the Aeronautical Chamber of Commerce aircraft register; this list is not complete, but it is more or less the bulk of the number that were built:

C-384; AKL-25A (# 1) Salmson AD-9
NC-484; AKL-25A (# 2) Salmson AD-9
C-485; AKL-25A (# 3) Salmson AD-9
C-587E; AKL-25A (# 4) Salmson AD-9
C-588E; AKL-25A (# 5) Salmson AD-9, #

6 had LeBlond 60.
NC-817E; AKL-25A (# 7) Salmson AD-9
C-9175; AKL-25A (# 8) Salmson AD-9
C-9176; AKL-25A (# 9) Salmson AD-9
C-9177; AKL-25A (# 10) Salmson AD-9
C-9178; AKL-25A (# 11) Salmson AD-9, not verified.
C-9179; AKL-25A (# 12) Salmson AD-9, not verified.
C-9180; AKL-25A (# 13) Salmson AD-9
NC-122H; AKL-25A (# 2-16) Salmson AD-9
NC-123H; AKL-25A (# 2-17) Salmson AD-9
C-162H; AKL-25A (# 2-20) Salmson AD-9
C-163H; AKL-25A (# 2-21) Salmson AD-9
NC-164H; AKL-25A (# 2-22) Salmson AD-9
-165H; AKL-25A (# 2-23) Salmson AD-9
NC-192H; AKL-25A (# 2-25) Salmson AD-9
-193H; AKL-25A (# 2-26) Salmson AD-9
-194H; AKL-25A (# 2-27) Salmson AD-9
-177M; AKL-25A (# 2-28) Salmson AD-9
NC-103M; AKL-25 A (# 2-34) Salmson AD-9
NC-160M; AKL-25A (# 2-35) Salmson AD-9
NC-195H; AKL-25A (# 2-39) Salmson AD-9
NC-104M; AKL-25A (# 2-42) Salmson AD-9
NC-178M; AKL-25 A (# 2-45) Salmson AD-9
NC-198M; AKL-25A (# 2-46) Salmson AD-9
-7482; AKL-25A (# 57) Salmson AD-9, next 8 a/c do not fit number pattern.
-7484; AKL-25A (# 83) Salmson AD-9
-7529; AKL-25A (# 92) Salmson AD-9
-7481; AKL-25A (# 93) Salmson AD-9
-435; AKL-25A (# 114) Salmson AD-9
-434; AKL-25A (# 115) Salmson AD-9
-10051; AKL-25A (# 116) Salmson AD-9

Fig. 65. The Fleet Model 1 was powered by a Warner "Scarab" engine; note heavy wing structure.

The "Fleet" sport-trainer definately shows it's parentage from the Consolidated PT and NY service-trainer series, and was a direct development from the Consolidated "Husky Junior", a similiar design that was built under A.T.C. # 84 (See U. S. CIVIL AIRCRAFT, Vol. 1). The first few Fleet aircraft were more or less re-named "Husky Juniors"; renamed in honor of Major Ruben Fleet, the guiding genius of "Consolidated" since the beginning way back in 1923. Considerable development went into the various models of the PT and NY trainer series, and much knowledge had been gained; the Fleet was certainly very fortunate to fall heir to such a background of ancestry. As pictured here, the Fleet posed as a brawny little ship that was extremely capable; well suited for the job intended it flew and handled well with a satisfying performance. A few of the early Fleet biplanes still had the one elongated cockpit, seating two in tandem as characteristic in the "Husky Junior", but this arrangement was soon modified to individual open cockpits, as shown. Certainly a stand-out feature on the sporty Fleet were the husky looking wing panels; wings of an unusually thick "Clark Y" airfoil section that had built in fortitude to spare. It was an advertised fact and a known fact that the wings of a Fleet biplane were stout enough to hold up the weight of two full grown elephants, or 13,125 pounds of sandbags! Proven beyond any doubt, then and since, the Fleet was an airplane that was eager to answer your every whim, and if you felt so-inclined, you could throw it all over the wild blue yonder in careless abandon, and not have fear of losing a single piece of the stout airframe. Casual observation will soon reveal that the Fleet was no great thing of beauty and lacked the near-feminine quality of the average airplane, but it was a "tomboy" that was trim and functional and a real satisfying joy to fly.

Typical of the earlier "Husky Junior", the Fleet Model 1 was powered with the smooth-running 7 cylinder Warner "Scarab" engine of 110 h.p. The type certificate number for the Fleet Model 1 was issued in May of 1929 and amended a month later to allow extra fuel in a "belly tank", and an increase in gross weight allowable of some 240 pounds. The Fleet biplane was manufactured by the Fleet Aircraft, Inc., a division of Consolidated Aircraft Corp. at Buffalo, New York where production had soon reached two a day. Lawrence D. "Larry" Bell was the president and general manager; Joseph M. Gwinn Jr. was chief engineer; and William "Bill" Wheatly was the chief pilot in charge of test and deve-

Fig. 66. The Fleet Model 1 was a development of the Consolidated "Husky Junior".

lopment.

Listed below are specifications and performance data for the Fleet Model 1 as powered with the 110 h.p. Warner "Scarab" engine; length overall 20'9"; hite overall 7'10"; wing span upper & lower 28'0"; wing chord both 45"; wing area upper 100 sq.ft.; wing area lower 95 sq.ft.; total wing area 195 sq.ft.; airfoil "Clark Y-15"; wt. empty 1022; useful load 558; payload with 24 gal. fuel was 232 lbs.; gross wt. 1580 lbs.; max. speed 113; cruising speed 95; landing speed 44; climb 780 ft. first min. at sea level; service ceiling 14,300 ft.; gas cap. 24 gal.; oil cap. 2.5 gal.; cruising range 360 miles; price at the factory was $5750. A reissue of the amended type certificate allowed extra fuel and a higher gross load, figures are as follows; wt. empty 1075; useful load 745; payload with 55 gal. fuel was 232 lbs.; gross wt. 1820 lbs.; maximum and cruising speeds were still comparable; landing speed jumped to 50 m.p.h.; climb 680 ft. first min. at sea level; service ceiling 13,020 ft.; total gas cap. 55 gal.; cruising range 750 miles. The fuselage framework was built up of welded chrome-moly steel tubing, lightly

Fig. 67. A few of the early Fleets had "bath-tub" cockpits.

Fig. 68. The Fleet was a versatile craft; it is shown here on skis.

faired to shape with wood fairing strips and fabric covered. The extra rugged wing framework was built up of heavy-sectioned solid spruce spars and heavy gauge stamped-out aluminum alloy wing ribs, also fabric covered. All wing and center-section struts were of heavy gauge chrome-moly steel tubing in a streamlined section, with interplane bracing wires of heavy gauge steel, also in streamlined section. The landing gear was of the robust "cross-axle" type with an "oleo strut" built into the lower end of each "vee". The upper wing was built into a one-piece section, and the gravity-feed fuel tank was mounted into the center portion. The fabric covered tail-group was built up of welded chrome-moly steel tube spar members and sheet steel ribs and formers. All tail surfaces were of

heavy cross-section, and the horizontal stabilizer was of a cambered "lifting section", a feature that was retained on Fleet models throughout the whole series. The fin was ground adjustable and the horizontal stabilizer was adjustable in flight; ailerons of the Freise offset-hinge type were on the lower panels only and were operated through a positive acting torque tube and bellcrank action. Like the earlier "Husky Junior", the "Fleet" was light and quick on the controls, nimble and eager, and it was capable of the complete retinue of aerobatic manuvers; manuvers that were only limited by the pilot's ability and his fortitude. The next development in the Fleet sport-trainer was the Kinner powered Model 2; see chapter for A.T.C. # 131 in this volume.

Fig. 69. The Waco "Taperwing" with Wright "Whirlwind" J5 engine, a fabulous craft of this period.

It is quite remarkable what a transformation took place in the docile "Waco 10" by just the replacement of wings; a set of tapered wing panels in the M-6 airfoil section turned the J5 powered "Waco Ten" into a truly fabulous airplane. Fast, eager, flashy, and highly manuverable, the "Taperwing" was the answer to a sportsman-pilots dream and prayer. Modified from the basic "Ten-W", the "Taperwing" (Ten-T) was also a three place open cockpit biplane powered with the Wright "Whirlwind" J5 engine of 220 h.p.; seating for three was in two open cockpits, but the front cockpit could be closed off with a metal cover to make it into a one-place airplane when used strictly for air-race work, stunt flying, or sport flying. As pictured here in various views, many characteristics of the basic "Model 10" are noticeable.

The "tapered wings", an inspiring feature that actually made this airplane such a rousing success, were conceived, designed, and developed, by Charlie Meyers, Advance Aircraft's gifted test-pilot who always bristled with new ideas. A prototype of the "Taperwing" was built for test early in 1928, and oddly enough, it was powered with the Curtiss OX-5 engine; there seemed to be some hesitancy in the front office about building a tapered wing airplane, especially with the fairly new M-6 airfoil section, but Charlie Meyers was allowed to go ahead with the development anyhow on a modest budget. The OX-5 powered prototype was a proven success, despite the misgivings, and the installation of a Wright J5 engine in the next "Taperwing", using this same set of wings, brought forth this airplane's full performance potential.

Quickly realizing what a star performer they now had, "Waco" entered three in the 1928 Transcontinental Air Derby from New York to Los Angeles; three lively craft that were spurred on by Johnnie Livingston, John P. Wood, and Charlie Meyers. They finished 1-3-5 (Class B), in the order listed above. At the National Air Races that were held in Los Angeles, they placed 2-3-4 in a civilian free-for-all over a 75 mile closed course; bested only by E. E. Ballough in his red-hot Laird "Speedwing" (LC-R). The "Taperwings" flown in the Air Derby had straight-axle landing gears to reduce weight, and were highly faired to reduce parasitic drag; later, these undercarriages were replaced with the standard "Waco" split-axle long leg oleo-spring gear. With it's performance and capabilities now very well proven, it excited the flying

Fig. 70. Freddie Lund flying the "Taperwing"; his stunt-flying was a work of art.

populace to no small degree; the "Taper-wing" began to sell in very good number, soon to be seen "stunting" at air-shows, winning air-races, or just pleasing the fancies and whims of some sportsman-pilot. We would usually hesitate to call an airplane fabulous, because the meaning of the word is so strong, but calling the Waco "Taperwing" a fabulous airplane can certainly be justified without hardly any misgivings; it was a real glamor-girl and probably the most romantic type to come off the "Waco" line. A design with a reputation and a personality that was equaled only by a very few airplanes.

Five "Taperwings" were sold to the government of China as training-craft early in 1929, and two (NS-24 and NS-25) were in service with the Department of Commerce, Aero. Branch; many others were sold to professional pilots for air-race and air-show work, and a good number were sold to noted sportsman-pilots who enjoyed flying and mastering a spirited airplane. The type certificate number for the J5 powered "Taperwing" was issued in March of 1929 and they were built by the Advance Aircraft Co. at Troy, Ohio. Shortly after this, the firm's name was changed to Waco Aircraft Co. to identify them better with their product; the plant however, remained in Troy, Ohio. The remarkable Clayton Bruckner was the president; E. E. Green was chief of engineering; and Chas. W. Meyers had been the chief pilot in charge

of test and development. Meyers left Waco late in 1928 for test-pilot chores with the newly formed Great Lakes Aircraft Corp.; he was replaced by Freddie Lund, who had been with the Gates Flying Circus for a number of years and was nationally known as "Fearless Freddie". A fine pilot, the diminutive Freddie Lund did much to make the "Taperwing" popular by way of his superb flying and great showman-ship; they say that he loved the craft immensely. Freddie Lund, flying the original J5 powered "Taperwing", that had the original set of tapered wings on it, performed the first outside-loop ever made by a commercial airplane. Other stand-out "Taperwing" pilots were James Hall, Art Davis, Stuart Auer, "Jock" Whitney, "Cliff" Durant, and "Tex" Rankin; years later there were many more added to this star-studded list.

Listed below are specifications and performance data for the "Waco Taperwing" (ATO) as powered with the Wright J5 engine; length overall 22'6"; hite overall 9'; wing span upper 30'3"; wing span lower 26'3"; chord both at root 62.5" total wing area 227 sq.ft.; airfoil M-6; wt. empty 1787; useful load 813; payload with 76 gal. fuel is 193 lbs.; payload with 100 gal. fuel is 50 lbs.; gross wt. 2600 lbs.; max. speed 135; cruising speed 110; landing speed 52; climb 1200 ft. first min. at sea level; service ceiling 19,000 ft.; gas cap. normal 76 gal.; gas cap. max.

Fig. 71. The prototype "Taperwing" had a Curtiss OX-5 engine.

100 gal.; oil cap. 5 gal.; cruising range 600-800 miles; price at the factory field was $8525. The fuselage framework was built up of welded chrome-moly steel tubing, faired to shape with wood fairing strips and fabric covered. The front cockpit had a door for easy entrance and exit, and a metal panel was available to close the front cockpit off when

Fig. 72. "Taperwing" flown by Johnnie Livingston to first place in the 1928 N.Y.-to-L.A. Air Derby.

Fig. 73.   Charlie Meyers, Waco test pilot and creator of the "Taperwing" configuration, is shown here with the first J5-powered version.

not in use; a small baggage compartment was located in the turtle-back section just behind the pilot. The wing framework was built up of heavy-sectioned solid spruce spars, with spruce and plywood built-up wing ribs that were closely spaced for added strength and to help preserve the true airfoil form; the completed panels. were fabric covered. The wings were tapered in plan-form and section, and there were 4 ailerons that were connected in pairs by a push-pull strut. The interplane struts were of heavy gauge chrome-moly steel tubing in a streamlined section, and the interplane bracing was of heavy gauge streamlined steel wire to withstand the stresses that would be imposed on an airplane of this type. Incidently, the "Taperwing" structure was stressed to take 450 h.p., so it's safety margin with the J5 engine was tremendous. The main fuel tank was in the fuselage just ahead of the front cockpit, and extra fuel was carried in a tank that was built into the center-section panel of the upper wing. The fabric covered tail-group was built up of welded chrome-moly steel tubing; the horizontal stabilizer was adjustable in flight. The landing gear was of the split-axle type, and was the typical long leg oleo-spring gear as used on other "Waco" models at this time; wheel tread was 78 inches. Standard equipment included metal propeller, dual controls, completely wired for lights, wheel brakes, and custom colors. An engine starter was optional equipment. For the next "Waco" development see chapter for A.T.C. # 168 in this volume, which discusses the BS-165 (BSO) that was powered with the 5 cylinder Wright engine.

A complete tabular listing of all the Wright J5 powered "Taperwing" (ATO) that were manufactured would be quite a task, so let it be sufficient to say that at least 53, or more, were built; then of course, there was the CTO with Wright J6-7-225 and a number of specials with Wright J6-9-300 and the Packard "Diesel".

Fig. 74.   The American Eagle Model A-129 with a 5 cylinder Kinner engine; the craft was gentle and easy to fly.

The popular "American Eagle" biplane as powered with the Curtiss OX-5 engine, had been built in good number·throughout 1928, and at least 300 of these craft were already flying in all parts of the country. The configuration had evolved from the early two-aileron job into the more familiar four-aileron job, and customers were actually lining up at the plant waiting for their delivery. In between production of the standard model, there· were various other versions of the "Eagle" that were powered with medium horsepower air-cooled radial engines; such interesting installations as the Quick Radial of 125 h.p.; the Anzani 120; the new Salmson 120: and the Ryan-Siemens of 125 h.p. The stock-pile of OX-5 engines certainly had been huge enough, but it was bound to run out soon and everyone knew it, so American Eagle was shopping around to find an engine that they could use as a suitable replacement. All of the above mentioned engines were satisfactory of course, but not quite the combination that they had hoped for. After tests with the "Eagle" powered by the Kinner engine, and an assurance from the manufacturer of it's continued supply, they felt the combination was right at last and decided to use this engine as the standard powerplant for the new series; the model A-129 for 1929 was the result.

The model A-129 was an all-purpose 3 place open cockpit biplane that was typical to the models just previous, in most all respects, except for the engine installation which was now the 5 cylinder Kinner K5 engine of 90-100 h.p. Because of the necessity to maintain proper balance, the little Kinner had to be stuck out there a good ways, and consequently, the "Kinner Eagle" wound up with a very long nose, and the result of this would not win any beauty contests to be sure. Before long, for reasons quite obvious, the A-129 had earned the nick-name of "Long Nose Eagle", "Ant-eater", and other such names, but regardless of this lack of good proportion the homely result was a good performer and a fine flying airplane. Though responsive and quite eager, the "Kinner Eagle" was also very stable and quite easy to fly; 'twas often said that it could actually feel the nervous hand of the student-pilot on the stick immediately, and sorta looked after him by overlooking mistakes. By virtue of this very commendable trait, many of these likeable craft were used for student pilot instruction.

The standard OX-5 powered "American Eagle" had been trimmed up somewhat for 1929 and was still in good production, and numerous old "Eagles" that had already seen many hours of service, had their OX-5 engines taken out at the factory and replaced with the 5 cyl. Kinner. Production of the A-129 took off in a flying start and held up in good number, some 100 had been built throughout the year, but towards the end of the year sales had been falling drastically; mainly due to the business recession. By the dawning of 1930, the "Kinner Eagle" was being offered with a $1500 flying course for the price of the airplane alone. With this offer came a little spurt of business, and the balance of the A-129 were cleared out to make way for new models, which the company felt would help create a new interest in buying and flying. The type certificate number for the model A-129 was issued in March of 1929 and it was manufactured by the American Eagle Aircraft Corp. at Fairfax Field, Kansas

City, Kansas. E. E. Porterfield Jr. was the president; Jack E. Foster was the chief engineer; and Larry Ruch was the chief pilot in charge of test and development. A "Kinner Eagle" (model A-129) was flown by the winsome Mae Haizlip to 23rd place in the National Air Tour for 1929; she didn't set any records, but they say she had a very enjoyable time. Altogether, we might sum this up by saying that though the American Eagle A-129 was almost ugly and sometimes laughed at, it did have a patient nature and upon closer association, it was very easy to like this airplane with affection and respect.

Listed below are specifications and performance data for the American Eagle model A-129 as powered with the 5 cyl. Kinner K5 engine of 90-100 h.p.; length overall 24'6"; hite overall 8'4"; wing span upper 30'6"; wing span lower 30'; wing chord both 62.5"; wing area upper 163 sq.ft.; wing area lower 140 sq.ft.; total wing area 303 sq.ft.; airfoil "Aeromarine 2A" Mod.; wt. empty 1220; useful load 800; payload 340; gross wt. 2020 lbs.; max. speed 105; cruising speed 90; landing speed 35; climb 620 ft. first min. at sea level; service ceiling 14,000 ft.; gas cap. 42 gal.; oil cap. 5 gal.; cruising range 6 hours; price at the factory field was $4895; during the early part of 1930, this price included a $1500 flying course. The fuselage framework was built up of welded chrome-moly steel tubing, faired to shape with wood fairing strips and fabric co-

Fig. 75. It is evident why the A-129 was called the "Anteater".

Fig. 76. The latest version of the OX-5-powered American Eagle was still selling in good numbers.

vered; cockpits were neatly upholstered and the front cockpit was provided with a door for ease of entry and exit. The wing framework was built up of solid spruce spar beams that were routed to an "I beam" section, with spruce and plywood built-up ribs; the completed framework was fabric covered. The upper wing was in two panels that were joined together at the center-line and fastened to a center-section cabane of inverted vee struts; there were four ailerons that were connected in pairs by a push-pull strut. The fuel supply of 42 gallons was carried in a tank that was mounted high in the fuselage, just ahead of the front cockpit; a direct-reading fuel gauge projected through the upper cowling. The landing gear was of the normal split-axle type and one spool of rubber shock-cord was used to take some of the jar out of bad bumps; the tail-skid was a chrome-moly steel tube that was also shock-cord sprung. Long fami-

liar on the "American Eagle" was the all-silver paint job with burnished swirls on all the metal cowling; the "Eagle" for 1929 was done up in a different dress. Standard color scheme was now a blue fuselage and yellow wings, or a red fuselage with grey wings; a wide accenting stripe in the lighter color was used to break up the expanse of the darker color on the fuselage. Being the cheapest Kinner-powered airplane that one could buy, the A-129 was equipped with a wooden propeller, and wheel brakes and engine starter were only available as special equipment. The next development in the "American Eagle" biplane was the "Phaeton" that was powered with the 5 cyl. Wright engine, this was the offering for 1930; an improved "Kinner Eagle", the model 201, was offered about a month or so later. These two models were approved on A.T.C. # 282 and A.T.C. # 293.

Fig. 77. The Swallow F28-AX was powered by an Axelson engine.

The Swallow Airplane Company of Wichita had been in continuous manufacture of "Swallow" airplanes for all of five years now; hailed and respected as America's first practical commercial airplane, and a design that spurred many others to follow suit, there were many hundreds of these flying in just about all parts of the country, and continued to be one of America's best and most popular all-purpose airplane of the open cockpit type. Improved progressively to keep pace with latest developments, the three-place Swallow was a sturdy and good flying craft that had already been done up in combination with the Curtiss OX-5 engine, the Hisso (Hispano-Suiza) A and E engines, and the popular Wright J5; now in a new development, the F28-AX, it was offered with the new 7 cylinder Axelson engine of 115-150 h.p. The Axelson was an air-cooled radial engine first known as the "Floco" that was manufactured by the Frank L. Odenbreidt Co.; it was soon to be seen mounted on models of various other manufacturers. It's rating was 150 horsepower at 1800 r.p.m., and it sold for $3250.

An occasional three-place Swallow with the OX-5 engine was still being built on order, as the design had not yet run it's course, but the bulk of production was in the new "Swallow TP" sport-trainer which was selling very good and kept the production lines quite busy. To compliment this, great plans and hopes were slated for the new F28-AX, but as it turned out, only one was known to be built and one other has been rumored; hence, this model must be considered as a very rare type. So rare in fact, that continuous research and prospecting had failed to dig up even one photograph of the type; so in description we'll think it sufficient to say that the F28-AX was a normal three-place Swallow biplane in just about every respect, except for it's engine installation. Whatever major changes did take place, were all ahead of the firewall. The type certificate number for the Axelson powered F28-AX was issued in March of 1929 and it was manufactured by the Swallow Airplane Company at Wichita, Kansas. W. M. Moore was the president; C. A. Noll was V.P.; George R. Bassett was secretary and general manager; M. D. Kirkpatrick was factory manager; Jay Sudowsky was the chief pilot; and Dan Lake was the chief engineer.

Fig. 78.   The Swallow F28-AX on twin floats.

Listed below are specifications and performance data for the Swallow F28-AX as powered with the 7 cyl. Axelson engine; length overall 23'10"; hite overall 8'11"; wing span upper 32'8"; wing span lower 32'3"; chord both 60"; wing area upper 156 sq.ft.; wing area lower 144 sq.ft.; total wing area 300 sq.ft.; airfoil U.S.A. 27; wt. empty 1574; useful load 923; payload 460-340; gross wt. 2497 lbs.; max. speed 110; cruising speed 93; landing speed 45; climb 640 ft. first min. at sea level; service ceiling 13,500 ft.; (the above performance figures were reported with Axelson of 115 h.p. rating); gas cap. 42-62 gal.; oil cap. 5.5 gal.; cruising range was 8 gal. per hour, or 450-630 miles; price at the factory field was $5850. The fuselage frame work was built up of welded chrome-moly steel tubing in a truss-type form; the framework was heavily faired to shape with wood fairing strips and fabric covered. The wing panel framework was built up of solid spruce spar beams and spruce and plywood built-up wing ribs, also fabric covered. The upper wing was in two panels that were joined together at the center-line and fastened to a center-section cabane of two "N type" struts; there were four ailerons that were connected together in pairs by a heavy stranded cable. The main fuel supply was carried in a tank that was mounted high in the fuselage, just ahead of the front cockpit; extra fuel was carried in two tanks that were built into the root end of each upper wing panel. The fabric

Fig. 79.   The Swallow biplane was well suited for general purpose service.

Fig. 80.   The Swallow F28-AX in Canadian "bush flying" service.

covered tail-group was built up of welded chrome-moly sheet formers and tubing members; the fin was ground adjustable and the horizontal stabilizer was adjustable in flight. The sturdy landing gear was of the normal split-axle type in an inverted tripod form, and used two spools of rubber shock-cord to take some of the jar out of the hard bumps; wheels were 28x4 or 30x5 with a 72 inch tread. Individual wheel brakes, a metal propeller, and a hand crank inertia-type engine starter were optional equipment. The standard color scheme for the Swallow biplane had long been a combination of a shiny black fuselage with bright orange-yellow wings, but by this time they were also using maroon-yellow and blue-yellow combinations; the model F28-AX was believed to be a combination of light green and a Diana cream, no doubt it looked quite handsome. The next development in the Swallow biplane was the Kinner powered TP-K sport-trainer, see the chapter for A.T.C. # 186 in this volume.

Listed below is the only known example of the Swallow model F28-AX as powered with the 7 cyl. Axelson engine:

C-8728; Swallow F28-AX (# 1038) Axelson 115-150, owned by B. G. Hall of Goleta, California.

Fig. 81. *The Sikorsky "Amphibion" Model S-38B; ship shown was the first craft with sloped windshield.*

The Sikorsky S-38B "Amphibion" might well be called the conqueror of the Caribbean Sea, a great body of water dotted with many large and small islands; some fairly close together like stepping stones and many were hundreds of miles apart. Pan American Airways, a pioneer in this part of the world, soon realized that to serve this area properly with air transportation, an airplane ""that could make it's own airport" was definately needed; this led to their extensive use of the Sikorsky "Amphibion". Now, practically any point in the Caribbean could be a port of call; Pan Am routes spread out from Cuba to Yucatan and down through Central America to the Panama Canal Zone. Another route extended south from Cuba to Puerto Rico and down through the Antilles to Dutch Guiana; from there the route bore westward back to the Canal Zone across the top of So. America, and Pan Am thus became a true airways system. By the end of 1929, Pan Am had grown from the original 90 mile hop from Key West to Cuba, into a multi-stop system of 13,000 miles; the original fleet of 3 airplanes had now swelled to 64, of which at least 20 were the S-38A & B "Amphibions". The choice of an amphibious aircraft for ser-

vice in this part of the world was only logical, because of it's ability to land in the heart of things, whether they be deep in the interior or on the shore-line of a body of water; only such utility was making air-travel worthwhile. It is quite interesting to note that people of the Caribbean and the Latin American countries, many of whom had never seen an airplane before, were patronizing air transport eagerly, making the transition from burro to airplane with remarkable nonchalance.

The Sikorsky "Amphibion" model S-38B was at first glance more or less typical to the earlier S-38A, however, several interesting modifications were incorporated into it's make-up. One of the most noticeable of the improvements was a change from the flat vertical windshield that was used on earlier models, to one that sloped back to a more streamlined shape; thereby gaining several miles per hour in cruising speed. Greater fuel capacity was also provided for a longer radius of action, which was now necessary on some of the Caribbean routes, and in other instances for business and private uses; with the fuel load the S-38B could now carry, a cruising range of 7 hours was entirely possible. The basic model S-38B was arranged as

Fig. 82. An early version of the S-38B with flat vertical windshield.

a 10 place transport with allowance for a bit more room, and a more plush interior arrangement; 8 of the seats could be quickly removed to provide cargo area. The S-38B was also arranged in 4 and 6· place "air yacht" versions, with highly plush interiors that often contained many special conveniences. Though there was no improvement apparent in the performance of the new model S-38B, it was nevertheless a much better craft in many ways; flight characteristics were excellent, manuverability was surprising for a ship of this size, and it could maintain flight and even climb with one engine dead, carrying a full gross load. The versatility and dependability of the Sikorsky "Amphibion" was a combination that was hard to beat, a majestic aircraft that had endeared itself to many and shall remain as one of the all-time "greats" in early aviation. The type certificate number for the model S-38B as powered with two Pratt & Whitney "Wasp" engines of 420-450 h.p. each, was issued in May of 1929 and some 75 or more examples of this model were

manufactured by the Sikorsky Aviation Corp. at Bridgeport, Connecticut; a division of the United Aircraft Corp. The incomparable Igor I. Sikorsky was a V.P. and the chief of engineering.

Listed below are specifications and performance data for the Sikorsky "Amphibion" model S-38B as powered with two "Wasp" engines of 425 h.p. each; length overall 40'3"; hite on wheels 13'10"; hite in water 10'2"; wing span upper 71'8"; wing span lower 36'0"; wing chord upper 100"; wing chord lower 59"; wing area upper 574 sq.ft.; wing area lower 146 sq.ft.; total wing area 720 sq.ft.; airfoil "Sikorsky GS-1"; wt. empty 6550 lbs.; useful load 3930; payload with 300 gal. fuel was 1735 lbs.; gross wt. 10,480 lbs.; max. speed 125; cruising speed 110; landing speed 55; climb 880 ft. first min. at sea level; climb in 10 min. was 7250 ft.; service ceiling 18,000 ft.; gas cap. max. 330 gal.; oil cap. 28 gal.; cruising range at 44 gal. per hour with 330 gal. fuel was 750 miles; basic price at the factory field was about $50,000.

Fig. 83. The S-38B pioneered with airlines in the Caribbean Sea.

Fig. 84. The S-38B was available with plush interior arrangements.

Fig. 85. An S-38B on the Twin Cities route of Northwest Airways.

The hull framework was built up of oak and ash wooden frame members that were reinforced with "dural" plates and gussets at the joints; the outer covering was of heavy gauge duralumin "Alclad" metal sheet that was riveted together, and fastened to the hull framework with screws. The hull was arranged into 6 water-tight compartments for greater safety in case of puncture; there was a compartment in the extreme nose·for anchor and mooring equipment, and a compartment just back of that for 200 lbs. of baggage. Entry into the main cabin was by way of a hatch and ladder in the rear section of the hull. The wing framework was built up of riveted and bolted duralumin girder-type spar beams and riveted duralumin truss-type wing ribs; the completed framework was fabric covered. All fuel tanks were in the upper wing for a maximum fuel load of 330 gal., and two oil tanks were also mounted in the upper wing. The unusual tailgroup was mounted on two booms that came aft from the upper wing, & were braced to the hull by streamlined struts; all surfaces were fabric covered and the rudders were aerodynamically balanced. The landing gear of 116 inch tread was of the retractable type, hydraulically operated to the extended or retracted position; wheels were 32x6 and brakes were available. Metal propellers, Eclipse inertia-type engine starters, navigation lights, fire extinguishers, anchor and mooring gear were standard equipment. The next development in the Sikorsky "Amphibion" was the 12 place model S-38C; for a discussion of this craft, see the chapter for A.T.C. # 158 in this volume.

Listed below is a partial list of S-38B en-

tries that were gleaned from various records; a complete listing would be quite extensive, so we will submit about 20 or so.

NC-9753; S-38B (# 114-1) 2 Wasp.
NC-9775; S-38B (# 114-2) 2 Wasp.
NC-9776; S-38B (# 114-3) 2 Wasp.
NC-9105; S-38B (# 114-4) 2 Wasp.
NC-9106; S-38B (# 114-5) 2 Wasp.
NC-9107; S-38B (# 114-6) 2 Wasp.
NC-9143; S-38B (# 114-7) 2 Wasp.
NC-9144; S-38B (# 114-8) 2 Wasp.
NC-9137; S-38B (# 114-9R) 2 Wasp.
NC-9151; S-38B (# 114-10) 2 Wasp.
NC-9138; S-38B (# 114-11) 2 Wasp.
NC-9139; S-38B (# 114-13) 2 Wasp.
NC-9140; S-38B (# 114-14) 2 Wasp.
NC-195H; S-38B (# 114-15) 2 Wasp.
NC-196H; S-38B (# 114-16) 2 Wasp.
NC-158H; S-38B (# 114-18) 2 Wasp.
NR-159H; S-38B (# 114-19) 2 Wasp.
NC-160H; S-38B (# 114-20) 2 Wasp.

The first model S-38B was serial # 114-1 and was known as 'Untin' Bowler, owned by Robert McCormick of the Chicago Tribune; serial # 114-7 and # 114-8 were 11 place craft on Group 2 approval 2-68; serial # 114-9R was a 4 place craft on Group 2 approval 2-69; serial # 114-11 was first S-38B with sloped windshield; serial # 114-12 was probably Army XC-6; serial # 114-14 was a 6 place craft on Group 2 approval 2-74; serial # 114-17 probably Navy PS-3; serial # 114-19 Sikorsky test-bed. The Sikorsky S-38 series were now built in blocks of 20; first block was 114-1 thru 114-20; second block was 214-1 thru 214-20; third block was 314-1 thru 314-20, etc., about 5 blocks were built in the various versions.

Fig. 86.   The "Cyclone"-powered Stearman "Speedmail" Model M-2 carried heavy loads with good performance.

Decidedly a typical Stearman biplane but almost twice as large, was the "Cyclone" powered "Speedmail" model M-2; a craft which was designed and developed especially to meet the increase in tonnage that was showing up occasionally on some portions of the air-mail system. The graceful "Speedmail" with it's cargo space of 91 cubic feet, that would carry a thousand-pound payload at some 125 m.p.h. cruise, was the answer to this problem at least for Varney Air Lines, who had 5 of them in regular scheduled service. Actually a big, beautiful airplane, the model M-2 was a typical Stearman mail-plane seating the pilot in an open cockpit, with two large mail and cargo compartments up in the forward section of the big fuselage. With most of the space in the forward fuselage taken up by cargo compartments, the fuel load more or less had to be carried in the large center-section panel of the upper wing. A good-sized baggage compartment for the pilot's personal effects was located in the fuselage, just in back of the pilot's seat; in case of an overflow load up front, a certain amount of cargo could also be carried in this compartment. The cockpit was piped for heat from an exhaust-manifold heater, but it couldn't have been too much comfort in cold winter weather. Rugged and well built, with the typical sure-footed performance of all Stearman biplanes, the model M-2 "Speedmail" was fast and efficient, was well liked by the pilots, and served the Varney Line faithfully for a good many years.

Born of plans laid out by Stearman, Short, and Varney, and developed in the latter part of 1928, the prototype "Speedmail" first flew on January 15th of 1929, and it's calculated performance and behavior was borne out almost down to the letter. Yet further proof of the designing genius that was inherent in the hearts and minds of both Lloyd Stearman and Mac Short; two of the most outstanding men in the business. Passing all required tests in very short order, the type certificate number for the Wright "Cyclone" powered "Speedmail" model M-2 was issued in March of 1929 and it was manufactured by the Stearman Aircraft Co. at Wichita, Kansas. The

Fig. 87. This "Speedmail" was the first of the type.

Stearman organization was a well-knit group that held together well; this in contrast to the constant upheaval that was taking place in other organizations, both large and small. Hard-working Lloyd Stearman was the president and general manager; faithful Mac Short was V.P. in charge of engineering; penny-pincher H. A. Dillon was treasurer; air-minded Walter Innes Jr. was secretary; enthusiastic J. E. Schaefer was in charge of sales; and pint-sized David P. "Deed" Levy was the chief pilot.

Listed below are specifications and performance data for the Stearman "Speedmail" model M-2 as powered with the Wright "Cyclone" engine of 525 h.p.; length overall 30'2";

hite overall 11'11"; wing span upper 46'; wing span lower 32'; chord upper 84"; chord lower 54"; wing area upper 311 sq.ft.; wing area lower 125 sq.ft.; total wing area 436 sq.ft.; airfoil "Goettingen 398"; wt. empty 3442; useful load 2136; payload 1026; gross wt. 5578 lbs.; max. speed 147; cruising speed 126; landing speed 55; climb 1050 ft. first min. at sea level; service ceiling 16,000 ft.; gas cap. 136 gal.; oil cap. 15 gal.; cruising range 730 miles. The fuselage framework was built up of welded chrome-moly steel tubing and faired to shape with wood formers and fairing strips; the portion from the rear of the pilot's cockpit, clear on up to the engine was covered in removable aluminum alloy panels, and the

Fig. 88. A three-seat "Speedmail" special built for Walter Varney.

Fig. 89. The grace of the "Speedmail" was inherent in Stearman design.

balance of the fuselage aft was covered in fabric. The two large hatch-covered cargo compartments were in the forward section of the fuselage and held a total of 91 cubic feet; a sturdy walk-way was provided on the inboard section of the left lower wing. The wing framework was built up of spruce and plywood box-type spar beams and spruce and plywood truss-type wing ribs; the completed framework was fabric covered. Ailerons of the self balancing offset-hinge type were in the upper wing panels, and were operated by a push-pull strut; the fuel supply was carried in a gravity-feed tank that was built into the center-section panel of the upper wing. The fabric covered tail-group was built up of welded chrome-moly steel tubing; the fin was ground adjustable and the horizontal stabilizer was adjustable in flight. The split-axle landing gear was of the typical Stearman outrigger type and used air-oil shock absorbing struts; instead of the typical tail-skid, the M-2, being a much heavier airplane, used a steerable tail-wheel which made ground manuvering much easier for a plane of this size. The distinctive collectorring on the front of the engine was connected to two long tailpipes running under the fuselage; this directed exhaust fumes well below the cockpit level, cut "engine roar" down to a minimum, and also provided manifold-heat to the pi-

lot's cockpit. Wheel brakes, a metal propeller, and an inertia-type engine starter were standard equipment; the "Speedmail" was completely equipped for night flying. The "Speedmail" configuration was used as the basis for design of the 5 place Stearman "Light Transport" model LT-1; for discussion of this model see chapter for A.T.C. # 187 in this volume. The next Stearman development, after the M-2 discussed here, was the one-place "pony express" mail-plane model C3MB, which is discussed in the chapter for A.T.C. # 137 in this volume.

Listed below are "Speedmail" entries that were gleaned from the Aeronautical Chamber of Commerce aircraft register; this list represents the total number of these aircraft:

C-9051; Speedmail M-2 (# 1001) Cyclone, Varney Air Lines

C-9052; Speedmail M-2 (# 1002) Cyclone, Varney Air Lines

C-9053; Speedmail M-2 (# 1003) Cyclone, Varney Air Lines

C9054; Speedmail M-2 (# 1004) Cyclone, Varney Air Lines

C-9055; Speedmail M-2 (# 1005) Cyclone, Varney Air Lines

-8199; Speedmail M-2 (#   ) Cyclone, 3 pl. for Walter Varney.

C-8827; Speedmail M-2 (# 1007) Hornet, 2 pl. for Cliff Durant.

Fig. 90.   Beautiful lines are evident in this view of the Buhl "Airsedan" Model CA-6.

The buxom Buhl "Airsedan" (CA-5 & CA-5A) of a year or so ago was slimmed and trimmed down into a rather sveldt looking lady of good looks and pleasant propertion. Now seating six with ample room and comfort, the "Standard Airsedan" model CA-6 was powered with the 9 cylinder Wright J6 series engine of 300 h.p., which made into a combination of excellent utility and very good performance. The Buhl "Standard Airsedan" model CA-6 was now of the true sesqui-plane type, an arrangement reputed, especially by Buhl, to have the distinct advantages of both the biplane and the monoplane; in fact, the sesqui-plane was a monoplane and a half. This type of wing cellule arrangement allowed for a shorter overall wing span in spite of the generous amount of lifting area provided, and allowed for a simple yet robust method of bracing for the wing truss. In short, the true "sesqui" enjoyed the structural rigidity of the biplane and was still able to enjoy most of the aerodynamic efficiency of the monoplane. The robust structure and the excellent performance of the CA-6 "Airsedan" in high altitudes and in close quarters, channeled much of it's popularity into the Pacific northwest where it served admirably without fan-

fare at all sorts of difficult chores. In fact, this popularity remained through the years and several of these staunch craft are believed to be in active service in this part of the country, even yet.

The capable Buhl "Airsedan" was an airplane of several outstanding accomplishments; first brought to public attention in 1927 during the hustle and bustle of the Dole Derby race to Hawaii. Though the flight of the "Miss Doran" in this race ended in tragedy, the remarkable thing apparent was the fact that the heavily-loaded "Airsedan" lifted off quite easily for the start of it's flight with 3 people on board, with their equipment, and 400 gallons of gasoline; a staggering load which amounted to a gross weight of over 5000 pounds. With the power of a 220 h.p. Wright J5 engine, this was more or less proof positive of the remarkable performance that was built into the Buhl "Airsedan" design. Another "Airsedan" which came into some prominence, for a short time at least, was the J5 powered "Angeleno" flown by Mendell and Reinhart when they set a refueled endurance record of over 246 hours in July of 1929. About a month later, a brand new Wright J6 powered model CA-6, named the "Sun God",

Fig. 91.   A veteran CA-6 which served many years in the Pacific Northwest.

was selected for a refueled endurance flight across the country from Spokane to New York, with a return flight to Spokane without landing; this was a flight that was arranged to prove the commercial possibilities of long non-stop flights carrying passengers across the length of the country, by refueling in mid-air while still headed for their destination. The Buhl "Sun God" flown by Nick B. Mamer, with Art Walker as co-pilot and hose-handling man, was a standard model CA-6 that was fitted with extra tanks for a 320 gallon fuel capacity. Leaving Spokane, Washington on the flight to New York City and return, they made 11 refueling contacts in the course of their flight and flew the 7200 mile circuit in 115 hours; a 5 day west-to-east and east-to-west tour at an average speed of just under 65 m.p.h. The high-light of this flight was a refueling contact at 10,000 feet over Wyoming, which caused a few anxious moments; Nick B. Mamer a staunch advocate of Buhl airplanes since 1927, had nothing but praise for the "Sun God" and the way she handled herself during the record-breaking flight.

Though not exceptionally fast, the model CA-6 "Airsedan" was nevertheless quite sprightly and very manuverable for a passenger-carrying craft of this type; certainly asking no favors from the pilot or the landing strip. Also available with twin-float gear as a seaplane, the CA-6 served regularly in passenger and air-cargo service in Canada, where terrain and the job at hand had a tendency to "separate the men from the boys", so to

speak. The type certificate number for the Buhl "Airsedan" model CA-6 as powered with the Wright J6 engine, was issued in May of 1929 and some 15 or more examples of this model were manufactured by the Buhl Aircraft Company at Marysville, Michigan. Lawrence D. Buhl was the president; A. H. Buhl was V.P; both were members of the prominent Buhl family of Detroit; Herbert Hughes was V.P. and general manager; Louis G. Meister was chief pilot and in charge of sales; and Ettienne Dormoy was chief of engineering, a brilliant man of great ability and tireless energy.

Listed below are specifications and performance data for the Buhl "Standard Airsedan" model CA-6 as powered with the 300 h.p. Wright J6 engine; length overall 29'8"; hite overall 8'7"; wing span upper 40'0"; wing span lower 26'1"; wing chord upper (constant) 81"; wing chord lower (tapered) 46" mean; wing area upper 239 sq.ft.; wing area lower 76 sq.ft.; total wing area 315 sq.ft.; wt. empty 2478; useful load 1722; payload with 100 gal. fuel was 890 lbs.; gross wt. 4200 lbs.; max. speed 140; cruising speed 120; landing speed 45; climb 900 ft. first min. at sea level; service ceiling 17,000 ft.; gas cap. 100 gal.; oil cap. 8 gal.; cruising range at 15 gal. per hour was 720 miles; price at the factory field was $13,500. The model CA-6 was also allowed an increase in useful load for extra fuel, figures are as follows; wt. empty 2478; useful load 1878; payload with 125 gal. fuel was 890 lbs.; gross wt. 4356 lbs.; performance was not noticeably affected and range

Fig. 92.   The famous "Spokane Sun God" made a memorable record flight.

was extended to some 950 miles.

The fuselage framework was built up of welded chrome-moly steel tubing, faired to shape with formers and fairing strips; the forward portion of the fuselage was covered with aluminum alloy panels and the rear portion from the entry door back, was covered in fabric. The cabin was comfortably and tastefully upholstered, and the cabin walls were sound-proofed and insulated; there was a large cabin entry door and a convenient step on each side. The stub section of the lower wing was built up of chrome-moly steel tube spar beams and wing ribs; the stub wing which was built integral with the fuselage was used as fastening points for the split-axle landing gear, a "tracking type" gear that used air-oil shock absorbing struts. The wing framework was built up of heavy-

sectioned spruce spar beams and spruce and plywood truss-type wing ribs; the completed framework was fabric covered. The ailerons in the upper wing panels were of the "balance horn" type, but the horns were inset from the tip and did not over-hang in a fashion popular just a few years back. There were two gravity-feed fuel tanks that were mounted in the upper wing, flanking each side of the fuselage. A large baggage compartment was aft of the cabin, and was accessible from inside or out; the cabin was provided with heating, lighting, and ventilation. A novel accessory for the pilot was an adjustable periscope for seeing back over the upper wing. The fabric covered tail-group was built up of welded chrome-moly steel tubing and both vertical and horizontal movable surfaces were aerodynamically balanced; the fin was ground

Fig. 93.   The Buhl CA-6 in Canadian service; the "Airsedan" operated well on floats.

Fig. 94.   The Buhl was noted for its efficient sesquiplane arrangement.

adjustable and the horizontal stabilizer was adjustable in flight. Individual wheel brakes and a steerable tail-wheel made for ease in ground manuvering. Wheel brakes, navigation lights, a metal propeller, and inertia-type engine starter were standard equipment; the engine was well muffled with a large exhaust collector-ring which allowed near-normal conversation at all times. The next development in the Buhl "Airsedan" series was the high-performance "Sport Airsedan" model CA-3D that was powered with the 300 h.p. Wright J6 engine, see the chapter for A.T.C. # 163 in this volume.

Listed below are Buhl model CA-6 entries that were gleaned from various records; this list may not be quite complete, but it does show the bulk of this model that were built.

C-9627; CA-6 (# 41) Wright J6.

C-9628; CA-6 (# 42) Wright J6.
C-9629; CA-6 (# 43) Wright J6.
C-9632; CA-6 (# 47) Wright J6.
NC-9633; CA-6 (# 48) Wright J6.
NC-9634; CA-6 (# 50) Wright J6.
NC-8446; CA-6 (#   ) Wright J6.
NC-8450; CA-6 (# 55) Wright J6.
C-8453; CA-6 (# 56) Wright J6.
C-8452; CA-6 (# 58) Wright J6.
NC-8448; CA-6 (# 60) Wright J6.
CF-AAY; CA-6 (#   ) Wright J6.

Model CA-6 serial # 43 was on Group 2 approval numbered 2-51 as a 4 place craft with Wright J6-9-300; model CA-6 serial # 48 was later on Group 2 approval numbered 2-133 as a 4 place craft with Wasp 420; model CA-6B serial # 61 (X-8455) was special 6 place craft with Wasp 420 engine.

Fig. 95. The Bellanca CH-300 "Pacemaker" was an example of airplane efficiency at its best.

Although Giuseppe Bellanca had experimented occasionally along somewhat more radical lines, he never turned completely away from the basic features of a great design, that had stood him in good stead for these past five years. Indeed, Bellanca knew by instinct and by past proof that he had a very good basic design that was flexible enough to allow occasional improvement and slight change, and there was surely not much point in discarding it just yet. Thus the model CH-300, which was also a "bow-legged" Bellanca, in it's early version, differs in only one major respect from the model just previous. A continuation of the basic CH design was developed into the CH-300 mainly by installation of the 9 cylinder Wright J6 engine of 300 h.p., in place of the J5 which developed only 220 h.p. Improved in some details but otherwise typical, the model CH-300 was also a high wing cabin monoplane with ample space for six in a roomy interior that offered comfort, style, and a generous payload. For a number of years the Bellanca had been a habitual winner of efficiency contests; contests which credit greatest speed with greatest load, it was not then unusual to see the CH-300 carry on this tradition. Two of the CH-300, flown by Geo. Haldeman and J. Wesley Smith, were

1-2 in two of the efficiency races at the 1929 National Air Races held in Cleveland, O., in one other contest, Haldeman came in first. In an out-and-out speed race for cabin aircraft at the 1929 National Air Races, the CH-300 came in 3 and 4, bested only by two speedy Wasp-Vegas. R. A. "Bob" Nagle flew a Bellanca CH-300 in the National Air Tour for 1929 and came in sixth amongst a strong field of contenders, which represented the best that the industry had to offer. Other records for altitude and endurance were taken in stride during the year. In short, it behooves one to admit that the efficient performance of the CH-300 had been proven in the field of contest, and it was ready to take it's place in the line-up of outstanding aircraft.

Business houses the country over were attracted to the Bellanca's utility, small air-lines and charter-flight operators, who had to watch their pennies closely, were attracted to it's efficiency, and the seaplane version was popular in the northern "bush country" for it's good short-field performance with a sizeable payload. Introduced about the first of the year, the type certificate number for the CH-300 as powered with the Wright J6-9 was issued in May of 1929 for both a landplane and a seaplane version; an amended

Fig. 96. The early version of the CH-300 still had bow-legged landing gear.

certificate was re-issued in Oct. of 1929 for both the landplane version and the seaplane version to carry extra fuel, to the amount of 113 lbs. Altogether some 35 or more of the model CH-300 were built and the last of these were blossoming into the "Pacemaker" series; all were manufactured by the Bellanca Aircraft Corp. at New Castle, Dela. Giuseppe M. Bellanca was the president and chief engineer; Andrew Bellanca was V.P. and secretary; H. G. Smith and S. Short were chief pilots, with George Haldeman and J. Wesley Smith doing much of the promotion.

Listed below are specifications and performance data for the Bellanca CH-300 as powered with the Wright J6 of 300 h.p.; length overall 27'9"; hite overall 8'4"; wing span 46'4"; wing chord 79"; total wing area 273 sq.ft.; lifting area in wing struts 47 sq.ft.; airfoil "Bellanca" (Mod. R.A.F.); wt. empty 2253 (2833); useful load 1797 (1772); payload 1067 (1042); gross wt. 4050 (4605); these wts. were issued 5-29-29; the figures in brackets are for seaplane; the following wts.

are for landplane as issued 10-11-29; wt. empty 2275 (2387); useful load 1800 (1913); payload 1070; gross wt. 4075 (4300) lbs.; figures in brackets are with extra fuel; the following wts. are for seaplane as issued 10-11-29; wt. empty 2810 (2922); useful load 1800 (1913); payload 1070; gross wt. 4610 (4835) lbs.; figures in brackets are with extra fuel; max. speed 140 (130); cruising speed 120 (110); landing speed 50 (55); climb 1100 (850) ft. first min. at sea level; service ceiling 18,000 (17,000) ft.; figures in brackets are for seaplane; gas cap. 86-105 gal.; oil cap. 7-9 gal.; range 675 (550) miles; price at the factory field was $14,950 for landplane and $17,400 for seaplane with Edo twin floats.

The fuselage framework was built up of welded chrome-moly steel tubing, faired to shape with wood fairing strips and fabric covered. The cabin walls were sound-proofed and insulated with blankets of Balsam-wool, and the cabin interior was tastefully upholstered; there was a large baggage compartment to the rear of the main cabin section, and a

Fig. 97. The Bellanca CH-300 was a terrific performer as a seaplane.

large sky-light was in the cabin roof of the pilot's section for overhead vision. The wing framework was built up of solid spruce spars that were routed to an "I beam" section, with spruce and plywood built-up ribs; completed framework was then fabric covered. The distinctive "lift struts" that braced the wing, were a framework built up of heavy chrome-moly steel tubing, with ribs spaced across the tube to maintain an airfoil section; the completed framework was fabric covered. The airfoiled lift-struts added 47 sq.ft. to the lifting area, and the extreme dihedral angle contributed greatly to the airplane's stability. The small strut cabane just inboard from the upper ends, were designed to stiffen this portion of the strut junction against the twist of heavy aileron loads. The fuel supply was carried in two tanks; one built into the root end of each wing. The fabric covered tail-group was built up of welded chrome-moly steel tubing, except for the horizontal stabilizer which was built up of spruce spars and plywood ribs; the horizontal stabilizer was adjustable in flight. The "bow-legged" cantilever landing gear as used on few of the early CH-300, was of chrome-moly steel construction and used rubber shock-cord to absorb the bumps; the later version of the CH-300 had a landing gear of much simpler design and used "Aerol" (air-oil) shock absorber struts. Wheels were 30x5 with a 74 inch tread, and wheel brakes were standard equipment. A metal propeller and engine starter were optional. The next development in the Bellanca "Pacemaker" series was the "Pacemaker" Freighter, which will be discussed in the chapter for A.T.C. # 245.

A short resume of Bellanca history was included in U. S. CIVIL AIRCRAFT, VOL. 1 (A.T.C. # 47) but we feel it appropriate to fill in some portions of this history that were left out. The name of Giuseppe Mario Bellanca has always been one of the foremost in aviation; it was known practically throughout the world, by reason of the many spectacular exploits of Bellanca airplanes in contest of every nature, and in endurance and trans-oceanic flights. The history of Bellanca, at least to the period of 1930, is interesting to follow because he really was one of the "early birds"; having built his first airplane in Italy during 1908. Since that date his entire life had been concentrated on the study and perfection of various phases in aeronautics. He received his training at the Royal

Fig. 98.   The tasteful interior of the Bellanca CH-300 "Pacemaker".

Institute of Milan, Italy and upon graduation, he embarked on a carreer as designer and engineer of machines to navigate the atmosphere. In 1911 he came to the U.S.A. with his parents, five brothers and a sister; settling in Brooklyn, New York. His first American production was an airplane built in the cellar of their home, the whole family assisting in it's construction. With this plane he taught himself to fly and soon opened a flying-school in Mineola. He operated the school for some years, but the urge to continue with airplane design and development was always with him. When the government took over all flying fields, prior to our entrance in World War 1, Bellanca gave up his school without regrets and became a designing-engineer with the Maryland Pressed Steel Co. of Hagerstown; while there he designed and developed two tractor biplanes which were each a notable advancement in the science of aeronautics. Shortly after, the Wright Aeronautical Co. (a builder of aircraft engines) entered into agreement with G.M.B. to build them an airplane to exploit their new J4 air-cooled engine; the famous Wright-Bellanca of 1925 was the result. The famous "Columbia" was a Wright-Bellanca built the next year; built in the Wright shop with Wright money, the second Wright-Bellanca was of course Wright property. Before Bellanca could carry out his dreamed-of plans, and come into possession of the airplane, it was necessary for him to find a backer with faith in Bellanca, and in the future of aviation, to put up money to buy the ship. Searching far and near, he found the man he sought in Charles A. Levine. In accordance to the program laid out by G.M.B., Clarence Chamberlin and "Bert" Acosta set a new endurance record in the

Fig. 99.   The Bellanca CH-300 was ideal for hunting and fishing in the "bush country".

"Columbia", and then Chamberlin with Levine as passenger, took-off in Chas. Lindbergh's wake and landed in Germany. Bellanca was finally coming into his own. He
earned world-wide acclaim and prominence;
the financial support that had been so long
in coming, was now offered to him on every
hand. Soon Bellanca found himself occupying
a new factory located on a 365 acre flying
field that bordered the Delaware River, just
south of Wilmington. In the summer of 1929,
the second trans-Atlantic flight by a "Bellanca" was successfully completed by Williams and Yancey when they took off from
Old Orchard, Maine and landed in Spain. It
had always been Bellanca's policy to enter
his aircraft in contests of all sorts, and they
have always come through with flying colors;

vindicating the faith placed in them by their
designer. Bellanca airplanes have won every
efficiency contest in which they have entered,
as well as acquitting themselves creditably
in air-races, endurance flights, and altitude
flights of all kinds. The complete list of Bellanca "firsts" would indeed be very impressive. As always, Bellanca spent most of his
time at the factory, with very little interest
in the world, outside of aeronautics. He
looked ahead to the future and was always
working on two or three new ideas, either to
improve his present product or to bring out a
new craft that would probably represent
another advancement in the science to which
he had already contributed so much. The next
few years was to see some of his finest designs.

Fig. 100. The stately Travel Air B-6000 had good performance plus comfort.

The buxom Travel Air monoplane had already been done up in a Wright "Whirlwind" J5 powered version as the Model 6000 (see chapter for A.T.C. # 100 in U. S. CIVIL AIRCRAFT, Vol. 1) which had but a short career, and soon after the monoplane was done up in a high-performance "Wasp" powered version as the model A-6000-A (see chapter for A.T.C. # 116 in this volume), but the most popular version by far was the Wright J6 powered model B-6000. The B-6000 was typical to the earlier 6000 in most all respects except for the engine installation which in this case was the 9 cylinder Wright J6 series engine of 300 h.p. The model B-6000, also referred to as the 6000-B, was a 6 place high wing cabin monoplane of generous proportions that had ample room and comfort for six, with amiable flight characteristics and a very good performance. Of an ideal size for the smaller feeder-line, it was used throughout the country on lines such as Southern Air Transport, and Texas Air Transport in Texas; Delta Air Lines in Louisiana, and several other lines in various other regions. Of a gentle and obedient nature, with a good sound structure, the 6000-B was around for many years and served patiently at all manner of uses. A number of the B-6000 found their way into service on routes in Mexico over treacherous mountain country, using miserable airports, if we could stretch a point and call these landing places airports, doing next to the impossible day after day, and the 6000-B stood up to it admirably without complaint.

The B-6000 had very good short-field characteristics and could hop a fair-sized load in and out with ease; such versatile utility was made to order for business-houses who had ports of call that were often far off the beaten path, and required operation from landing fields that were little more than backlot pastures. For some time previous, men of business were forced to use the high-powered open cockpit biplane for making these calls, now they could do it in cabin comfort. Several owners used the 6000-B to good advantage as a personal air-taxi to out-of-way places, one such was Victor Fleming, a prominent Hollywood film director; the complete roster of satisfied owners surely would be an impressive one. The type certificate number for the Travel Air monoplane model 6000-B (B-6000) as powered with the Wright J6-9 engine, was issued in March of 1929 as a landplane, and amended in July of 1929 to include the seaplane version (S-6000-B), as

Fig. 101.   An S-6000-B in Canadian service.

mounted on twin-float gear. The Travel Air monoplane was manufactured by the Travel Air Company at Wichita, Kansas and some 55 or more of this model were built. The iniminable Walter H. Beech was the president; Thad C. Carver and J. H. Turner were V.P.; Wm. R. Snook was secretary; C. G. Yankey was treasurer; the brilliant Herb Rawdon was chief of engineering; and Clarence E. Clark was chief pilot in charge of test and development. Lending itself nicely to modification, the basic 6000 design was changed very little to produce the 6000-B type; as a consequence, neary all of the J5 powered model 6000 were converted to the 6000-B or the S-6000-B version.

Listed below are specifications and perfor-

mance data for the Travel Air monoplane model S-6000-B as powered with the 300 h.p. Wright J6 series engine; length overall 30′10″; hite overall 9′3″; wing span 48′7″; wing chord 78″; total wing area 282 sq.ft.; airfoil "Clark Y-15"; wt. empty 2608; useful load 1622; payload with 82 gal. fuel was 900 lbs.; gross wt. 4230 lbs.; max. speed 130; cruising speed 110; landing speed 60; climb 800 ft. first min. at sea level; service ceiling 16,000 ft.; gas cap. 82 gal.; oil cap. 7 gal.; cruising range 550 miles; price at the factory field was $13,000 for earlier versions and later raised to $13,500. The following figures are for the seaplane version as mounted on twin Edo floats; wt. empty 3030; useful load 1590; payload with 82 gal. fuel was 870 lbs.; gross wt. 4620 lbs.;

Fig. 102.   Victor Fleming, the Hollywood film director, was the proud owner of a B-6000 used for pleasure flying.

Fig. 103.   The S-6000-B was a good seaplane.

max. speed 120; cruising speed 100; landing speed 65; climb 680 ft. first min. at sea level; service ceiling 15,000 ft.; all other dimensions and data remain the same.

The fuselage framework was built up of welded chrome-moly steel tubing, the portion from the nose end to the rear of the cabin was braced in a truss form with steel tubing, and from there back was braced with steel tie-rods in an X truss form; the framework was faired to shape with formers and wood fairing strips and fabric covered. From the front wing strut forward, the fuselage was covered with metal panels. The cabin walls were sound-proofed and insulated, and all windows were of shatterproof glass; the seats were comfortable and well upholstered, and could be removed quickly for hauling bulky freight loads. The wing framework of semi-cantilever design, was built up with box-type spar beams of spruce flanges and plywood webs, with spruce and plywood girder-type wing ribs; the leading edges were covered with "dural" sheet and the completed framework was fabric covered. Two gravity-feed fuel tanks were mounted in the wing flanking the fuselage; long narrow chord ailerons were of the Freise offset-hinge type and were operated

differentially. The wing bracing struts and the landing gear struts were of heavy gauge chrome-moly steel tubing and were encased in balsa-wood fairings. The fabric covered tailgroup was built up of welded chrome-moly steel tubing; the fin was ground adjustable and the horizontal stabilizer was adjustable in flight. The split-axle landing gear of 9 foot tread was of the outrigger type and used "Aerol" shock absorber struts; wheel brakes and a steerable tail-wheel made ground manuvering a fairly easy chore. The engine was fitted with a large volume collector-ring that muffled exhaust noises to a great degree and permitted conversation in the cabin at all times. Bendix wheels and brakes (32x6), a metal propeller, navigation lights, and inertia-type engine starter were standard equipment. After a year or so, the S-6000-B was developed into the new 6-B which was approved under A.T.C. # 352. There were also a number of the 6000-B in a 7 place version that were licensed under a Group 2 approval numbered 2-138. The next Travel Air development was a biplane, the Wright J5 powered model B-4000 which is discussed here in the chapter for A.T.C. # 146 in this volume.

Fig. 104.   The utility of the Travel Air monoplane was greatly enhanced by twin-float gear; here is a fine view of an S-6000-B.

Fig. 105.   The Fleet Model 2 with 100 h.p. Kinner K5 engine.

The Fleet Model 2 was a companion offering to the Fleet Model 1, and was just about typical in all respects except for the engine installation which in this case was the 5 cylinder Kinner K5 engine of 100 h.p. Performance of this twin to the Model 1 was more or less typical, and it shared all of the inherent qualities and characteristics that made the Fleet sport-trainer such a great favorite. Of the 350 or so that were built into 1930, perhaps well over one-half of this number were of the Model 2 version, which was produced in the greatest quantity and was undoubtedly the most popular. Designed primarily for training purposes, the Fleet Model 2 was being used extensively by flying schools all over the country, and had trained thousands of pilots. Never bragging of speed as one of it's cardinal virtues, the Fleet nevertheless cruised along happily at about 90 m.p.h., with an eagerness coursing through it that was easily felt by any pilot. The ability to perform any aerobatic manuver then known was it's outstanding quality, and it's rugged structure was well able to withstand the abuse of excessive air loads. Performing outside loops was a mania that developed during 1929 and 1930, the outside loop was an abnormal

manuver both to airplane and to pilot, that was often used to prove to one and all the excellent design and sound structure of an airplane. Some manufacturers cried loud and long if their product performed one, or maybe two, and before long, the record stood at 19; Paul Mantz, stunt-pilot that he was, decided to get in the act with his Fleet 2 and racked up a record of 46 outside loops in 1930. Though capable, the Fleet was also quite friendly and gentle, and remained quite a favorite with the private-owner for just plain fun flying. Due to the strong character of the Fleet, and in view of the large number that were built, it is not unusual that a very good number have been restored to active life and are flying yet, to this day.

During the course of continuous development, the Fleet biplane was tried out with several of various powerplants; the Fleet Model 3 was a version that was powered with the 5 cylinder Wright J6 series engine of 165 h.p. and two were built as special-purpose aircraft, but the version never reached production; the Fleet Model 4 was a version that was powered with the 6 cylinder Curtiss "Challenger" engine of 170 h.p. for test; the Fleet Model 5 was a version powered with the 6

Fig. 106.   The brawny Fleet was not much on speed, but it performed with eagerness.

cylinder Brownback C-400 engine of 90 h.p. and was to have been slated for limited production but further details beyond the one example built are unknown; a special version of the Fleet was built for testing hook-on pick-ups onto a trapeze affair that projected below the belly of a dirigible, this model is believed to be the Fleet 6; among other versions tested was a Fleet 1 Special that was powered with the new 4 cylinder inverted Menasco "Pirate" engine of 95 h.p.; during the course of all this experimentation with the various combinations, normal production of about two planes per day, both in the Fleet 1 and Fleet 2, continued and by the end of 1930, some 350 examples had already been built. The type certificate number for the Fleet Model 2 as powered with the Kinner K5 engine, was issued in June of 1929 for a landplane version and amended in October of 1929 to include the seaplane version with twin-float gear; both versions could be equipped with a "belly tank" for a total fuel load of 55 gallons. The Fleet biplane was manufactured by the Fleet Aircraft, Inc. at Buffalo, New York. Lawrence D. "Larry" Bell

Fig. 107.   An ideal trainer, the Fleet Model 2 is shown here with blind-flying hood.

was the president and general manager; Joseph M. Gwinn Jr. was the chief engineer; and William B. "Bill" Wheatly was the chief pilot in charge of test and development.

Listed below are specifications and performance data for the Fleet Model 2 as powered with the 100 h.p. Kinner K5 engine; length overall 20'9"; hite overall 7'10"; wing span upper & lower 28'0"; wing chord both 45"; wing area upper 100 sq.ft.; wing area lower 95 sq.ft.; total wing area 195 sq.ft.; airfoil "Clark Y-15"; wt. empty 1010 (1063); useful load 565 (757); payload with 24 gal. fuel was 239 lbs.; payload with 55 gal. fuel was 237 lbs.; gross wt. 1575 (1820) lbs.; max. speed 113; cruising speed 95; landing speed 44 (50); climb 780 (680) ft. first min. at sea level; service ceiling 14,300 (13,020) ft.; gas cap. 24 (55) gal.; oil cap. 2.5 gal.; cruising range 360 (750) miles; price at the factory was $5500; the figures in brackets represent data for the landplane with "belly tank" and max. fuel load of 55 gallons. The following figures are for the seaplane version as mounted on twin floats; wt. empty 1183 (1223); useful load 543 (728); payload with 24 gal. fuel was 217 lbs.; payload with 55 gal. fuel was 208 lbs.; gross wt. 1726 (1951) lbs.; max. speed 103; cruising speed 88; landing speed 48 (52); climb 700 (620) ft. first min. at sea level; service ceiling 13,000 (12,500) ft.; the figures in brackets represent data for the seaplane version with "belly tank" and a max.

fuel load of 55 gallons.

The fuselage framework was built up of welded chrome-moly steel tubing, lightly faired to shape with wood fairing strips and fabric covered; the fuselage frame was a rugged structure of exceptional rigidity and weighed only 87 pounds. The bucket seats were formed of aluminum sheet and had deep wells for holding a parachute pack; quite unusual were the little rectangular windshields which were a carry-over from the Army PT series. The extra rugged wing framework was built up of heavy-sectioned solid spruce spar beams and heavy gauge stamped-out aluminum alloy wing ribs; the leading edges were covered back to the front spar to preserve the airfoil form and the completed framework was fabric covered. The fabric covering was fastened to the wing ribs with self-threading metal screws that were covered with fabric tape. The upper wing was built into a one-piece section, and the gravity-feed fuel tank was mounted in the center portion; the upper wing also had a large trailing edge cut-out and handgrips for ease of entry and exit to the front cockpit. All wing and center-section struts were of heavy gauge chrome-moly steel tubing in a streamlined section, with interplane bracing wires of heavy gauge steel, also in a streamlined section. The landing gear was of the robust "cross-axle" type with an "oleo strut" built into the lower end of each "vee"; wheels were 24x4. The fabric covered tail-

Fig. 108. A Fleet Model 2 in Canadian service; the Fleet was popular for training and sport.

Fig. 109. The Fleet Model 2 traded its wheels for skis in winter flying.

group was built up of welded chrome-moly steel tube spar members and sheet steel ribs and formers. All tail surfaces were of heavy cross-section, and the horizontal stabilizer was of a cambered "lifting section", a feature that was retained on Fleet models throughout the whole series. The fin was ground adjustable and the horizontal stabilizer was adjustable in flight; ailerons of the Freise offset-hinge type were on the lower panels only and were operated through a positive acting torque tube and bellcrank action. Like the earlier "Husky Junior", the "Fleet" was light and quick on the controls, nimble and eager, and it was capable of the complete retinue of aerobatic manuvers; manuvers that were only limited by the pilot's ability or his fortitude. The next development in the standard Fleet biplane was the Model 7, that was powered with the 5 cylinder Kinner B5 engine of 125 h.p., and introduced in the latter part of 1930; this model will be discussed in the chapter for A.T.C. # 374.

FORD "TRIMOTOR", 4-AT-E

Fig. 110. The Ford Trimotor Model 4-AT-E was an improved version of the "Tin Goose".

The Ford "Tri-Motor" certainly had become a significant part of American aviation by this time, and a sight of majestic form that was becoming quite familiar to people in all walks of life across the length and breadth of our country. Already serving on the nation's airways with such carriers as National Air Transport, Rapid Air Lines, Maddux Air Lines, Gray Goose Air Lines, Stout Air Service, Colonial-Western Airways, and several others, it's services were flung to all corners of our nation and many thousands had received their first experience in air-travel by way of the old "Tin Goose". One of the first transport airplanes to be used regularly in business-flying, it's users was an impressive roster that included such well-known firms as the Standard Oil Co., Royal Typewriter Co., Texaco Oil Co., Monarch Foods, Firestone Rubber Co., and several others. The Ford "Tri-Motor" was initially conceived to deliver efficient transport in good performance, with a dependability that would prove attractive and essential to those realizing the importance of air-travel; that it had fulfilled the promises extended is well reflected in the words related above.

The model 4-AT had progressed through continuous improvement and various stages of development, and had become a better air-

plane each time; with progress in aviation came a faster pace and a constant clamor for better performance, the next logical step in 4-AT development was an increase in horsepower. The model 4-AT-E, which was basically a model 4-AT-B with more power, was in answer to this and mounted three of the new 9 cylinder Wright J6 engines of 300 h.p. each; a combination which boosted the performance a good deal and allowed operators to offer faster schedules from point to point, with a utility and dependability that was not often matched. The model 4-AT-E seated 10 or 11 passengers with ample room and good comfort in a cabin that was sound-proofed and insulated, heated and ventilated, and neatly trimmed in good taste. A pilot and his co-pilot, who was more often just a mechanic, were seated in the forward compartment which now contained several refinements and aids that made the pilot's job a good bit easier. At this stage, one would tend to notice that though the original Ford "Tri-Motor" (4-AT-1) had been conceived less than three years ago, the "Tin Goose" had been developed into a graceful lady of impressive accomplishments; accomplishments that have been a fine example of good, sound, engineering. The type certificate number for the model 4-AT-E as powered with three Wright

J6 engines was issued in March of 1929 and some 17 or more of this model were manufactured by the Stout Metal Airplane Co., a division of the Ford Motor Co. at Dearborn, Michigan. Edsel B. Ford was the president; William B. Mayo was V.P.; William B. Stout was V.P. in charge of engineering; and Leroy Manning was the chief pilot.

Listed below are specifications and performance data for the Ford "Tri-Motor" model 4-AT-E as powered with three Wright J6 engines of 300 h.p. each; length overall 49'10"; hite overall 12'8"; wing span 74'; wing chord at root 156"; wing chord at tip 92"; total wing area 785 sq.ft.; airfoil "Goettingen 386"; wt. empty 6500-6696; useful load 3630-3434; payload 1725-1675; gross wt. 10,130 lbs.; empty wt. and useful load will vary with cabin arrangement and fuel load; max. speed 132; cruising speed 112; landing speed 58; climb 920 ft. first min. at sea level; climb in 10 min. was 7200 ft.; service ceiling 14,000 ft.; gas cap. 235 gal.; oil cap. 24 gal.; cruising range 570 miles; price at the factory field was $42,000. The fuselage framework was built up of duralumin open channel sections that were riveted together in a truss-type form; an "Alclad" aluminum alloy metal skin that was corrugated for added strength and stiffness, was riveted to the fuselage framework. All unexposed parts of the duralumin structure were treated against corrosion, and all exposed metal surfaces were covered with "Alclad", which was an aluminum alloy sheet stock that was sandwiched with two coats of pure aluminum to resist corrosion. The cabin was spacious and neatly appointed in good taste; removable seats were of aluminum alloy framing and covered in real leather, cabin windows were large and well placed, and of shatter-proof glass; the cabin walls were soundproofed and insulated. The cabin section was heated, there were lavatory facilities, and a 30 cubic foot baggage compartment. The pilot's compartment up forward was separated from the main cabin by a bulkhead, with a door leading to and from. The wing framework in three panels, was built up of duralumin spar beams that were riveted together from open section members into a truss-type beam, with wing ribs that were more or less constructed in the same manner; the completed framework was covered in corrugated Alclad sheet. The center-section panel of the wing was of constant width and section that was nearly 32 inches thick and it was built integral to the fuselage; to it were mounted the outboard engine nacelles and in it were housed the three fuel tanks of 235 gallon capacity in total. The outer wing panels were tapered in plan-form and section and were bolted to the center panel by six large bolts; ailerons were of the offset-hinge type and were operated by stranded cable. The tail-

Fig. 111. The C-9 was the 4-AT-E version in Army dress.

Fig. 112. A veteran "Tin Goose" in this 4-AT-E, serving as late as 1956.

group was metal framed and built up in a manner much the same as the wings, and it was also covered in corrugated Alclad sheet; the horizontal stabilizer was adjustable in flight. The out-rigger landing gear of nearly 17 foot tread was fastened to the fuselage and under each outer nacelle; rubber "donut" rings in compression were used as shock absorbers, and the earlier tail-skid was now replaced with a swiveling tail-wheel. Standard equipment included complete wiring for navigation lights, wheel brakes, metal propellers, and inertia-type engine starters.

Listed below are 4-AT-E entries that were gleaned from the Aeronautical Chamber of Commerce aircraft register, and checked against the listing that is presented in the "Ford Story" by Wm. T. Larkins, which is a complete history of the Ford "Tri-Motor" and is highly recommended to those desiring more complete information on this subject.

NX-7864; 4-AT-E (# 4-AT-49) 3 Wright J6-9-300

NC-9611; 4-AT-E (# 4-AT-54) 3 Wright J6-9-300

NC-9612; 4-AT-E (# 4-AT-55) 3 Wright J6-9-300

NC-9613; 4-AT-E (# 4-AT-56) 3 Wright J6-9-300

NC-9614; 4-AT-E (# 4-AT-57) 3 Wright J6-9-300

NC-9642; 4-AT-E (# 4-AT-58) 3 Wright J6-9-300

A8273; 4-AT-E (# 4-AT-59) 3 Wright J6-9-300, U.S.Marines

A8274; 4-AT-E (# 4-AT-60) 3 Wright J6-9-300, U.S.Navy

NC-9678; 4-AT-E (# 4-AT-61) 3 Wright J6-9-300

NC-8400; 4-AT-E (# 4-AT-62) 3 Wright J6-9-300

NC-8401; 4-AT-E (# 4-AT-63) 3 Wright J6-9-300

NC-8402; 4-AT-E (# 4-AT-64) 3 Wright J6-9-300

NC-8403; 4-AT-E (# 4-AT-65) 3 Wright J6-9-300

NC-8404; 4-AT-E (# 4-AT-66) 3 Wright J6-9-300

NC-8405; 4-AT-E (# 4-AT-67) 3 Wright J6-9-300

NC-8406; 4-AT-E (# 4-AT-68) 3 Wright J6-9-300

NC-8407; 4-AT-E (# 4-AT-69) 3 Wright J6-9-300

NC-8408; 4-AT-E (# 4-AT-70) 3 Wright J6-9-300

Several of the 4-AT-B models were later converted to 4-AT-E models by installation of the Wright J6-9-300 engines; some of these conversions taking place in 1930-1931. As newer and faster equipment was replacing the 4-AT-E, it was relegated to serve on the smaller feeder-lines, with air-taxi and charter operators, several were exported to foreign duty, and as the years went by, some were hauling joy-riders on scenic flights, and a number were steadily employed in the bush-country as work-a-day transports that hauled just about anything that would go in the cabin. Some of these are probably flying, even yet!

Fig. 113. The fabulous Boeing 100 was one of aviation's all-time greats.

The Boeing "Model 100", as shown here, was a handsome and dashing single-seater airplane of exciting qualities and a terrific performance; a very famous design that was the civil-commercial version of the sensational P-12 and F4B fighter series as used by this country's military and naval air services. The prototype of this whole series of airplanes apeared in 1928 as the "Design 83" and "Design 89"; two designs which were a private venture developed by Boeing engineers as a carrier-based fighter type they had hoped to sell to the Navy air arm. The Navy accepted this basic design as the F4B, and the Army accepted this design also, as the P-12. The reputation of this series of fighter airplanes is truly a lasting tribute to a great design; probably the best-known and best-liked military airplanes of the 1929-1939 decade.

The "Model 100" was developed as a single place special-purpose airplane and was powered with the 9 cylinder Pratt & Whitney "Wasp" engine of 450 h.p. Though not exactly a great deviation from the generally accepted practice, the configuration of the

"Boeing 100" was eye-catching and rather exciting to say the least. The performance of this airplane was certainly bordering on the sensational. The first example of the "Model 100" was delivered to the Department of Commerce (Airways Div.); it was registered as NS-21 and had manufacturers serial # 1142. The second airplane of this type was delivered to Pratt & Whitney Aircraft Div. as a flying test-bed for engine development; this craft was registered as NX-872H and had serial # 1143. This airplane was eventually acquired by the inimitable Milo Burcham, a daring and very famous stunt-pilot of the "thirties". The third "Model 100" was used by the Boeing company as a factory demonstrator; this craft was registered as NC-873H and had serial # 1144. This particular airplane led an active life and was eventually acquired by Paul Mantz, a famous Hollywood stunt-pilot who is very active in movie-making; incidently, this is the only example of the "100" that is left to this day. The fourth and last airplane of this type was NC-874H (serial # 1145) and it was sold in Japan.

Fig. 114.   The Boeing Model 89 was the basis for the design of the Boeing 100.

The type certificate number for the "Model 100" was issued in April of 1929, and a revised certificate for a higher empty weight was issued in June of 1929. Only four examples of this particular type were built and they were manufactured by the Boeing Airplane Company of Seattle, Wash. W. E. "Bill" Boeing was chairman of the board; Philip G. Johnson was the president; C. N. Monteith was chief of engineering, and L. R. Tower was the chief pilot.

Listed below are specifications and performance data of the Boeing "Model 100" as powered with the Pratt & Whitney "Wasp" engine of 450 h.p.; length overall 20'1"; hite overall 9'7"; span upper 30'; span lower 26'4"; chord upper 60"; chord lower 45"; wing area upper 141.5 sq.ft.; wing area lower 86 sq.ft.; total wing area 227.5 sq.ft.; airfoil "Boeing 106"; empty wt. 1805-1882; useful load 895-817; payload would be 280-202 lbs. minus the weight of the pilot; gross wt. 2700 lbs.;

Fig. 115.   This Boeing 100 was a test-bed for Pratt & Whitney engines.

Fig. 116. Capt. Ira Eaker flying the Boeing 100 on a record run; belly-tank carried extra fuel.

weights listed are as first approved 4-29, and as later revised in 6-29; max. speed 166; cruising speed 140; landing speed 56; climb 2400 ft. first min. at sea level; service ceiling 24,000 ft.; gas cap. 95 gal.; oil cap. 6 gal.; approx. range 600 miles; price at the factory field was $20,000.

The construction method of the "100" was very interesting and contained many innovations. The fuselage framework was built up in a combination of welded chrome-moly steel tubing and plate in the highly stressed portions and the balance of the fuselage structure was built up of square-sectioned aluminum alloy tubing that was bolted together; the framework was faired to shape and the forward portion up to the pilot's cockpit was covered in aluminum alloy panels, the portion aft was covered with fabric. The unusual cylinder fairings which looked so well on the "100", were an attempt to streamline the airflow around the engine cylinders which had a tendency to disturb and scatter the airstream to a great degree, and cause quite a bit of parasitic drag. This method of engine fairing was used to some extent, both here and abroad, but it actually did not justify the extra expense that it involved. A much better and cheaper method of streamlining or straightening-out the air-flows around the engine cylinders was the "Townend ring" type of low-drag engine cowling; a simple cowling that was almost universally put to use in the next year or so. The upper and lower wing panels of the "100" were built up of box-type spars that were fabricated of spruce

beams and mahagony plywood webs, wing ribs that were very closely spaced to preserve the airfoil section were built up of spruce members reinforced with mahagony plywood in a girder-type form; both upper and lower wing panels were one-piece of continuous section and were fabric covered. A push-pull strut, operated by torque tubes and bell-cranks in the lower wing, operated the ailerons which were in the upper wing panel; these were of metal semi-monococque construction and were covered with corrugated stressed-skin Alclad sheet. The thick-sectioned tail-group was also built up of metal semi-monococque construction and was covered with stressed-skin corrugated Alclad sheet; the horizontal stabilizer was adjustable in flight. The landing gear, with a tread of 76 inches, was of the split-axle type and used the cross-axle configuration that was fairly typical on other Boeing types; shock-absorbers were Boeing "oleo-struts" that were first put into use by the "Model 100". Wheels were 30x5 and wheel brakes were operated from the rudder pedals; the steerable tail-skid was mounted on the very end of the fuselage to help offset the short-couple as much as possible. As it turned out, the "100" was a "ground-looping dilly" and the pilot had to be strictly heads-up when landing this craft.

An interesting modification of the Boeing "100" type was the two-place model "100A" that was built for millionaire Howard Hughes; this model was certificated on a Group 2 approval numbered 2-83.

A.T.C. #134
(4-29)
BOURDON "KITTY HAWK", B-2

Fig. 117. The Bourdon "Kitty Hawk" Model B-2 with a Siemens-Halske engine.

Allen Bourdon's "Kitty Hawk", designed and engineered by John E. Summers, was a compact and delightful little airplane; a 3 place open cockpit biplane of basically conventional lines, but well arranged and done up into a rather pert machine. Though quite plain and simple, the "Kitty Hawk" had a compatible character and a lovable, quiet, feminine charm. The Bourdon "Kitty Hawk" model B-2, as shown here, was powered with a 7 cylinder Yankee-Siemens (SH-14) engine of 105-113 h.p. which made it into a quiet and easy-going combination that delivered a sprightly and satisfying performance. Dainty and compact, the B-2 was designed to perform all of the duties to be expected from a general-purpose airplane of this type but with the thought in mind that it was not necessary to haul a lot of extra airframe bulk around, in performance of these varied duties; consequently, the "Kitty Hawk" was an efficient airplane that performed in good economy. The type certificate number for the model B-2 as powered with the Yankee-Siemens engine of 105 h.p., was issued in April of 1929 and re-

issued for the B-2 with Yankee-Siemens of 113 h.p. and some change in weights, in September of 1929. In between these times, the B-2 (serial # 2 thru 7) were also on a Group 2 approval numbered 2-56. The "Kitty Hawk" model B-2 was manufactured by the Bourdon Aircraft Corp. at Hillsgrove, Rhode Island. Allen P. Bourdon was the president, general manager, and spare time chief pilot; Roderick Makepeace secretary, treasurer and purchasing agent; and John E. Summers was chief of design and engineering.

The first Bourdon "Kitty Hawk", powered with a 7 cylinder Ryan-Siemens engine of 97 h.p., was out in the early part of 1928; it was tested successfully but it's potential career was greatly shortened by a hangar fire in which it was destroyed. The second airplane of this type was powered with the rare "Hallett" engine for a short time; the third airplane was powered with a Ryan-Siemens engine, and the next four airplanes of this type were powered with the Yankee-Siemens engines. These engines were quite similiar to the earlier Ryan-Siemens except for a few minor changes which generally centered around the valve-action, and were being distributed here in the U.S.A. by K. G. Frank. Airplane number 8 in the "Kitty Hawk" series was powered with the 5 cylinder Kinner K5 engine of 90-100 h.p. and was designated the model B-4; see chapter for A.T.C. # 166 in this volume.

Listed below are specifications and performance data for the Bourdon "Kitty Hawk" model B-2 as powered with the Yankee-Siemens engine of 105-113 h.p.; length overall 22'11"; hite overall 8'6"; span upper and lower 28'; wing chord both 54"; wing area upper 123 sq.ft.; wing area lower 110 sq.ft.; total wing area 233 sq.ft.; airfoil "USA-27"; wt. empty 1142-1139; useful load 808-760; payload 398-350; gross wt. 1950-1899 lbs.; max. speed 105-110; cruising speed 90-93; landing speed 42-40; climb 610-730 ft. first min. at sea level; service ceiling 14,000-14,500 ft.; gas cap. 35 gal.; oil cap. 3.5 gal.; range at 6 gal. per hour was 450 miles; price at the factory field was $5200. The double figures shown for weights, were as issued 4-29 and 9-29. The fuselage framework was built up of welded 1025 steel tubing; faired to shape with wood fairing strips and fabric covered. A front cockpit entrance door was provided on the left side, and a convenient step was provided for the pilot. The wing framework was built up of solid spruce spars that were routed to an "I beam" section; wing ribs were built into a Warren truss of spruce members and plywood gussets, and the completed framework was also fabric covered. The upper wing was built in two panels that were joined together at the center-line, and fastened to a center-section cabane of inverted vee struts. Ailerons were on all four panels, and were connected in pairs with a streamlined push-pull strut. The fabric covered tail-group was built up of 1025 welded steel tubing; the fin was ground adjustable and the horizontal stabilizer was adjustable in flight. The split-axle landing gear had long telescoping legs of a fairly wide tread (88"); shock absorbers were of rubber discs in compression and later were of laced rubber shock-cord. Wheels were 26x4, without brakes, and the tail-skid was of the spring-leaf type with a hardened removable shoe. The next development in this series was the Kinner powered "Kitty Hawk" model B-4, see chapter for A.T.C. # 166 in this volume.

Listed below are Bourdon "Kitty Hawk" model B-2 entries gleaned from the Aeronau-

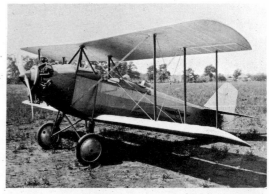

Fig. 118. An early version of the Bourdon "Kitty Hawk"; this may have been the prototype with the 97 h.p. Ryan-Siemens engine.

tical Chamber of Commerce aircraft register:

X-5598; Bourdon B-2 (# 1) Ryan-Siemens, destroyed in fire.

-7479; Bourdon B-2 (# 2) Hallett engine, used only 10 hrs.

C-7627; Bourdon B-2 (# 3) Ryan-Siemens.

C-7928; Bourdon B-2 (# 4) Yankee-Siemens

C-336E; Bourdon B-2 (# 5) Yankee-Siemens

NC-191H; Bourdon B-2 (# 6) Yankee-Siemens

NC-479 ; Bourdon B-2 (# 7) Yankee-Siemens

Discussions in this chapter generally center around April of 1929; therefore, it might be appropriate to mention the doings at the second annual Aircraft Show at Detroit, Michigan which was held in April. From the stand-point of attendance, the show was once again a record-breaker; there were more than 100 airplanes on the floor of Convention Hall, and nearly everyone of them was a riot of color. Everything from the huge transport to the open cockpit sport-plane was represented, and actual business was quite brisk; many manufacturers actually had more orders than they could hope to fill. Financiers were taking aviation seriously and many deals were consummated and planned for the expected boom yet to come. Judging from models on the floor, the greatest trend was towards the small cabin monoplane, but the small open biplane was far from being a dead issue, as could be plainly seen by the colorful and varied offerings that were on exhibit. More attention to streamlining was very evident, better finishes, and neater interiors; manufacturers had much to offer for the coming year and the outlook was indeed very excellent.

Fig. 119. The Butler "Blackhawk" was a handsome craft patterned after familiar lines.

The capable-looking "Blackhawk" was a 3 place open cockpit biplane that had a nice eager stance, a plane of many qualities that would appeal to the sportsman-pilot; powered with the Wright J5 engine of 220 h.p., it had a snappy performance with payload and range enough for extensive cross-country travels. Foremost among the "Blackhawk's" many features was a rugged construction of exceptional rigidity, which would allow the average pilot to vent his exuberance by man-handling this craft in the air as much as he dared and not have to worry about it's coming apart. The configuration of the "Blackhawk" was fairly typical to the normal, and one might say, even quite familiar. Discussions about this airplane had always brought to the fore, it's striking resemblance to the "Stearman", which might be accounted for in the fact that the "Blackhawk" was designed by a Stearman, but it was Waverly M. and not Lloyd. Waverly Stearman was Lloyd Stearman's brother, and with E. M. "Matty" Laird, the trio is credited with the designing of America's first commercial airplane (Swallow). Young Waverly had been designing airplanes since he

was 17 years old, and was considered one of the most outstanding in this field. The Butler "Blackhawk" which actually first came out as the "Skyway", was designed by Waverly M. Stearman early in 1928 and the first ship was rolled out of the shop in October of that year. Wilton M. Briney, who tested this first example, exhibited such a trust in this new design that he flew the test-hop without a parachute.

The first few models of this type were manufactured in the shops of the Butler Manufacturing Co. of which the Butler Aircraft Corp. was a subsidiary. Upon completion of first tests, the design was slightly modified into a somewhat smaller craft and the name "Skyway" was dropped; all succeeding craft were designated the "Blackhawk". The well-known Col. Art Goebel of "Dole Derby" fame, flew one of the first production models on cross-country tests from Kansas City to San Antonio, Texas; upon his return from this jaunt, he exhibited such exuberance about the airplane that he bought it for his personal use. Numerous other shake-downs and demonstration flights brought many praises from the pilots on it's stability and sure-footed ease of handling; truly a plane designed with the sportsman-pilot in mind. "Hoot" Gibson, well-known "cowboy" movie-actor for years, loved a good, spirited airplane and bought one of the first "Blackhawk" to come off the line. A good half-dozen or so were quickly sold, but the "bank scare" and the resulting crash in the stock markets of 1929 practically nullified this plane's sales potential and consequently, not very many were built. The Butler company also developed and built a four place cabin biplane they called the "Coach", and later tried a last-ditch stand with a sport biplane called the "Blackhawk Sport", but neither of these made any great impression on the market and eventually all manufacturing activities were suspended. Though the "Blackhawk" was few and rare, one or two of these were being used as "crop dusters" even into the latter "thirties".

The type certificate number for the "Blackhawk" as it appeared in production was issued in April of 1929 and it was manufactured by the Butler Aircraft Corp., a subsidiary of the Butler Manufacturing Co. at Kansas City, Missouri. E. E. Norquist was the president; Wm. A. Knapp and Roy S. Kemp were V.P.; Oscar D. Nelson was treasurer; Waverly M. Stearman was chief of design and engineering; Milton C. Bauman was assistant engineer, and Wilton M. Briney was chief pilot and in charge of sales.

Listed below are specifications and performance data for the Butler "Blackhawk" as powered with the Wright J5 engine of 220 h.p.; length overall 24'; hite overall 9'; span upper 34'; span lower 28'6"; chord upper 64"; chord lower 54"; wing area upper 180 sq.ft.; wing area lower 115 sq.ft.; total wing area 295 sq.ft.; empty wt. 1885; useful load 1015; payload with 70 gal. fuel 360 lb.,; payload with 45 gal. fuel 510 lb.; gross wt. 2900 lbs.; max. speed 130; cruising speed 110; landing speed 47; climb 1,000 ft. first min. at sea level; service ceiling 14,000 ft.; gas cap. 45-70 gal.; oil cap. 8-10 gal.; max. cruising range at 11.5 gal. per hour was 650 miles; price at the factory was $7995. The fuselage framework was built up of welded steel tubing in varying grades to suit the strain that would be imposed, with allowance for ample safety factor; the fuselage frame was faired to shape and fabric covered. The steel tubing throughout the structure was treated internally and externally for protection against rust and corrosion. The cockpits were upholstered in real leather, the pilot's seat was adjustable and had a well to fit a parachute pack. Both right-hand and left-hand throttles were provided for the pilot so that he could fly with either hand, and individually operated wheel brakes were connected to brake pedals adjacent to the rudder pedals. A metal panel could be quickly attached over the front cockpit when not being used; covering the gaping front cockpit could increase the cruising speed by as much as 5 m.p.h. The wing framework was built up

Fig. 120. Designed by Waverly Stearman, the prototype was called "Skyway".

Fig. 121. This "Blackhawk" was owned by Art Goebel, who took a fancy to its performance.

of solid spruce spars and spruce and plywood built-up ribs, also fabric covered; ailerons were on the lower panels only. The main fuel tank was in the fuselage, but there were two fuel tanks, one in the root section of each upper wing for extra fuel supply. The upper wing was built in two sections and joined together at the center-line; there being no separate center-section panel as would usually be the case. The fabric covered tail-group was built up of welded chrome-moly steel tubing; the fin was ground adjustable and the horizontal stabilizer was adjustable in flight. The split-axle landing gear was of the cross-axle configuration and was of fairly wide tread, it was built up of chrome-moly steel tubing in streamlined section; wheels were

Fig. 122. The 4-place Butler "Coach" was a planned version, but built only as a prototype.

30x5 and wheel brakes were standard equipment. A baggage compartment with a door on the side, was located in the fuselage behind the pilot's cockpit. A metal propeller, exhaust collector-ring, and hand crank inertia-type engine starter were also offered as standard equipment.

Listed below are "Butler" entries gleaned from the Aeronautical Chamber of Commerce aircraft register; this may not account for all that were built, but it is more or less the bulk of them:

X-7857; Skyway (# 100) J5, prototype.
        (#101) unaccounted for.
X-146E; Coach (# 102) J5, 4 place cabin.
NC-521; Blackhawk (# 103) J5, Art Goebel, Grp. 2-49.
C-896E; Blackhawk (# 104) J5
NC-730K; Blackhawk (# 105) J5, "Hoot" Gibson.
NC-599H; Blackhawk (# 106) J5
NC-598H; Blackhawk (# 107) J5
NC-597H; Blackhawk (# 108) J5
        (# 109) unaccounted for.
NC-593H; Blackhawk (# 110) J5
X-299N; Blackhawk (# 111) J6-5-165, "Sport".

Fig. 123. The Stinson "Detroiter" Model SM-1F had a background of illustrious achievements.

The illustrious Stinson "Detroiter" monoplane had an enviable background of achievements not very often equalled by any one airplane; born in an age of trans-oceanic flights, endurance flights, and all manner of attempts to seek world-wide recognition, and to give proof of capabilities in contest, the wandering "Detroiter" had finally settled down, so to speak, had taken it's bows and now was content to take it's place alongside others as a vehicle of commerce. Changing slightly and improving progressively in the past year or so, the "Detroiter" evolved into the SM-1D series (see chapters for A.T.C. # 74-76-77-78 in U. S. CIVIL AIRCRAFT, Vol. 1) which laid the foundation for the development of the SM-1F. The model SM-1F was a high wing cabin monoplane of typical Stinson configuration that seated six with ample room and comfort, and was powered with the 9 cylinder Wright J6 engine of 300 h.p. Whereas the SM-1D series were rather underpowered with the 220 h.p. Wright J5, and were soon converted to the SM-1D300 which mounted the 300 h.p. Wright J6 engine, the lessons learned in this venture brought about the development of the new SM-1F. which had performance to spare and certainly asked no favors. Being of ideal size for the small feederline, it was used to transport cargo and passengers from outlying areas into stations of the main airway systems, used by business men as an air-taxi to points of call not served by regular transports, or hauling loads of cargo into areas that were best served by the airplane. Two of this model were serving in Alaska, and four were being used to haul mail in China. The SM-1F was designed to work and was doing it admirably.

The type certificate number for the model SM-1F as powered with the 300 h.p. Wright J6 series engine, was issued in April of 1929 and some 26 or more of this model were built within a year's time. The SM-1F was manufactured by the Stinson Aircraft Corp. at Wayne, Michigan in their new plant that was taking shape into one of the finest aircraft producing installations in the land. Edward A. Stinson was the president; Harvey J. Campbell was the V.P.; William A. Mara was secretary and in charge of sales; Wm. C. Naylor and Kenneth M. Ronan were engineers; and Randolph Page was the chief pilot in charge of test and development. Though "Eddie" Stinson was the president, he did not conduct much of his duties from behind a desk, he was an ambassador-at-large and more at home behind the control wheel of a Stinson airplane; it was not unusual for him to test-hop an experimental airplane after following it through all stages of it's development and manufacture, or fly clear across the country to set up plans for a distributing organization. If paper-work had piled up while he had been busy elsewhere, his loyal staff was always eager to give him a helping hand to get out from under.

Listed below are specifications and performance data for the Stinson "Detroiter" model SM-1F as powered with the Wright J6 engine

Fig. 124. The 300 h.p. Wright J6 engine powered the SM-1F "Detroiter" and gave better performance.

of 300 h.p.; length overall 32'8"; hite overall 9'0"; wing span 46'8"; wing chord 84"; total wing area 292 sq.ft.; airfoil "Modified M-6"; wt. empty 2614; useful load 1686; payload with 100 gal. fuel was 874 lbs.; gross wt. 4300 lbs.; max. speed 132; cruising speed 113; landing speed 56; climb 850 ft. first min. at sea level; service ceiling 16,000 ft.; gas cap. 100 gal.; oil cap. 6 gal.; cruising range at 15 gal. per hour was 680 miles; price at the factory field was $13,500. The fuselage framework was built up of welded chrome-moly steel tubing and was gusseted at every joint with chrome-moly plate stock to make a structure of exceptional strength and rigidity; the framework was faired to shape with formers and fairing strips and fabric covered. The cabin walls were sound-proofed and insulated, and upholstered in tasteful combinations of mohair; the cabin was heated by hot air coming off the exhaust manifold of the engine. All windows were of shatter-proof glass and could be rolled up or down; there was a door and a convenient step on each side for exit or entry into the cabin. The engine was well muffled by a collector-ring and long tail-pipes; this allowed near-normal conversation at all times. The wing framework was built up of heavy-sectioned solid spruce spars; the front spar was routed to an "I-beam" section and the wing ribs were built up of spruce and plywood in a truss-type form. The leading edge was covered with duralumin sheet back to the front spar to preserve the airfoil form, and the completed framework was fabric covered. The ailerons were of a welded steel tube framework and were fabric covered; the fuel supply was carried in two tanks that were mounted in the wing, flanking each side of the fuselage. The fabric covered tail-group was built up of welded chrome-moly steel tubing; the fin was ground adjust-

Fig. 125. The SM-1F "Detroiter" hauled mail on Chinese routes; preparation for the initial flight is shown here.

Fig. 126. A later version of the SM-1F had a low-drag engine cowling for added speed.

able and the horizontal stabilizer was adjustable in flight. The wing bracing struts were of heavy gauge chrome-moly steel tubes that were encased in balsa-wood fairings; these fairings were shaped to airfoil section to give added lift and stability. The wide-tread landing gear was of the outrigger type and used air-oil shock absorber struts; wheel brakes and a steerable tail-wheel made ground manuvering an easy task. The wheels were 32x6 and wheel tread was 110 inches. A metal propeller and an inertia-type engine starter were standard equipment. The next development in the "Detroiter" monoplane was the SM-1FS which was typical but was mounted on twin-float seaplane gear; this version will be discussed in the chapter for A.T.C. # 212. The next development in the Stinson monoplane, following the SM-1F as described here, was the "Junior" model SM-2AA; see chapter for A.T.C. # 145 in this volume.

Listed below are SM-1F entries that were gleaned from various records; this list is not complete, but it does show the bulk of this model that were built.

NC-9691; SM-1F (# 500) Wright J6-9-300.
NC-9693; SM-1F (# 501) Wright J6-9-300.
NC-9692; SM-1F (# 502) Wright J6-9-300.
NC-9698; SM-1F (# 503) Wright J6-9-300.
C-8420; SM-1F (# 504) Wright J6-9-300.
C-8421; SM-1F (# 505) Wright J6-9-300.
C-8424; SM-1F (# 506) Wright J6-9-300.
C-8429; SM-1F (# 507) Wright J6-9-300.
C-8430; SM-1F (# 508) Wright J6-9-300.
      ; SM-1F (# 509) Wright J6-9-300.
      ; SM-1F (# 510) Wright J6-9-300.
      ; SM-1F (# 511) Wright J6-9-300.
      ; SM-1F (# 512) Wright J6-9-300.
C-8468; SM-1F (# 513) Wright J6-9-300.
C-8436; SM-1F (# 514) Wright J6-9-300.
C-8469; SM-1F (# 515) Wright J6-9-300.
NC-445H; SM-1F (# 516) Wright J6-9-300.
NC-448H; SM-1F (# 517) Wright J6-9-300.
      ; SM-1F (# 518) Wright J6-9-300.
NR-487H; SM-1F (# 519) Wright J6-9-300.
NC-404M; SM-1F (# 520) Wright J6-9-300.
NC-413M; SM-1F (# 521) Wright J6-9-300.
NC-910W; SM-1F (# 522) Wright J6-9-300.
      ; SM-1F (# 523) Wright J6-9-300.
NC-961W; SM-1F (# 524) Wright J6-9-300.
NC-966W; SM-1F (# 525) Wright J6-9-300.
Serial # 509-510-511-512 believed to have been delivered to Chinese government for airmail route.

Fig. 127. The Stearman C3MB was a single-seater with covered cargo hold for carrying mail.

As the chapters of bygone days in aviation are written, the "Stearman" biplane without any doubt will be remembered as a very capable and very proud airplane; proud of it's exceptional ability and proud also of it's heritage. Altogether, it has had a very useful and very commendable existence, and must surely be regarded as one of our all-time greats in the make-up of early commercial aviation; a budding industry that stood on the threshold of a wonderful future, back there in 1927. Well planned and well arranged, it is known fact that the basic design was so excellent from the very outset, that it remained the basis for every "Stearman" airplane that was ever built. Like the thoroughbred that it was, every Stearman had that visible evidence of true class and good breeding; a craft that was blessed with many inborn attributes that held up admirably throughout all of the models that were produced in the following years. By nature, the Stearman biplane was extremely rugged in character and the unfailing dependability, especially "when the chips were down", was a by-word long known among the folks that fly. As time went on, it's use and proven success as a "Pony Express" on many of the

short haul feeder-lines that fed the main transcontinental airmail system, was more than likely it's greatest claim to fame and fond remembrance; as a work-a-day airplane the performance and utility of the Wright J5 powered C3B biplane was surely among the very best. They were a charm to fly, with spirit and sure-footed determination, yet were obedient and well-behaved.

The Stearman model C3MB as discussed here, was a development from the model C3B and was typical, in most all respects except that the front cockpit, which normally held two passengers, was now converted into a cargo carrying compartment of some 33 cubic feet that was covered with a hinged metal hatch to protect it's load. Used strictly for carrying air-mail and express cargo, the model C3MB was a one-place airplane and was still powered with the 9 cylinder Wright "Whirlwind" J5 engine of 220 h.p. The C3MB version was not actually a new design, but more or less a conversion, and many of the existing C3B type and even the earlier C2B type, that were being used daily on various routes across the country, were modified to conform to this new specification. All of the Stearman biplanes of the C2B and

C3B type were eligible for this conversion to the new specification. Among the carriers that modified their equipment to conform to this version, or ordered new equipment in this version were: National Air Transport; Varney Air Lines; Western Air Express; Texas Air Transport; Interstate Air Lines; National Parks Airways; and Continental Air Lines. In all, some 18 or more examples of the C3MB type were serving regularly until replaced some years later by heavier equipment. The type certificate number for the model C3MB as powered with the Wright J5 engine was issued in April of 1929. The Stearman biplane was manufactured by the Stearman Aircraft Company at Wichita, Kansas. Lloyd Stearman was the president and general manager; Mac Short was V.P. and chief engineer; H. A. Dillon was treasurer; Walter Innes Jr. was secretary; J. E. Schaefer was sales manager; and David P. "Deed" Levy was the chief pilot. Working for Stearman at this time was a quiet and studious young chap named Hall L. Hibbard, that only 4 or 5 years later became the chief engineer at Lockheed Aircraft Co.

Listed below are specifications and performance data for the Stearman biplane model C3MB as powered with the 220 h.p. Wright J5 engine; length overall 24'0" (23'6" without prop spinner); hite overall 9'0"; wing span upper 35'0"; wing span lower 28'0"; wing

Fig. 128. The original Stearman mailplane for Varney Air Lines was a C2MB.

chord upper 66"; wing chord lower 54"; wing area upper 187 sq.ft.; wing area lower 109 sq.ft.; total wing area 296 sq.ft.; airfoil "Stearman # 2"; wt. empty 1895; useful load 935; payload with 68 gal. fuel was 330 lbs.; gross wt. 2830 lbs.; max. speed 130; cruising speed 112; landing speed 45; climb 1000 ft. first min. at sea level; service ceiling 17,500 ft.; gas cap. 68 gal.; oil cap. 8 gal.; cruising range 560 miles; price at the factory field was $8730. The fuselage framework was built up of welded chrome-moly steel tubing, faired to shape with wood fairing strips and fabric covered; the space normally used as a front cockpit was converted into a 33 cubic foot cargo compartment for hauling air-mail and express cargo. The wing framework was built up of solid spruce spar beams and spruce and plywood built-up wing ribs, the

Fig. 129. This handsome C3B had the second cockpit covered when not carrying a passenger.

Fig. 130. The simple grace of the Stearman biplane was a familiar sight on early airways.

completed framework was fabric covered. The main fuel tank was mounted high in the fuselage, just ahead of the cargo compartment, and a gravity-feed fuel tank for extra fuel load was mounted in the center-section panel of the upper wing. Ailerons were on the upper panels only and were operated by push-pull tubes that came out of the pilot's cockpit and up into the center-section panel of the upper wing, where they were connected to torque tubes and bellcranks for smooth and positive action. The fabric covered tail-group was built up of welded chrome-moly steel tubing, the fin was ground adjustable and the horizontal stabilizer was adjustable in flight. The landing gear of 90 inch tread was of the outrigger type and used a combination of "oleo" shock absorber struts that were snubbed by rubber shock-cord; wheel brakes and 30x5 wheels were standard equipment. The C3MB versions and conversions did not all look alike; some had prop spinners and no head-rest, and some had head-rest and no propeller hub "spinner", this would account for different times of manufacture. A metal propeller, inertia-type engine starter, and a

full complement of night flying equipment were available for the C3MB. The next Stearman development was the 5 place light transport model LT-1, this "Hornet" powered craft will be discussed in the chapter for A.T.C. # 187 in this volume.

Listed below are Stearman C3MB entries that were gleaned from various records; this list may not be complete, but it does show the bulk of this model that were in operation on the various air-lines mentioned, during 1929 and 1930.

C-1598; C3MB (# 103) Wright J4, modified from C2B.

C-3863; C3MB (# 106) Wright J5, modified from C2B.

C-4011; C3MB (# 108) Wright J5, modified from C2B.

C-4273; C3MB (# 111) Wright J4B, modified from C2B.

C-5500; C3MB (# 114) Wright J4B, modified from C2B.

C-4552; C3MB (# 115) Wright J5, modified from C3B.

C-7171; C3MB (# 138) Wright J5, modified from C3B.

C-6486; C3MB (# 172) Wright J5, modified from C3B.

C-6436; C3MB (# 176) Wright J5, modified from C3B.

C-6487; C3MB (# 178) Wright J5.

C-6490; C3MB (# 188) Wright J5

C-9058; C3MB (# 201) Wright J5.

C-9057; C3MB (# 205) Wright J5.

C-6496; C3MB (# 207) Wright J5.

C-9059; C3MB (# 210) Wright J5.

C-9060; C3MB (# 211) Wright J5.

C-9061; C3MB (# 212) Wright J5.

C-8836; C3MB (# 242) Wright J5, modified from C3B.

A.T.C. #138
(4-29)
STAR "CAVALIER", B

Fig. 131. The Star "Cavalier" Model B had a 55 h.p. Velie engine; the "Cavalier" seated two side-by-side.

The Star "Cavalier" was an interesting little cabin monoplane of normal configuration that seated two side-by-side in chummy comfort, and was powered with the 5 cylinder Velie engine of 55 h.p. Arranged so as to appeal to the small fixed-base operator or private-owner, it had good flight characteristics and ample performance for an airplane of this type; a simple and rugged structure offered care-free flying with a minimum amount of repair and maintenance. Though possessing many desirable features, and a very pleasant nature, the "Cavalier" was destined to come up with a "tough row to hoe"! Being of a similiar type, the little "Cavalier" monoplane had to compete for sales against the sky-rocketing popularity of the "Monocoupe"; right then and there it was at a decided disadvantage because most everyone seems to support a winner, and the pixie-like "Monocoupe" was surely the winner in this case. Several of the Velie-powered "Cavalier" (Model B) were built, and a good number of various other models too, that were powered with engines such as the 5 cyl. LeBlond 60, the 5 cyl. Lambert 90, the 5 cyl. Armstrong-Siddeley "Genet" 80, and 2 or 3 other versions. In all, some 55 of the various "Cavalier" type were reported built in a production period of 4 or 5 years. The type cer-

tificate number for the Velie powered "Cavalier" (Model B) monoplane, was issued in April of 1929 and it was manufactured by the Star Aircraft Company at Bartlesville, Oklahoma. Wm. D. Parker was general manager and in charge of sales; E. A. Riggs was chief of design and engineering. Designed by E. A. "Gus" Riggs and Wm. D. "Billy" Parker, a combination of great talents and experience, the "Cavalier" was formally announced in September of 1928 at the Los Angeles Aircraft Exposition, held in connection with the National Air Races. Further development and certification of the first model was taking time and it was not until the latter part of 1929 that "Cavaliers" began making an appearance here and there.

The well-known "Billy" Parker launched his aviation carreer in 1912; circumstances prevailing forced him to build his own airplane, and then he had to learn to fly it, by himself. From that time on, Parker barnstormed throughout the northwestern part of the country during the summer months, and his winters were occupied with the task of building new airplanes for the following season. The performance of Parker-built airplanes was soon recognized amongst the flying fraternity and he was obliged to build a few for other barn-storming pilots. Following

Fig. 132. The "Cavalier" B seated two in comfort; it was rugged and easy to fly.

World War 1, where he served as a captain in the British R.F.C., and as chief-instructor for a flying-school here in the U.S.A., "Billy" Parker became a test pilot where he became acquainted with the problems and struggles of aircraft design, development, and manufacture, in the early "twenties". E. A. "Gus" Riggs was one of America's earliest aeronautical engineers; designing and building some advanced tractor-type biplanes (engine in front) before World War 1 for some of the country's most noted exhibition pilots. His teaming up with "Billy" Parker led to the design and development of the "Cavalier" monoplane series.

Listed below are specifications and performance data for the "Cavalier" model B as powered with the Velie engine; length overall 19'11"; hite overall 6'3"; wing span 31'6"; wing chord 61"; total wing area 157 sq.ft.; airfoil "Clark Y"; wt. empty 862; useful load 538; payload 199; gross wt. 1400 lbs.; max. speed 100; cruising speed 85; landing speed

38; climb 575 ft. first min. at sea level; service ceiling 10,000 ft.; gas cap. 25 gal.; oil cap. 3 gal.; cruising range 500 miles; price at the factory was $2985. According to the following figures, the first "Cavalier" must have been a light-weight with a rather snappy performance; wt. empty 575; useful load 475; payload 375; gross wt. 1050 lbs.; max. speed 105; cruising speed 87; landing speed 35; climb 700 ft. first min. at sea level; service ceiling 12,000 ft.; gas cap. 13 gal.; oil cap. 2 gal.; cruising range at 20 m.p.g. was 200 miles. Dimensions of this early version were as listed above, but, as is most always the case, production airplanes are always much heavier than the prototype and performance suffers in direct proportion. The fuselage framework was built up of welded chrome-moly steel tubing in a Warren truss form; the framework was faired to shape with wood fairing strips and fabric covered. The wing framework was built up of solid spruce spars that were routed to an "I beam" section,

Fig. 133. The "Cavalier" was designed for low-cost flying pleasure.

Fig. 134. A "Cavalier" B, still flying in the late 1950's.

with spruce and plywood truss-type wing ribs, also fabric covered; the fuel supply was carried in two tanks that were mounted in the wing, flanking the fuselage. The cabin was trimmed and upholstered neatly, with large cabin windows and an overhead sky-light for good visibility; there was a baggage shelf in back of the seat. The fabric covered tail-group was built up of welded chrome-moly steel tubing; the fin was ground adjustable and the horizontal stabilizer was adjustable in flight. The landing gear was of the normal split-axle type and employed two spools of rubber shock-cord to absorb taxiing loads and landing shock; wheel tread was 66 inches and wheels were 26x4. There was a "Cavalier" version with the 4 cyl. upright "Gipsy" engine that was entered in the 1929 National Air Races held at Cleveland, by Ennis L. Stewart, but other details about this particular craft are unknown. The next development in the "Cavalier" series was the LeBlond 60 powered Model C; this version will be dis-

cussed in the chapter for A.T.C. # 255.

Listed below are "Cavalier" B entries that were gleaned from the Aeronautical Chamber of Commerce aircraft register; this list may not be complete, but it does show the bulk of this model that were built.

C-7239; Cavalier B (# 101) Velie
C-7240; Cavaier B (1# 102) Velie
C-453; Cavalier B (# 103) Velie
C-451; Cavalier B (# 104) Velie
C-450; Cavalier B (# 105) Velie, unverified.
NC-331H; Cavalier B (# 107) Velie
NC-263K; Cavalier B (# 108) Velie
C-990H; Cavalier B (# 109) Velie
C-960H; Cavalier B (# 111) Velie
NC-350M; Cavalier B (# 112) Velie
NC-351M; Cavalier B (# 113) Velie
NC-24N; Cavalier B (# 114) Velie
NC-941E; Cavalier B (# 117) Velie
Some of the unlisted serial numbers represent models with LeBlond 60, Gipsy, Genet, and Lambert 90 engines.

Fig. 135. The Alexander "Eaglerock" Model A-12 had a "Comet" engine.

The new A-series "Eaglerock" biplane which was introduced about mid-year in 1928, had been selling in very good number throughout the year and was being produced in four different models; the A-1, A-2, A-3, and A-4. In between the regular production run, there was a constant experimentation with various air-cooled powerplants in an effort to develop new models to fortify the line. The hairy-chested model A-5 with the Menasco-Salmson engine of 230-260 h.p., was developed in time to be flown in the National Air Tour for 1928 by Benny O. Howard, but this combination, though a sky-rocketing performer, was not entirely satisfactory. Several more versions were tried in rapid succession, with powerplants such as the Ryan-Siemens (Siemens-Halske), the Anzani 120, the Hallett, the Warner "Scarab", the Floco (Axelson), and several unverified versions that one would hear about from time to time, but none of these were ear-marked for regular production. The Ryan-Siemens powered A-7 showed good promise, and one example of this model was built, but various circumstances that cropped up, nipped this project in the bud and further development was not carried out. In the meantime, the "Comet" powered version was tested satisfactorily and added to the line as the model A-12.

The "Eaglerock" model A-12 as powered with the 7 cylinder "Comet" radial air-cooled engine of 130-150 h.p., was a large three place open cockpit biplane of a configuration that was surely quite familliar to the flying-folk by this time; upwards of 200 in the A-series "Eaglerock" had already been built in several different models, and they were flying in all parts of the country. Added to this of course, were about 400 or so of the old Combo-Wing and Long-Wing type. The "Comet" powered A-12 was a good flying airplane of pleasant character, and shared all the other inherent qualities that made the "Eaglerock" biplane such a great favorite. The type certificate number for the model A-12 was issued in April of 1929 and some nine or more of this model were manufactured by the Alexander Aircraft Co. at Colorado Springs, Colorado; a subsidiary division of Alexander Industries. J. Don Alexander was the president; D. M. Alexander was V.P.; J. A. McInaney was V.P. in charge of sales; Al W. Mooney was chief of engineering; and Cloyd Clevenger was chief pilot in charge of test and development. Daniel Noonan, who had designed the very first "Eaglerock", had left the firm and was now in charge of production for American Eagle Aircraft. The 7 cylinder "Comet" engine was first introduced with

115 h.p., but was soon re-rated for 130 h.p. Further development and refinement boosted horsepower to 150, and the latest version of 1929-30 was putting out 165 horsepower.

Listed below are specifications and performance data for the "Eaglerock" model A-12 as powered with the "Comet" engine of 130 h.p.; length overall 24'11"; hite overall 9'10"; wing span upper 36'8"; wing span lower 32'8"; wing chord both 60"; wing area upper 183 sq.ft.; wing area lower 153 sq.ft.; total wing area 336 sq.ft.; airfoil "Clark Y"; wt. empty 1627; useful load 953; payload 370; gross wt. 2580 lbs.; max. speed 112; cruising speed 94; landing speed 40; climb 720 ft. first min. at sea level; service ceiling 11,800 ft.; gas cap. 59 gal.; oil cap. 7 gal.; cruising range 650 miles; price at the factory field was $5847. Performance figures with the "Comet" of 150 h.p., and 165 h.p. were proportionately better. The fuselage framework was built up of welded chrome-moly steel tubing in a truss form, and was heavily faired to shape with wood fairing strips and fabric covered. The cockpits were deep and the occupants were well protected; there was a small baggage compartment in the turtle-back section of the fuselage, just behind the pilot's cockpit. The wing panels were built up of solid spruce spar beams that were routed to an "I beam" section, with spruce and plywood built-up wing ribs; the completed panels were fabric covered. There were four ailerons that were connected in pairs by a streamlined push-pull strut. The fabric covered tail-group was also built up of welded chrome-moly steel tubing and was braced with streamlined tie-rods; the fin was ground adjustable and the horizontal stabilizer was adjustable in flight. Dual controls were available, and the set in the front cockpit were quickly removable when carrying passengers. The robust landing gear was of the split-axle type and used two spools of shock-cord to absorb the bumps; wheels were 28x4. The wings were wired for navigation lights as standard equipment, but wheel brakes, metal propeller, and engine starter were optional equipment. The next development in the A-series "Eaglerock" biplane was the Curtiss "Challenger" powered model A-13; see chapter for A.T.C. # 141 in this volume.

Listed below are "Eaglerock" A-12 entries that were gleaned from the Aeronautical Chamber of Commerce aircraft register; this list may not be complete, but it does show the bulk of this model that were built:

C-504E; Eaglerock A-12 (# 763) Comet
C-8217; Eaglerock A-12 (# 785) Comet
C-8248; Eaglerock A-12 (# 787) Comet
C-8270; Eaglerock A-12 (# 788) Comet
C-8277; Eaglerock A-12 (# 843) Comet
C-8257; Eaglerock A-12 (# 844) Comet
NC-733H; Eaglerock A-12 (# 860) Comet
NC-752H; Eaglerock A-12 (# 861) Comet
NC-308W; Eaglerock A-12 (# 957) Comet

Fig. 136. A later version of the A-12 had a 165 h.p. "Comet" engine.

Fig. 137. *A rare and brawny version was the 260 h.p. Model A-5.*

Fig. 138. The Lockheed "Vega" 2 was powered by a 300 h.p. Wright J6 engine.

The well known Lockheed "Vega" was a man-made star that shone brightly and consistently in these days of aviation; performing feats and accomplishments of record-breaking nature that kept the name blazed in head-lines all over the nation, whereby it earned the respect and admiration from people in all walks of life. The "Vega" was like a magic combination of frame and power which sought the thrills and romance that lay just beyond the horizon, and it even looked like a fitting steed for an impatient globe-trotter that was ever-ready to be off on some expedition or another. Without question, the Lockheed "Vega" was a design of compatible marriage and pure symmetry, which stands well to be remembered as one of the most outstanding developments in aviation from this day and age.

First produced as the Vega 1 with the Wright "Whirlwind" J5 engine, and shortly after as the Vega 5 with the thundering "Wasp" engine, the Vega was now offered in a version powered with the 300 h.p. Wright J6 series engine as the Vega 2. The Vega 2 was also a five place high wing cabin monoplane with the typical internally braced cantilever wing and the familiar cigar-shaped

monococque fuselage; an uncluttered combination that translated into a useful efficiency and a bonus in speed. Though endowed with speed, the Vega was no temperamental "racer" by any means, and served usefully, efficiently, and faithfully on several air-lines or as a personal transport; a few even finding their way into service in Canada. The Model 2 with the J6-300 engine, also known as the "Whirlwind-Vega", was sort of an in-between model in the Vega line that was offered by Lockheed, and was built in rather small number; no example of this model had any great claim to fame but they were a worthwhile addition to the series. The type certificate number for the Vega Model 2 as powered with the 300 h.p. Wright J6 engine, was issued in April of 1929 and some four or more examples of this model were manufactured by the Lockheed Aircraft Company at Burbank, Calif. The venerable Eddie Bellandi had long been the chief test-pilot for Lockheed, and since then, both Lee Schoenhair and Wiley Post had taken a turn at this chore during this period of time.

Listed below are specifications and performance data for the Lockheed Vega 2 as powered with the 300 h.p. Wright J6 engine;

Fig. 139.   The "Vega" 2 carried five and could top 155 m.p.h.

length overall 27'6"; hite overall 8'4"; wing span 41'0"; wing chord at root 102"; wing chord at tip 63"; M.A.C. 80"; total wing area 275 sq.ft.; airfoil at root Clark Y-18; airfoil at tip Clark Y-9.5; wt. empty 2140; useful load 1713; payload with 100 gal. fuel was 860 lbs.; gross wt. 3853 lbs.; max. speed 155; cruising speed 133; landing speed 50; climb 950 ft. first min. at sea level; service ceiling 17,000 ft.; gas cap. 100 gal.; oil cap. 10 gal.; cruising range 800 miles; price at the factory was about $14,000.

The monocoque fuselage was of 5/32 inch spruce veneer glued in layers and then formed to shape in a mold called a "concrete tub" under 150 pounds pressure; the two shell halves were then assembled over circular laminated spruce formers that were held in line by a few horizontal wood stringers. After the shells were joined, necessary cut-outs were formed for the windows, entry door, wing attach, and the pilot's cockpit. The cantilever wing framework was built up of spruce and plywood box-type spar beams and the wing ribs were built up of laminated ply-

wood webs with spruce cap-strips; the completed framework was covered with 3/32 inch laminated spruce plywood veneer. The cantilever tail-group was all-wood and of a similiar construction as the wing; the horizontal stabilizer was adjustable in flight. The two fuel tanks of 16 gauge aluminum were mounted in the wing for gravity-feed; one tank was each side of the center-line. The split-axle landing gear of 96 inch tread was of two long telescopic legs with oleo shock absorbers; wheels were 30x5 and brakes were standard equipment. The engine was well muffled by a large-volume collector-ring which kept the noise level down and enabled near-normal conversation in the cabin at all times. A metal propeller and an electric engine starter were standard equipment. The next development in the Lockheed "Vega" series was the six place model 5-A powered with the Pratt & Whitney "Wasp" engine, see the chapter for A.T.C. # 169 in this volume.

Listed below are Vega 2 entries that were gleaned from various records; this list may not be quite complete, but it does show the bulk of this model that were built.
NC-7895; Vega 2 (# 30) Wright J6.
NC-623E; Vega 2 (# 58) Wright J6.
NC-2875; Vega 2 (# 60) Wright J6.
NC-857E; Vega 2 (# 64) Wright J6.
Vega 2 serial # 30 was first a Vega 1 with J5 engine, after modification into Vega 2 it served in Canada as CF-AAL; Vega 2 serial # 60 and # 64 with Canadian-American Air Lines; Vega 2 serial # 58 originally sold to Schlee-Brock.

Fig. 140.   A "Vega" 2 in Canadian service; the "Vega" was popular in many countries.

Fig. 141.   The Alexander "Eaglerock" Model A-13 with 6 cylinder Curtiss "Challenger" engine.

In an effort to capture every bit of business that was available, many manufacturers of the all-purpose three place open cockpit biplane were lashing out in all directions to produce the models that would please or suit as many buyers as was possible. Conformance to this sort of thing was rather contagious and Alexander, of course, was deemed to follow suit; they were already producing five models regularly in various engine combinations from 90 to 220 h.p., and now, the A-13 was introduced as the sixth. One would probably stop to question the rationality of all this, but it must be borne in mind that not everyone could be pleased with only a limited offering, so as the saying goes, variety was the spice of life. The models A-2, A-3, and A-4, were powered with water-cooled engines from 90 to 180 h.p.; the model A-1 was powered with an air-cooled engine, but it's power stood at 220. There seemed to be a need for a combination somewhere in between. The A-12 was an in-between model that sported 130 horse, and the A-13 with it's Curtiss "Challenger" engine was now another that had 170 h.p. Whatever your purse or purpose, there was ample opportunity to take your pick.

The model A-13 as powered with the famous 6 cylinder Curtiss "Challenger" engine of 170 h.p., was a very compatible combination that delivered a good performance, with exceptional reliability. Much the same as any other "Eaglerock", the A-13 was docile and obedient, but could be spurred into some lively action upon the command of a good pilot. The type certificate number for the "Challenger" powered model A-13 was issued in May of 1929 and some 6 or more of this model were manufactured by the Alexander Aircraft Co. at Colorado Springs, Colorado. It might be interesting and revealing to note that Alexander's "factory field" was the highest factory airport in the world; towering some 6200 feet above sea level. This is well over a mile high, and this is where the "Eaglerocks" were born and reared to make their way to all parts of our country, and even into Canada, Alaska, and Mexico. Thus it is certainly no wonder that the "Eaglerock" performed so well at high altitudes, and in the rugged, mountainous portions of our west. Buying airplanes on the installment plan was introduced to the industry by Alexander, way back in 1926; Alexander's present plan was 40 per-cent as a down payment, and the balance to be paid in 20 semi-monthly installments with a 10 per-cent financing charge on the balance. This made buying airplanes a bit easier, and an airplane could

Fig. 142.   The prototype version of the "Eaglerock" A-13.

actually pay for itself and still render the owner-buyer a bit of profit. The capable Al W. Mooney had been chief of design and engineering at Alexander since early in 1928, but was replaced about this time, by Ludwig Muther when Al Mooney left with plans for his own interests. Ludwig Muther had been an engineer with Junkers and Fokker in Europe, and with Keystone Aircraft here in the U.S.A.; he was co-designer with Mooney and Max Munk on the development of the Alexander "Bullet".

Listed below are specifications and performance data for the "Eaglerock" model A-13 as powered with the Curtiss "Challenger" engine of 170 h.p.; length overall 24'6; hite overall 9'10"; wing span upper 36'8"; wing span lower 32'8"; wing chord both 60"; wing area upper 183 sq.ft.; wing area lower 153 sq.ft.; total wing area 336 sq.ft.; airfoil "Clark Y"; wt. empty 1705; useful load 945; payload 365; gross wt. 2650 lbs.; max. speed 117; cruising speed 100; landing speed 40; climb 840 ft. first min. at sea level; service ceiling 14,700 ft.; gas cap. 59 gal.; oil cap. 7 gal.; cruising range 600 miles; price at the factory field was $6096. The fuselage framework was built up of welded chrome-moly steel tubing in a truss form, and was heavily faired to shape with wood fairing strips; the upper half of the fuselage from the engine firewall, back through the cockpit area, was covered with aluminum cowling panels and the balance of the fuselage was covered in fabric. The cockpits were roomy and comfortable, and the occupants were well protected; there was a small baggage compart-

ment in the turtle-back section of the fuselage, just in back of the pilot's cockpit. The wing panels were built up of solid spruce spars that were routed to an "I beam" section, with spruce and plywood truss-type wing ribs; the completed panels were fabric covered. There were four ailerons that were connected together in pairs by a streamlined push-pull strut. The fabric covered tail-group was also built up of welded chrome-moly steel tubing and was externally braced by streamlined steel tie-rods; the fin was ground adjustable and the horizontal stabilizer was adjustable in flight. The robust landing gear was of the split-axle type and used two spools of shock-cord to absorb the bumps; wheels were 28x4. The wings were wired for navigation lights, an exhaust collector-ring, and a metal propeller were offered as standard equipment; wheel brakes and an inertia-type engine starter were optional equipment. The next development in the A-series "Eaglerock" biplane was the Kinner powered model A-15, see chapter for A.T.C. # 190 in this volume.

Listed below are "Eaglerock" A-13 entries that were gleaned from the Aeronautical Chamber of Commerce aircraft register; this list may not be complete, but it does show the bulk of this model that were built.

C-6360; Eaglerock A-13 (# 733) Challenger

C-8230; Eaglerock A-13 (# 784) Challenger

C-8265; Eaglerock A-13 (# 823) Challenger

C-8246; Eaglerock A-13 (# 824) Challenger

Fig. 143. An A-13 at Alexander Airport. Some 6200 feet above sea level, this field was the highest factory field in the world.

NC-704H; Eaglerock A-13 (# 848) Challenger

NC-338W; Eaglerock A-13 (# 958) Challenger

Other "Eaglerock" entries of considerable interest were as follows.

X-4570; Eaglerock A-7 (# 451) Ryan-Siemens

-6846; Eaglerock A-1 (# 569) Wright J4

X-7187; Eaglerock A-? (# 610) Curtiss C-6A

-6326; Eaglerock A-? (# 633) Floco (Axelson)

X-6315; Eaglerock A-11 (# 649) Warner 110

X-6355; Eaglerock A-? (# 680) Hallett

X7981; Eaglerock A-? (# 696) Hallett

Other versions in the A-series "Eaglerock", though unverified, were believed to mount engines such as the Kimball "Beetle", Anzani 120, Fairchild-Caminez, Jacobs & Fisher, Brownback 90, and the MacClatchie "Panther".

A.T.C. #142
(6-29)
RYAN "BROUGHAM", B-5

Fig. 144. The Ryan "Brougham" Model B-5 with a 300 h.p. Wright J6 engine.

Each year the Ryan "Brougham", like the fair lady she was, kept getting more buxom, more mellow, and more beautiful; the passing years were kind and seemed to add pride and confidence with the knowledge of a job well done. Tranquil in nature and of thoroughly predictable character, the "Brougham" had a devoted legion of owners that were it's staunchest boosters, and upon this was built a good stable foundation of trust and popularity. It had been over two years now since "Lindy" flew the Atlantic Ocean on that memorable day, in his famous "Spirit of St. Louis" (Ryan NYP), and the Ryan "Brougham" was still being tagged as a sister-ship, but this was not merely promotion, it was yet an honor well to be proud of. The "Brougham" had come a long way since then; over 150 had already been built and the record of service that was piled up by this craft all across the nation, was surely a testimonial to the validity of the promises extended by the proud manufacturer. The genial B. F. "Frank" Mahoney, due to illness and a desire to relax from constant business pressures, sold out to a group of financiers and the "Brougham" was now being built in new quarters at St. Louis, Mo. The name of Mahoney was soon

dropped from the mast-head and the new operation became known as the Ryan Aircraft Corp. Facilities for quantity production were soon installed and quickly expanded; an average of two airplanes per week were coming off the line.

Big business was taking to the skies, and the Ryan B-5 was doing it's share to prove the efficiency and the pay-off that was possible in conducting business calls by airplane; among the satisfied converts were Prest-O-Lite; Fleet-Wing Oil; Pampa Refineries; Haynes Drilling Co. and several others. Small feeder-type airlines and charter service operators were also attracted to the "Brougham's" performance and utility; air lines such as Pickwick Airways, and Pikes Peak Air Lines, to name just a few. They have often said that "all work and no play makes Jack a dull boy", so the B-5 took time off occasionally to shoot at a record or two; Marvel Crosson, capable aviatrix from California flew a Ryan B-5 to 24,000 feet or better, which set a new record for women pilots. Russell Young flew a B-5 in the National Air Tour for 1929 and came to 19th place amongst a determined field of strong competitors; but time out from work came seldom, so the B-5 was more at work than at play. Nine of the Ryan B-5 were sold to China for air-line service; at least two were mounted on floats to operate from the numerous rivers.

The Ryan "Brougham" model B-5 was a high wing cabin monoplane with ample room and comfort for six, and was powered with the 300 h.p. Wright J6 series engine. Neat and trim with good proportion, the B-5 was quite a deft and capable airplane with a sprightly performance; fully loaded with six people and 100 gallons of fuel, take-off was accomplished consistently in less than 300 feet, and a climb-out to 1000 feet in a minute and 35 seconds from point of take-off, was the type of get-up-and-go that made the "Brougham" B-5 popular for high altitudes and close quarters. Also approved with twin-float seaplane gear; the switch from one to the other could be made in less than four hours. The type certificate number for the Ryan "Brougham" B-5 as powered with the Wright J6 engine, was issued in June of 1929 for both a landplane and a seaplane version; some 60 or more examples of this model were manufactured by the Ryan Aircraft Corp. at St. Louis (Anglum), Missouri. Merger into the Detroit Aircraft Corp. was approved in June of 1929 and "Mahoney" was dropped from the trade-name. Edward S. Evans was the president; James Work was the V.P.; John C. Nulsen was general manager; W. A. "Bill" Mankey was chief engineer; and J. J. "Red" Harrigan was the chief pilot.

Listed below are specifications and performance data for the Ryan "Brougham" model

B-5 as powered with the 300 h.p. Wright J6 engine; length overall 28'4"; hite overall 8'10"; wing span 42'4"; wing chord 84"; airfoil Clark Y; total wing area 280 sq.ft.; wt. empty 2251 (2269); useful load 1749 (1731); payload with 100 gal. fuel was 924 (906) lbs.; gross wt. 4000 lbs.; weights in brackets are for a/c fitted with "Townend ring" low-drag engine cowling; max. speed 140; cruising speed 120; landing speed 50; climb 1000 ft. first min. at sea level; service ceiling 18,000 ft.; gas cap. 100 gal.; oil cap. 8 gal.; cruising range 720 miles; price at the factory field was $13,250 and later lowered to $12,500. The following figures are for the B-5 on Edo twin-float seaplane gear; wt. empty 2582 (2600); useful load 1518 (1500); payload with 100 gal. fuel was 693 (675) lbs.; gross wt. 4100 lbs.; weights in brackets are for B-5 seaplane fitted with "Townend ring" low-drag engine cowling; max. speed 128; cruising speed 110; landing speed 50; climb 950 ft. first min. at sea level; service ceiling 17,5000 ft.; cruising range 650 miles; all other dimensions and data remained the same; price at the factory was $14,350.

The fuselage framework was built up of welded chrome-moly steel tubing in a rigid truss form, faired to shape with wood fairing strips and fabric covered. The walls of the cabin were sound-proofed and insulated and the interior was upholstered in blue mohair with mahogany trim; a baggage compartment was aft of the cabin section and was accessible from inside or out. The cabin was heated and ventilated, and large windows

Fig. 145. Big business was taking to the air, and the Ryan "Brougham" Model B-5 was proving the efficiency of business calls by airplane.

Fig. 146. The Ryan B-5 performed well as a seaplane; the prototype version is shown here during tests.

offered excellent visibility to passengers and pilot. The wing framework was built up of 3-ply laminated spruce spars that were routed to an "I beam" section, with wing ribs built up of spruce and plywood in a Warren type truss; the completed framework was fabric covered. The ailerons were of a fabric covered steel tube framework, and were of the offset-hinge type that operated differentially; normal turns were possible by ailerons alone. The wing bracing struts were of heavy gauge chrome-moly steel tubing and were encased in balsa-wood fairings that were shaped to an airfoil form for added lift and stability; the two fuel tanks were mounted in the wing for gravity feed, one each side of the fuselage. The split-axle landing gear was of the outrigger type and was formed into a truss with the front wing struts; "Aerol" shock absorbing (air-oil) and wheel brakes were standard equipment. The fabric covered tail-group was built up of welded chrome-moly steel tubing; the fin was ground adjustable and the horizontal stabilizer was adjustable in flight, through a range of 9 degrees. A novel oil-cooling radiator core was mounted on the engine firewall; cool air intake came in the scoops at the right and vented out from louvers on the left, adjustment of intake was provided for temperature control. The engine was well muffled by a large collector-ring for near-normal conversation in the cabin at all times. A ground-adjustable metal propeller, wiring for navigation lights, swiveling tail-wheel, and inertia-type engine starter were standard equipment. The next development in the Ryan "Brougham" series was the P & W "Wasp" powered model B-7, this craft will be dis-

cussed in the chapter for A.T.C. # 262.

Listed below are Ryan B-5 entries that were gleaned from various records; this list is not a complete tally, but it does show a good portion of the number that were built.

X-8321; Ryan B-5 (# 187) Wright J6.
C-9230; Ryan B-5 (# 188) Wright J6.
C-9231; Ryan B-5 (# 189) Wright J6.
C-9232; Ryan B-5 (# 190) Wright J6.
NC-9233; Ryan B-5 (# 191) Wright J6.
NC-9234; Ryan B-5 (# 192) Wright J6.
C-9235; Ryan B-5 (# 193) Wright J6.
C-9236; Ryan B-5 (# 194) Wright J6.
C-9237; Ryan B-5 (# 195) Wright J6.
NC-9238; Ryan B-5 (# 196) Wright J6.
NC-9239; Ryan B-5 (# 197) Wright J6.
       Ryan B-5 (# 198) Wright J6.
NC-14H; Ryan B-5 (# 199) Wright J6.
NC-15H; Ryan B-5 (# 200) Wright J6.
NC-16H; Ryan B-5 (# 201) Wright J6.
NC-17H; Ryan B-5 (# 202) Wright J6.
NC-18H; Ryan B-5 (# 203) Wright J6.
NC-8320; Ryan B-5 (# 204) Wright J6.
NC-306K; Ryan B-5 (# 205) Wright J6.
NC-307K; Ryan B-5 (# 206) Wright J6.
NC-308K; Ryan B-5 (# 207) Wright J6.
NC-128W; Ryan B-5 (# 208) Wright J6.
       Ryan B-5 (# 209) Wright J6.
NC-131W; Ryan B-5 (# 211) Wright J6.
NC-313K; Ryan B-5 (# 212) Wright J6.
NC-314K; Ryan B-5 (# 213) Wright J6.
NC-315K; Ryan B-5 (# 214) Wright J6.
NC-378K; Ryan B-5 (# 215) Wright J6.

The next 30 serial numbers at least were Ryan B-5 a/c; the Ryan model B-7 started at serial # 249 or thereabouts; serial # 210 was a J5 powered model B-3A.

Fig. 147. A Curtiss-Robertson "Robin" Model C-1, one of the most popular airplanes of this period.

Since the first Curtiss "Robin" was introduced back in March of 1928, Curtiss-Robertson had built some 300 of the OX-5 powered models (Robin B) and some 50 or more of the "Challenger" powered models (Robin C) up to this time, which was about a year or so later. Consequently, there were enough of these craft out in service now to give the Curtiss engineers and the Curtiss-Robertson people a fair idea of how things were measuring up with the "Robin", and what would be needed in the way of changes and improvements. The "Robin" being such a well thought out and excellent design at the outset, there were actually very little changes warranted or necessary, but one of the major changes in this new model was to incorporate fittings into the structure to adapt it for the installation of twin-float seaplane gear; this was asked for by several operators. With the added improvements, allowance for extra equipment, and the necessary structural changes, the new "Robin" wound up just a little bit heavier when empty; some addition was allowed in the total useful load and the gross weight now stood at some 150 pounds heavier. Altogether, the new "Robin" was somewhat better for it, and the performance remained pretty much the same.

The new Curtiss-Robertson "Robin" model C-1, as shown here in various views, was also a 3 place high wing cabin monoplane of typical configuration and was also powered with the 6 cylinder Curtiss "Challenger" engine of 170 h.p.; the engine was somewhat modified and was re-rated to 185 h.p. towards the end of 1929, and this brought about noticeable performance increases in some instances. Literally hundreds of "Robins" were now flying all about the country-side as air-taxis for business people, as pleasure-craft for the private owner, in charter service, hauling enthusiasts for airplane rides and the like with numerous fixed-base operators, and used for flight instruction too with a good number of flying schools. The "Robin" was certainly well known, a familiar sight in all parts of the country, and one of the most popular cabin airplanes of this particular time. Probably the best known "Robin" of all time was the "St. Louis Robin", a model C-1 that was flown to fame by Jackson and O'Brine in 1929.

During the hustle and bustle of 1929, there were a rash of endurance flights, flights that were prolonged way beyond normal by being refueled in mid-air; the planes and the crews remained aloft and took their fuel and other needs from a "tanker" that trailed a refueling

Fig. 148.   Dale "Red" Jackson servicing his engine during a record endurance flight.

hose from it's bottom, or lowered supplies to the crew below on the end of a cable. The endurance plane usually remained aloft until something forced them down, be whatever it may. The Army Air Corp's "Question Mark" sparked off this craze in January of 1929, by staying aloft for over 150 hours; Reinhardt and Mendell then bettered this record some time later, and so it went on. Dale Jackson and Forrest O'Brine toyed with the idea of giving the boys a real record to shoot at, and did just that in their "St. Louis Robin", by staying aloft for over 420 hours. Their "Robin" was a model C-1, as shown here, and was powered with the 6 cylinder "Challenger" engine which did not miss a beat for over 17 days. Refueling was carried out by lowering the filler hose from the "tanker" above through a trap-door in the roof of the cabin, and there were narrow steel-tube cat-walks on either side of the fuselage, where they crawled out (in mid-air) to make adjustments! Jackson and O'Brine had certainly hung up a good one, and their record stood intact until mid-1930 when the Hunter brothers flew their Stinson "Detroiter" monoplane for over 553 hours before coming down. The type certificate number for the "Robin" model C-1 as powered with the Curtiss "Challenger" engine, was issued in June of 1929 for both a landplane and a seaplane version, and some 200 or more of this model were manufactured by the Curtiss-Robertson Airplane Mfg. Corp. at Anglum, St. Louis County, Missouri; a division of the Curtiss

Airplane & Motor Co. Major William B. Robertson was the president of the St. Louis division; engineering was handled by the Curtiss corps of engineers; and Dale "Red" Jackson was the chief pilot in charge of test and development.

Listed below are specifications and performance data for the Curtiss-Robertson "Robin" model C-1 as powered with the 170-185 h.p. Curtiss "Challenger" engine; length overall 24'1"; hite overall 8'0"; wing span 41'0"; wing chord 72"; total wing area 243 sq.ft.; airfoil "Curtiss C-72"; wt. empty 1638; useful load 962; payload with 50 gal. fuel was 452 lbs.; gross wt. (landplane) 2600 lbs.; max. speed 120; cruising speed 102; landing speed 48; climb 640 ft. first min. at sea level; service ceiling 12,500 ft.; gas cap. 50 gal.; oil cap. 5 gal.; cruising range 500 miles; price at the factory field was $7500. The following figures are for the model C-1 as fitted with Edo DeLuxe floats; wt. empty 1885; useful load 775; payload with 50 gal. fuel was 165 lbs.; gross wt. 2660 lbs.; max. speed 113; cruising speed 97; landing speed 45; climb 640 ft. first min. at sea level; service ceiling 12,000 ft.; cruising range 475 miles; all other dimensions and data remain the same. The following figures are for the model C-1 as fitted with Edo model P floats which were bigger; wt. empty 2017; useful load 893; payload with 50 gal. fuel was 383 lbs.; gross wt. 2910 lbs.; max. speed 110; cruising speed 95; landing speed 52; climb 540 ft. first min. at sea level; service ceiling 11,500 ft.; cruising range 450

Fig. 149. The "Robin" was boxy-looking, but its efficiency was well proven.

miles; all other dimensions and data remain the same.

The fuselage framework was built up of welded chrome-moly steel tubing of square and round section that was built up in truss form; the framework was lightly faired to shape and fabric covered. The pilot sat in a bucket type seat up front, and the passengers sat in back on a split bench-type seat that was adjustable fore and aft separately, to stagger the seats for more elbow room. The semi-cantilever wing framework was built up of solid spruce spar beams and stamped-out Alclad aluminum alloy wing ribs; the completed framework was fabric covered. There were two gravity-feed fuel tanks, one in the root end of each wing, flanking the fuselage. The fabric covered tail-group was built up of welded chrome-moly steel tubing; the fin was

ground adjustable and the horizontal stabilizer was adjustable in flight. The split-axle landing gear of 96 inch tread, was of the outrigger type and built into the front wing struts in a truss form; wheel brakes were standard equipment and the tail-skid was steerable to help in ground manuvering. The two cabin entry doors were on the right side, and a convenient step was provided for both doors to lessen the awkward stretch that would be necessary to get in and out. The model C-1 was equipped with a Curtiss-Reed forged metal propeller, wired for navigation lights, 30x5 wheels and brakes, steerable tail-skid, and inertia-type hand crank engine starter. The next development in the Curtiss-Robertson "Robin" series was the model C-2, see the chapter for A.T.C. # 144 in this volume.

Fig. 150. The "Robin" on Edo twin-float gear.

A.T.C. #144
(5-29)
CURTISS-ROBERTSON "ROBIN", C-2

Fig. 151. The Curtiss-Robertson "Robin" Model C-2 was a rare version of which only a few were built.

The first St.Louis made "Robin" took to the air on August 7th of 1928 with Maj. Wm. B. Robertson at the controls; after a short flight he pronounced it satisfactory and turned it over to Dan Robertson. Young Danny Robertson then took it up for further testing that same day, with orders to "break it up if you can"; witnesses of the testing will report that he tried his very best to break it up, but she hung together quite well. Jubilant over their new product, quantity production was soon commenced in a 3 unit plant that covered some 45,000 sq. ft. In less than a year's time, over 350 "Robins" were manufactured and plant space was being added from time to time to handle the volume of orders that kept pouring in; by mid-1929 Curtiss-Robertson Div. was rolling out 17 completed airplanes each week. Well over 300 of the OX-5 powered "Robin B" had been produced by now, also 50 or more of the "Challenger" powered "Robin C"; the new "Robin C-1" was already introduced into production and started off in good number, and along with this came a development called the 'Robin C-2".

There are times when a historian comes up against a blank wall, and not any normal amount of searching will reveal a way out; histories of certain aircraft disappear quickly and are now deeply buried in the secrets of the past, secrets that can be likened to an insurmountable blank wall. From the attempts made to discover accurate information, it seems that the Curtiss-Robertson "Robin" model C-2, in discussion here, has turned out to be one of these dark secrets of the past. After spending a considerable time in research through the available records, we have yet to come up with a why-for to explain the "Robin C-2". One can only surmise from various bits of data that were dug up here and there, and this is about how it looks from here.

The "Robin" model C-2 from all indications, was typical to the model C-1 described previously, but it was not eligible for certification as a float-mounted seaplane, and was usually registered in the restricted category as a special-purpose airplane. The model C-2 was also powered with the 6 cylinder Curtiss "Challenger" engine, but was always listed with 170 h.p. and never with the later improved "Challenger" of 185 h.p.; this is a point to ponder on. Only a few of the model C-2 were built and nearly all of these were in operation with the Curtiss Flying Service in different parts of the country; this is another point to ponder, but it serves no clue. Operating at the same gross weight as the model C-1, the model C-2 was some 29 lbs. heavier when empty, and one would guess that it should be the other way around. Some data informs us that "Robin" aircraft from serial # 249 and up, are eligible as the model C-2;

this cannot help matters either, so we will just admit that this one has us and many others completely stumped. The type certificate number for the "Robin" model C-2 as powered with the 170 h.p. "Challenger" engine, was issued in May of 1929 and some 6 examples of this model were manufactured by the Curtiss-Robertson Airplane Corp. at Anglum (St.Louis), Missouri; a division of the Curtiss Aeroplane & Motor Co.

Listed below are specifications and performance data for the "Robin" model C-2 as powered with the 170 h.p. Curtiss "Challenger" engine; length overall 25'1"; hite overall 8'0"; wing span 41'0"; wing chord 72"; total wing area 223 sq.ft.; airfoil "Curtiss C-72"; wt. empty 1667; useful load 933; payload with 50 gal. fuel was 418 lbs.; gross wt. 2600 lbs.; max. speed 120; cruising speed 102; landing speed 47; stall speed 56; climb 640 ft. first min. at sea level; climb to 5000 ft. was 7.7 min.; climb to 10,000 ft. was 18.7 min.; service ceiling 12,700 ft.; gas cap. 50 gal.; oil cap. 6 gal.; cruising range at 10 gal. per hour was 4.9 hours or 500 miles; price at the factory field was approx. $7500.

The fuselage framework was built up of welded chrome-moly steel tubing, lightly faired to shape and fabric covered. The pilot sat in a bucket-type seat in front, and the two passengers sat on a split bench-type seat that was adjustable fore and aft for more shoulder room, by a staggering of the occupants; a baggage compartment of 12.5 cubic feet was to the rear of the cabin section. The wing framework was built up of solid spruce spar beams, with stamped-out Alclad wing ribs; the completed framework was fabric covered. The stamped-out "Alclad" wing ribs weighed .59 lbs. each and would take a load of 535 lbs. without any yield or distortion; metal wing ribs are quite common in aircraft of today, but in 1928-1929 they were quite an innovation. The fuel tanks were mounted in the wing, flanking the fuselage; the oil tank

was mounted on the fire-wall in the engine section. The fabric covered tail-group was built up of welded chrome-moly steel tubing; the fin was ground adjustable and the horizontal stabilizer was adjustable in flight. The split-axle landing gear of 96 inch tread was of the out-rigger type and used oleo-spring shock absorbing struts; wheels were 28x4 and brakes were standard equipment. The tail-skid was steerable for better ground manuvering. The "Robin" was equipped with a Curtiss-Reed metal propeller that was of forged duralumin alloy; the familiar bent-slab Curtiss-Reed prop was still available in limited number, but only used for 150 h.p. or less. Navigation lights and inertia-type engine starter were available as optional equipment. Some 750 or more "Robins" were manufactured in all, and some 10 years later at least 400 were still in active operation; a good testimonial of the "Robin's" rugged nature and continued popularity. The next Curtiss development was the "Challenger" powered "Thrush" which could be best described as a 6 place "Robin"; for discussion of the "Thrush", see the chapter for A.T.C. # 159 in this volume.

Listed below are Curtiss-Robertson "Robin" model C-2 entries that were gleaned from various records; this list may not be quite complete, but it does show the bulk of this model that were registered.

NC-9273; Robin C-2 (# 338) Challenger.
NC-55H; Robin C-2 (# 420) Challenger.
NR-323K; Robin C-2 (# 480) Challenger.
NR-324K; Robin C-2 (# 482) Challenger.
NR-325K; Robin C-2 (# 484) Challenger.
NC-373K; Robin C-2 (# 536) Challenger.

All craft listed were registered with the Curtiss Flying Service initially. Curtiss-Robertson "Robin" from serial # 249 and up, were eligible as the model C-2. "Robins" of serial # 228 and up, were allowed 2600 lb. gross weight with the "Challenger" engine.

Fig. 152. A Stinson "Junior" Model SM-2AA with a Wright J6 engine of 165 h.p.

The Stinson "Junior" was the "baby" of the familiar Stinson monoplane line and had been formally introduced somewhat more than a year ago. Since that time, a good number had been built, and it was taking it's place in various phases of commercial aviation. In it's basic form, the "Junior" was a 3-4 place strut-braced high wing cabin monoplane and was typically a "Stinson" type, but only in a scaled-down fashion. The original SM-2 of a year or so ago was powered with the 7 cylinder Warner engine of 110 h.p., and though performance was quite good considering the low horsepower available, it was somewhat shy in the punch that would make it more attractive to the small business man who would sometimes ask and expect the impossible in airplane utility and performance. It was originally Ed Stinson's intent to offer this light enclosed monoplane as a personal-type airplane with just adequate performance for the average private-owner, or the small business-man who would have a bent to travel occasionally in conducting certain phases of his business. The "Junior", also called the "Detroiter Jr.", was not exactly a dainty little sprite to begin with, and successive power increases in the interests of better performance caused it to become somewhat bigger and bulkier, and a good deal heavier. It was not too long before this "baby" of the Stinson Monoplane line grew right out of the light plane class.

The introduction of the model SM-2AA was the first step in this direction. The SM-2AA was still quite typical and hardly any changes were introduced except the installation of the new 5 cylinder Wright J6 series engine of 150-165 h.p., and seating was now definately established at 4 places, plus baggage. Noticeable improvement in performance was gained in some respects, and it was felt that this combination would be better suited to the requirements as demanded by potential customers for an airplane of this particular type. The type certificate number for the Stinson "Junior" model SM-2AA as powered with the 5 cylinder Wright engine, was issued in May of 1929 and some 22 or more of this model were built by the Stinson Aircraft Corp. at their new plant located in Wayne, Michigan. A Group 2 approval numbered 2-73 was issued 5-31-29 for the first five airplanes. The famous Edward A. Stinson was the president; the honorable Wm. A. Mara was secretary; former engineer Wm. C. Naylor was replaced by Kenneth M. Ronan as chief engineer; and old-time mail pilot Randolph Page was in charge of test and development. Sales of the new "Junior" model (SM-2AA) spurted off to a good start and soon they were to be seen here and there about the country. Several were operated by private owners as pleasure craft, some by business men, a few were used by flying-schools to teach cabin monoplane techniques, and at least one or two had found their way into Alaskan service. There seemed to be a good future for the "Junior" line, and modifications continued, but the development seemed to revolve around increases in horsepower.

Listed below are specifications and performance data for the Stinson "Junior" model SM-2AA as powered with the 5 cyl. Wright engine of 165 h.p.; length overall 26'4"; hite overall 7'6"; wing span 41'6"; wing chord 75"; total wing area 238 sq.ft.; airfoil "Clark Y"; wt. empty 1972; useful load 1180; payload 610; gross wt. 3152 lbs.; max. speed 115; cruising speed 97; landing speed 47; climb 610 ft. first min. at sea level; service ceiling 12,000 ft.; gas cap. 60 gal.; oil cap. 5 gal.; cruising range 580 miles; price at the factory field was $8500. The fuselage framework was built up of welded chrome-moly steel tubing and was gusseted in all major stress points; faired to shape with plywood formers and wood fairing strips and fabric covered. The whole forward section of the fuselage up to and including the metal doors, was covered in aluminum panels. The cabin was upholstered in blue mohair with matching trim, the walls were sound-proofed against noise and insulated against temperature changes, cabin heating and ventilation were provided, and a large baggage compartment was located behind the rear seats. The wing framework was built up of heavy-sectioned solid spruce spars, with spruce and plywood reinforced wing ribs; the completed framework was fabric covered. Two fuel tanks of 30 gal. capacity each were built into the wing and flanked each side of the fuselage. The fabric covered tail-group

was built up of welded chrome-moly (molybdenum) sheet stock and steel tubing; the horizontal stabilizer was adjustable in flight. Flight controls on the big Stinson monoplane were of the "Dep" wheel type, but flight controls on the "Junior" were of the normal joy-stick type; dual controls were available. The landing gear was of the outrigger type and tied into the wing brace struts, shock absorbers were of the air-oil type and wheels were 30x5; the wheel tread was unusually wide at 110 inches. Wheel brakes were standard equipment, and this convenience in connection with the swivelling tail-wheel made ground manuvering a rather simple chore. A metal propeller, and inertia-type engine starter were optional equipment. When Continental Motors introduced their model A-70 engine of 165 h.p., they selected a "Junior" of this type to act as a test-bed for demonstration and further development; however, this combination was rare and never went beyond this one example. Another version of the "Junior" was tried with the 6 cyl. Curtiss "Challenger" engine of 170 h.p., and two examples were built, but never went beyond the test stage; later modified to SM-2AA & SM-2AB. The next development in the Stinson "Junior" was the earth-pawing SM-2AB that was powered with the Wright J5 engine of 220 h.p., see chapter for A.T.C. # 161 in this volume.

Listed below are Stinson "Junior" SM-2AA entries that were gleaned from the Aeronautical Chamber of Commerce aircraft register; this list may not be complete, but it does show the bulk of this model that were

Fig. 153. The Stinson "Junior" was also tried out with the Curtiss "Challenger" engine in early testing.

Fig. 154. The "Junior" was a typical Stinson in scaled-down form.

built.

NC-9695; SM-2AA (# 1046) Wright 5, Grp. 2-73.

C-8422; SM-2AA (# 1049) Wright 5, Grp. 2-73.

NC-8431; SM-2AA (# 1051) Wright 5, Grp. 2-73.

NC-8425; SM-2AA (# 1052) Wright 5, Grp. 2-73.

NC-8435; SM-2AA (# 1054) Wright 5, Grp. 2-73.

NC-8437; SM-2AA (# 1057) Wright 5, A.T.C. # 145.

NC-8433; SM-2AA (# 1058) Wright 5, A.T.C. # 145.

NC-8440; SM-2AA (# 1060) Wright 5, A.T.C. # 145.

NC-8441; SM-2AA (# 1061) Wright 5, A.T.C. # 145.

C-8465; SM-2AA (# 1066) Wright 5, A.T.C. # 145.

C-8466; SM-2AA (# 1067) Wright 5, A.T.C. # 145.

C-8467; SM-2AA (# 1068) Wright 5, A.T.C. # 145.

C-8471; SM-2AA (# 1069) Wright 5, A.T.C. # 145.

C-8472; SM-2AA (# 1071) Wright 5, A.T.C. # 145.

C-8481; SM-2AA (# 1075) Wright 5, A.T.C. # 145.

NC-447H; SM-2AA (# · 1077) Wright 5, A.T.C. # 145.

NC-457H; SM-2AA (# 1079) Wright 5, A.T.C. # 145.

NC-465H; SM-2AA (# 1101) Wright 5, A.T.C. # 145.

X-472H; SM-2AA (# 1107) Continental A-70, test.

NC-475H; SM-2AA (# 1110) Wright 5, A.T.C. # 145.

NC-476H; SM-2AA (# 1111) Wright 5, A.T.C. # 145.

NC-400M; SM-2AA (# 1126) Wright 5, A.T.C. # 145.

There were 14 unlisted aircraft, some may have been SM-2AA, others were SM-2K; SM-2AB; and SM-2AC.

Fig. 155. The B-4000 was an improved version of the familiar J5 Travel Air.

The Travel Air model 4000 prototype was introduced with great pride in 1926, and since that time had achieved great popularity and rounded out an outstanding career of service and performance. Like all airplanes in the past few years, it was continuously being modified for improvement of structure, improvements in utility for a greater number of uses, and being tailored to newer concepts of performance. Subsequent variations of the "J5 Travel Air" were then becoming somewhat heavier, carried substantial increases in payload, and therefore operated in normal service at higher gross loads; performance in some certain instances suffered accordingly, but that was to be expected. Unlike a few years back, it was not now sufficient to say, "gentlemen, here we have a craft of 220 horses that will take off from a dime, climb like a home-sick angel, and go like the blazes". There were now those that wanted a little more payload, and perhaps a baggage compartment that was somewhat bigger than the usual hat-box, they wanted more instruments in the cockpit, and there were those that cried out for more cruising range, which also meant additional pounds of weight and more stresses on the structure; all these things had to be dealt with of course, and consequently the "4000" was

getting stronger, bulkier, and heavier, and some of the "old flash" had to be sacrificed to make way for new requirements.

As can be surmised from the statements above, the "J5 Travel Air" was always leveled at a clientele that had special demands; like the playful sportsman-pilot who was prone to ask the very most from any airplane, or the business-houses who had to have a swift, good-performing air taxi that would stand in readiness and could go just about anywhere on demand. That the "4000" had always catered to this type of owner-service was well known the country over, and the model B-4000 was now being introduced as the last and perhaps the finest example of this great series. The model B-4000 was also a 3 place open cockpit biplane and was still very typical in general appearance, but evolution caused numerous noticeable changes. The dear old "elephant ears" that were so characteristic of the Travel Air biplane for years, were now gone; a change made supposedly in the interests of greater speed. With the increases in gross load and the almost inevitable increase in stresses, the wings had to be beefed-up a good deal with heavier spars and interplane bracing of heavier gauge; even the airfoil in the upper wing was changed to one of less

Fig. 156. The Traval Air B-4000 featured stronger wings and a sturdy undercarriage.

drag and somewhat greater lift. The familiar shock-cord sprung landing gear was also changed to the heavier outrigger type which employed oleo-spring shock absorber struts, and was more suitable to the increased gross weight of this airplane. Of course, wheel brakes, metal propeller, navigation lights, and inertia-type engine starter were standard equipment.

If one could study the cross-section of B-4000 owners, it would give some indication of the varied duties that were asked of this craft, and they ran the gamut of sportsman-pilots, charter operators, flying schools, oil companies, and government service with the Dept. of Agriculture and the Aero. Branch of the Dept. of Commerce. Though small in number and one of the lesser known models of the Travel Air biplane series, during these rapidly changing times, the B-4000 was a new concept that laid the foundation for new models yet to come; models such as the BC-4000, the B9-4000, and the popular 4D. The type certificate number for the model B-4000 as powered with the 9 cylinder Wright J5 engine of 220 h.p., was issued in June of 1929 and at least 20 examples of this model were manufactured by the Travel Air Company at Wichita, Kansas.

Listed below are specifications and performance data for the Travel Air biplane model B-4000 as powered with the 220 h.p. Wright J5 engine; length overall 23'4"; hite overall 9'1"; wing span upper 33'0"; wing span lower 28'10"; chord upper 66"; chord lower 56"; wing area upper 171 sq.ft.; wing area lower 118 sq.ft.; total wing area 289 sq.ft.; airfoil T.A.# 2 upper & T.A.# 1 lower; wt. empty 1893; useful load 1007; payload with 67 gal. fuel was 392 lbs.; gross wt. 2900 lbs.; max. speed 128; cruising speed 110; landing speed 52; climb 950 ft. first min. at sea level; service ceiling 14,000 ft.; gas cap. 67 gal.; oil cap. 6 gal.; cruising range 530 miles; price at the factory field was $ 8500.

The fuselage framework was built up of chrome-moly steel tubing, faired to shape with wood fairing strips and fabric covered. The roomy cockpits were well upholstered and there was a good-sized baggage compartment in the lower fuselage, just in back of the rear cockpit. The wing framework was built up of solid spruce spars that were routed to an "I beam" section, with spruce and plywood built-up wing ribs; the completed framework was fabric covered. The main fuel tank was mounted high in the fuselage just ahead of the front cockpit, and extra fuel was carried in a tank that was mounted in the center-section panel of the upper wing. The fabric covered tail-group was built up of welded chrome-moly steel tubing; the vertical fin was ground adjustable and the horizontal stabilizer was adjustable in flight. The robust landing gear of 85 inch tread was of the outrigger type and employed oleo-spring shock-absorber struts; wheels were 30x5 and brakes were standard equipment. Navigation lights, metal propeller, and inertia-type engine starter were also standard equipment. The next development in the Travel Air biplane was the single place BM-4000 that was powered with the Wright J5 engine; see the chapter for A.T.C. # 147 in this volume.

Listed below are B-4000 entries that were gleaned from various records; this list may not be quite complete, but it does show the bulk of this model that were built.

NC-8136; B-4000 (# 897) Wright J5.
NC-8716; B-4000 (# 1000) Wright J5.
NC-8888; B-4000 (# 1011) Wright J5.
NC-8891; B-4000 (# 1014) Wright J5.
C-9822; B-4000 (# 1042) Wright J5.
C-9904; B-4000 (# 1064) Wright J5.
NC-9921; B-4000 (# 1065) Wright J5.
NC-601H; B-4000 (# 1168) Wright J5.
NC-602H; B-4000 (# 1169) Wright J5.
NC-603H; B-4000 (# 1170) Wright J5.
NC-604H; B-4000 (# 1171) Wright J5.
NC-605H; B-4000 (# 1172) Wright J5.
NC-630H; B-4000 (# 1176) Wright J5.
NC-631H; B-4000 (# 1177) Wright J5.
NC-632H; B-4000 (# 1178) Wright J5.
NC-657H; B-4000 (# 1242) Wright J5.
NS-16; B-4000 (# 1261) Wright J5.
NS-5; B-4000 (# 1262) Wright J5.
NS-174V; B-4000 (# 1365) Wright J5.

Fig. 157. The Travel Air BM-4000 was a B-4000 with front cockpit converted into a cargo hold.

It may seem repetitious to say that the "J5 Travel Air" has had a varied career, as varied as one could possibly imagine, but carrying the air-mail was certainly one of it's lesser known accomplishments. Some air lines, mostly of the feeder-type that shuttled mail into and out of the main transcontinental system, used the small medium-powered open cockpit biplane as sort of a "Pony Express" for the lighter loads that were carried on a more frequent schedule. The Stearman biplane and the Pitcairn "Mailwing" were better known by everyone for handling this type of service, but the Travel Air biplane was in there too for it's fair share on several of these lines. Among some of the operators that were using the "J5 Travel Air" for mail-carrying duties was the Interstate Air Lines on a route from Atlanta to St. Louis; the Pacific Air Transport on a route from Los Angeles to Seattle; Robertson Air Lines on a route from St. Louis to Chicago; and the Continental Air Lines on a route from Cleveland to Louisville, Kentucky. This is not to imply that the Travel Air biplane had built up any great reputation as a mail-carrier, but only to point

out that this craft took this sort of thing in stride as one of it's many duties in a versatile career of useful servitude.

The model BM-4000 that is under discussion here is indeed a rare and mystifying airplane, and it cannot be determined from any of the meagre records left behind, if any were ever built as such, or if possibly this was a certificate of approval that was issued to allow modification of existing "4000" types to conform to this specification. The answer lies somewhere to be sure, but we have not found it to this point. The Stearman biplane of the C3MB type (see the chapter for A.T.C. # 137 in this volume) was covered by a certificate of approval that allowed for modification of the front cockpit, which normally held two passengers, into a hatch-covered compartment used only for transporting air-mail and air-cargo. We can speculate that it would be no great effort at all to modify an existing "J5 Travel Air", into an aircraft with a covered compartment that would be used only for the transport of cargo, whether it be air-mail or whatever. If the designation BM-4000 refers to aircraft only of the B-4000 type, we can

Fig. 158. These mail-carrying Travel Airs operated with Continental Air Lines; note modified rudders.

then guess that it would be but a simple matter to modify a B-4000 to the new specification and come up with a model BM-4000.

The model BM-4000 was a single-place open cockpit biplane that was powered with the 9 cylinder Wright J5 engine of 220 h.p.; payload for the craft was some 500 pounds, and it's high empty weight in comparison to typical craft in the "4000" series, would suggest extra equipment that probably even included a full complement of night-flying gear. No BM-4000 even appears in any registration records, and the nearest example that could be judged as a craft of the BM-4000 type, was a Travel Air B-4000 that was operated by the Michigan Air Express, but it may even have been of the 3 place open cockpit version as described in the previous chapter. The type certificate number for the model BM-4000 as powered with the Wright J5 engine, was issued in May of 1929 to the Travel Air Company at Wichita, Kansas.

Listed below are specifications and performance data for the Travel Air biplane model BM-4000; length averall 23'4"; hite overall 9'1"; wing span upper 33'0"; wing span lower 28'10"; wing chord upper 66"; wing chord lower 56"; wing area upper 171 sq.ft.; wing area lower 118 sq.ft.; total wing area 289 sq.ft.; airfoil "Travel Air"; wt. empty 1928; useful load 1072; payload with 67 gal. fuel was 500 lbs.; gross wt. 3000 lbs.; max. speed 130; cruising speed 112; landing speed 55; climb 900 ft. first min. at sea level; service ceiling 13,500 ft.; gas cap. 67 gal.; oil cap. 6 gal.; cruising range 550 miles; price at the factory field probably averaged at $9000.

The fuselage framework was built up of welded chrome-moly steel tubing, faired to shape with wood fairing strips and fabric covered. The front cockpit which normally sea-

ted two passengers, was converted to a hatch-covered compartment for transporting airmail and air-cargo; usually a compartment of this type would be lined with laminated plywood or metal panels. The wing framework was built up of heavy-sectioned solid spruce spars that were routed to an "I beam" section, with spruce and plywood built up wing ribs; the completed framework was fabric covered. The main fuel tank was mounted high in the fuselage just behind the engine firewall; extra fuel was carried in a tank that was mounted in the center-section panel of the upper wing. Some cargo-carrying versions of the Travel Air had all the fuel in the upper wing; 3 tanks that were mounted one in the center-section panel and one in the root end of each wing. This provided more fuselage area for a bigger compartment, a compartment that was able to handle bulkier loads. The fabric covered tail-group was built up of welded chrome-moly steel tubing and sheet steel former ribs; the vertical fin was ground adjustable and the horizontal stabilizer was adjustable in flight. The landing gear on the B-4000 type was of the outrigger configuration that used oleo-spring shock absorber struts; a landing gear of this type was more suitable to the higher gross loads and the resulting increases in stress and shock during take-off runs and in landing. Wheels were 30x5 and brakes were standard equipment. For air-mail service, it is most likely that this craft would be equipped with a full complement of night-flying gear, that would include landing lights and parachute flares. The next development in the Travel Air biplane was the model A-4000 as powered with the Axelson engine; see the chapter for A.T.C. # 148 in this volume.

Fig. 159.   *The prototype of the A-4000 was a handsome craft; note the neat installation of the Axelson engine.*

The popular "Travel Air" biplane, like any other 3 place open cockpit biplane of the general-purpose type, could change it's entire personality and pattern of behavior just by the installation of a different engine. After all, the engine was the heart and life of the airplane that would motivate it into action; how the airplane would act or the pilot react, was entirely dependent on the spirit and the muscle that was available in the powerplant. From the many types that were available on the market, one would perhaps tend to wonder why all the variation in powerplants would be justifiable, but then, there were many things to take into consideration when selecting a certain airplane-powerplant combination. It could be the type of duties that were planned for this certain airplane, or perhaps the price-tag had some bearing on the selection, or it may even have been personal preference for an engine of a certain type. It is well known fact that pilots pledged devotion to an airplane mounting a certain engine, with the same cussed fervor they might show for a certain brand of car they had been buying for years; it is one of the

quirks of man-kind and best left alone or catered to, rather than trying to change someone's mind and gain a convert.

The model A-4000 was a 3 place open cockpit biplane of the popular general-purpose type, and was quite typical to the standard Travel Air biplane that was coming off the line at this particular time. The powerplant for the model A-4000 was the 7 cylinder air-cooled Axelson engine that made up into a rather pleasant combination with the compatible qualities of the Travel Air biplane, offering good performance with good utility and a fair economy of operation. The first few examples of this model had the old-style wings with the familiar and lovable "elephant ear" ailerons, but later examples were done up with new wing panels that had rounded tips, with the ailerons inset from the end. The new wing panels were somewhat "beefier" with heavier spars and what-not, and weighed about 40 pounds more per set; added equipment boosted the weight still higher and therefore, this model operated at a rather high gross load. The trend at this particular time seemed to be forming in a desire to sacrifice some performance for the addition of wheel-brakes, engine starter, and other extra equipment to make things easier or a little simpler; this would not be the case a few years back, when performance was paramount and above all else. The type certificate number for the model A-4000 as powered with the Axelson engine, was issued in May of 1929 and some 9 or more examples of this model were manufactured by the Travel Air Company at Wichita, Kansas.

Listed below are specifications and performance data for the Travel Air biplane model A-4000 as powered with the 115-150 h.p. Axelson engine; length overall 24'6"; hite overall 8'11"; wing span upper 34'8" (33'0"); wing span lower 28'10"; wing chord upper 66"; wing chord lower 56; wing area upper 178 (171) sq.ft.; wing area lower 118 sq.ft.; total wing area 296 (289) sq.ft.; figures in brackets are for wings without "ears"; wt. empty 1655; useful load 995; payload with 60 gal. fuel was 425 lbs.; gross wt. 2650 lbs.; following figures based on 115 h.p.; max. speed 108; cruising speed 92; landing speed 47; climb 525 ft. first min. at sea level; service ceiling 10,000 ft.; gas cap. 60 gal.; oil cap. 6 gal.; cruising range 550 miles; following figures based on 150 h.p.; max. speed 115; cruising speed 97; landing speed 47; climb 620 ft. first min. at sea level; service ceiling 12,000 ft.; all other dimensions and data remained the same; price at the factory field was $5750.

The fuselage framework was built up of welded chrome-moly steel tubing, faired to shape with wood fairing strips and fabric covered. The cockpits were comfortable and well appointed; a baggage compartment with access from the left side, was in the lower fuselage just behind the rear seat. The wing framework was built up of solid spruce spars that were routed to an "I beam" section, with spruce and plywood built-up wing ribs; the completed framework was fabric covered. The main fuel tank was mounted high in the fuselage just ahead of the front cockpit; extra fuel was carried in a tank that was mounted in the center-section panel of the upper wing. The fabric covered tail-group was built up of welded chrome-moly steel tubing and sheet steel former ribs; the fin was ground adjustable and the horizontal stabilizer was adjustable in flight. The landing gear of 78 inch tread was of the typical split-axle type and used two spools of rubber shock-cord to snub the bumps; wheels were 30x5 and brakes were standard equipment. A metal propeller, navigation lights, and inertia-type engine starter were available as optional equipment. The next development in the Travel Air biplane was the Curtiss "Challenger" powered model C-4000; for discussion of this model see the chapter for A.T.C. # 149 in this volume.

Listed below are A-4000 entries that were gleaned from various records; this list may not be quite complete, but it does show the bulk of this model that were built.

X-9017; A-4000 (# 818) Axelson.
 C-8703; A-4000 (# 979) Axelson.
 C-8842; A-4000 (# 1006) Axelson.
NC-9820; A-4000 (# 1019) Axelson.
NC-9939; A-4000 (# 1123) Axelson.
NC-9988; A-4000 (# 1164) Axelson.
NC-679H; A-4000 (# 1236) Axelson.
NC-443N; A-4000 (# 1344) Axelson.
NC-454N; A-4000 (# 1370) Axelson.

Fig. 160. The Travel Air A-4000 had a 7 cylinder Axelson engine.

Fig. 161. The Travel Air C-4000 with a 170 h.p. Curtiss "Challenger" engine.

As pictured here in various views, the model C-4000 was the compatible mating of the popular Travel Air biplane with the reliable Curtiss "Challenger" engine. A combination of rather good performance that shared all the inherent qualities and flight characteristics that made the Travel Air biplane a popular choice, a popular choice with just about any engine combination. The 6 cylinder Curtiss "Challenger" engine was a two-row "staggered radial" that could be likened to a pair of 3 cylinder engines put together; this was a powerplant of exceptional reliability that gained world-wide fame and recognition by propelling the "St.Louis Robin" to a new endurance record of over 420 hours. First rated for 170 h.p. at 1800 r.p.m. and later re-rated to 185 h.p. at 2000 r.p.m., the "Challenger" was a short-stroke engine of fairly small diameter that presented less-than-average frontal drag area, and faired in nicely into the average installation.

First developed in the latter part of 1928 as a special-purpose airplane, under the watchful eyes of Curtiss engineers, the model C-4000 prototype (X-6413) was indeed a handsome airplane with a performance well on a par with it's good looks. The standard production version of the model C-4000, though blessed with good looks of less exciting character, differed in many respects; it was a 3 place open cockpit biplane and planned as a general-purpose airplane, a craft of the type that would best serve the interests of the penny-concious flying service operator. Most of the earlier examples in the C-4000 series were fitted with the old style wings which had the familiar "elephant ear" ailerons; all the later examples were fitted with the new style wings which discarded the balance-horns and had rounded wing tips. The new wing panels were beefed-up a bit with heavier construction, and this added some 40 pounds to the empty weight of the airplane. Though operating with a fairly high gross weight and able to carry a substantial payload, the model C-4000 performed very well under all types of service. The type certificate number for the model C-4000 as powered with the Curtiss "Challenger" engine, was issued in May of 1929 and some 19 or more examples of this model were manufactured by the Travel Air Company at Wichita, Kansas.

Listed below are specifications and performance data for the Travel Air biplane model C-4000 as powered with the 170 h.p. "Challenger" engine; length overall 24'6"; hite overall 8'11"; wing span upper 34'8" (33'0"); wing span lower 28'10"; wing chord upper 66"; wing chord lower 56"; wing area upper 178 (171) sq.ft.; wing area lower 118 sq.ft.; total wing area 296 (289) sq.ft.; airfoil T.A. #1; figures in brackets are for new wings without "ears"; wt. empty (old wings) 1590;

useful load 1007; payload with 67 gal. fuel was 392 lbs.; gross wt. 2597 lbs.; wt. empty (new wings) 1630; useful load 1007; payload with 67 gal. fuel was 392 lbs.; gross wt. 2637 lbs.; max. speed 124; cruising speed 105; landing speed 48; climb 800 ft. first min. at sea level; service ceiling 13,000 ft.; gas cap. 67 gal.; oil 6 gal.; cruising range at 10 gal. per hour was 630 miles; price at the factory field was $6275.

The fuselage framework was built up of welded chrome-moly steel tubing, faired to shape with wood fairing strips and fabric covered; a baggage compartment of 9 cubic foot capacity was in back of the rear cockpit with an access door on the left side. The wing framework was built up of solid spruce spars that were routed for lightness, with spruce and plywood built-up wing ribs; the completed framework was fabric covered. The main fuel tank was mounted high in the fuselage just ahead of the front cockpit, and extra fuel was carried in a tank that was mounted in the center-section panel of the upper wing. The fabric covered tail-group was built up of welded chrome-moly steel tubing and sheet steel formers; the fin was ground adjustable and the horizontal stabilizer was adjustable in flight. The landing gear was of the normal split-axle type with two spools of rubber shock-cord to snub the shocks; tread was 78 inches, with Bendix wheels (30x5) and brakes. A Curtiss-Reed metal propeller of forged aluminum alloy, and inertia-type engine starter were also available. The next Travel Air de-velopment was the "Wasp" powered cabin monoplane that was mounted on twin-float seaplane gear; this was the model SA-6000-A which is discussed in the chapter for A.T.C. # 175 in this volume.

Listed below are model C-4000 entries that were gleaned from various records; this list may not be quite complete, but it does show the bulk of this model that were built.

X-6413; C-4000 (# 754) Challenger.
NC-8156; C-4000 (# 971) Challenger.
NC-8179; C-4000 (# 972) Challenger.
C-8890; C-4000 (# 1013) Challenger.
CF-ABH; C-4000 (# 1015) Challenger.
C-9824; C-4000 (# 1044) Challenger.
C-9825; C-4000 (# 1045) Challenger.
C-9920; C-4000 (# 1110) Challenger.
NC-380M; C-4000 (# 1301) Challenger.
NC-381M; C-400 (# 1302) Challenger.
NC-687K; C-4000 (# 1306) Challenger.
NC-389M; C-4000 (# 1309) Challenger.
NC-390M; C-4000 (# 1310) Challenger.
NC-391M; C-4000 (# 1311) Challenger.
NC-392M; C-4000 (# 1312) Challenger.
NC-396M; C-4000 (# 1315) Challenger.
NC-398M; C-4000 (# 1318) Challenger.
NC-404N; C-4000 (# 1335) Challenger.
NC-436N; C-4000 (# 1350) Challenger.
Model C-4000 serial # 1015 was registered in Canada; model C-4000 serial # 1110 was converted from a model 2000; model C-4000 serial # 1310 was originally a model E-4000; model C-4000 serial # 1335 was originally a model E-4000; model C-4000 serial # 754 also registered as NC-6413.

Fig. 162. The prototype of the Traval Air C-4000 is shown here during early tests.

Fig. 163. This Travel Air C-4000 was originally a 2000 with OX-5 engine; more modern engines replaced the OX-5.

Fig. 164. The Command-Aire 3C3-T trainer with a Curtiss OX-5 engine.

With the gradual and mounting increase in pilot training, over that of a year or so ago, there was developed a great need and a good bit of clamor among the flying school operators for a two place training craft that was built especially for the job on the flight line. Consequently, several of the manufacturers of the open cockpit biplane, were contemplating on modifying their standard three place models to conform to this new demand; which in some cases would only take a little bit of doing. The Command-Aire trainer however, was planned and developed as a new design and though it was typical to the standard 3C3 in outward appearance, it was arranged to handle the job of student training much better, and calculated to instill in the student the feeling that he was flying a craft that was designed especially to teach him the art of flying.

The Command-Aire 3C3-T trainer was a two place open cockpit biplane seating two in tandem, of a typical Command-Aire configuration, except that the fuselage was narrowed down somewhat to conform to the new seating arrangement, and used the one elongated cockpit commonly called the "bath tub".

There was some variation in this cockpit arrangement over that as introduced on the prototype version, but in production, it more or less remained as pictured here. The Curtiss OX-5 engine was selected as still the most economical and most logical choice for a craft of this type, so the new model 3C3-T was powered with the 8 cylinder OX-5 engine of 90 h.p. Performance of this two place craft was quite good and the inherent flight characteristics were excellent, but it was considered a mite too easy to fly; this should not be argued, as it is a matter of personal opinion. Some instructors thought it better to use an airplane for primary training that was a bit harder to fly, others did not. Kept bare of most frills and extra fancy appointments, the 3C3-T trainer was designed to do a certain job with the least amount of lost time from the flight line, where it was often seen from dawn until dusk. The type certificate number for the model 3C3-T trainer as powered with the Curtiss OX-5 engine, was issued in May of 1929 and some 30 or more of this model were manufactured by Command-Aire, Inc. at Little Rock, Arkansas.

Listed below are specifications and perfor-

Fig. 165. The Command-Aire 3C3-T was an ideal trainer of gentle nature and rugged structure.

Fig. 166. Flight characteristics of the 3C3-T were ideal for sport flying.

mance data for the Command-Aire biplane trainer model 3C3-T as powered with the 90 h.p. Curtiss OX-5 engine; length overall 24'6"; hite overall 8'4"; wing span upper and lower 31'6"; wing chord both 60"; wing area upper 169 sq.ft.; wing area lower 134 sq.ft.; total wing area 303 sq.ft.; airfoil "Aeromarine 2A"; wt. empty 1290; useful load 700; payload with 40 gal. fuel was 250 lbs.; gross wt. 1990 lbs.; figures for a later version were as follows; wt. empty 1439; useful load 670; payload with 40 gal. fuel was 220 lbs.; gross wt. 2109 lbs.; max. speed 100; cruising speed 85; landing speed 35; climb 550 ft. first min. at sea level; service ceiling 10,000 ft.; gas cap. 40 gal.; oil cap. 4 gal.; cruising range 450 miles; price at the factory field with new OX-5 engine, when available, was $3350; also sold less engine and propeller. The price with OX-5 engine was reduced to $2250 in 1930.

The fuselage framework was built up of welded chrome-moly steel tubing, lightly faired to shape with fairing strips and fabric covered. Quite novel was the one-piece metal turtle-back, which was quickly removable for inspection or maintenance to components in the rear section of the fuselage. The wing framework was built up of solid spruce spar beams with spruce and plywood built-up wing ribs; the completed framework was fabric covered. The upper wing was in two panels that were joined together on the center line and fastened to the center-section cabane struts; there was no separate center-section panel in the upper wing as would normally be the case. The fuel tank was mounted high in the fuselage, just ahead of the front seat of the cockpit with a direct reading fuel gage that projected through the cowling. Instrumentation was adequate but held down to only bare necessities. The split-axle landing gear of 78 inch tread was of the normal Com-

mand-Aire layout and used two spools of rubber shock-cord to snub the bumps; the tail-skid was of the spring-leaf type and wheels were 26x4. The engine coolant radiator was of the nose-section type and was faired into the engine cowling, which was profusely louvered to vent the engine section; a propeller "spinner" was optional. The next development in the Command-Aire trainer series was the Warner powered 3C3-AT as described in the chapter for A.T.C. # 151 in this volume.

Listed below are 3C3-T trainer entries that were gleaned from various records; this list may not be quite complete, but it does show the bulk of this model that were built.

C-10012; 3C3-T (# 533) OX-5
C-237E; 3C3-T (# 551) OX-5
C-351E; 3C3-T (# 569) OX-5
C-437E; 3C3-T (# 572) OX-5
C-445E; 3C3-T (# 580) OX-5
C-475E; 3C3-T (# 585) OX-5
NC-476E; 3C3-T (# 586) OX-5
C-477E; 3C3-T (# 587) OX-5
C-478E; 3C3-T (# 588) OX-5
C-581E; 3C3-T (# 605) OX-5
NC-582E; 3C3-T (# 606) OX-5
NC-583E; 3C3-T (# 607) OX-5
NC-584E; 3C3-T (# 608) OX-5
NC-585E; 3C3-T (# 609) OX-5
C-907E; 3C3-T (# 611) OX-5
NC-904E; 3C3-T (# 612) OX-5
C-902E; 3C3-T (# 613) OX-5
NC-901E; 3C3-T (# 614) OX-5
C-903E; 3C3-T (# 615) OX-5
C-908E; 3C3-T (# 617) OX-5
NC-909E; 3C3-T (# 618) OX-5
NC-917E; 3C3-T (# 620) OX-5
NC-966E; 3C3-T (# 636) OX-5
NC-967E; 3C3-T (# 637) OX-5
NC-968E; 3C3-T (# 638) OX-5
NC-969E; 3C3-T (# 639) OX-5

A.T.C. #151
(5-29)
COMMAND-AIRE 3C3-AT

Fig. 167.   The Command-Aire trainer Model 3C3-AT had a 110 h.p. Warner engine; docile in behavior, it was a good training craft.

To help satisfy the need and still the clamor from flying school operators for a two place training craft, designed especially for pilot training, Command-Aire had developed the OX-5 powered 3C3-T biplane which was described here previously. Though an excellent trainer for the primary phases of pilot instruction, the model 3C3-T was a bit lacking in certain instances for training in the secondary phases, which required more difficult manuvers and more advanced techniques. For this purpose, Command-Aire developed the model 3C3-AT which was similiar to the OX-5 powered trainer, but was powered with the 7 cylinder Warner "Scarab" engine of 110 h.p. A combination that gave the new development a little more verve and much better manuverability; a better aptitude for executing the more advanced techniques in pilotage and the more difficult manuvers that were required for a pilot's license in the higher category, a license such as the "limited commercial" or the "transport". It was also anticipated that this type of craft would be suitable too as a sport-craft for the private-owner. Though not a fully aerobatic airplane in the true sense, the 3C3-AT was extremely capable, with performance and flight characteristics calculated to please and delight the average sport-flyer.

The Command-Aire 3C3-AT trainer, as pictured here, was a two place open cockpit biplane seating two in tandem, of a normal Command-Aire configuration that provided a seating arrangement for two in the one elongated cockpit, a cockpit that was commonly called the "bath tub" type. Performance of this two place craft was quite good and the inherent flight characteristics were excellent, especially in the lower speed ranges, though inherently docile in behavior, it could be prompted into some spirited action if so desired. The rather high useful load that was available, was calculated to allow at least 200 pounds each for both the instructor and the student, who many times would be dressed in heavy winter-weight flying clothes and wearing parachute packs, during the course of flight instruction. Also kept bare of frills and most fancy appointments, the 3C3-AT trainer was designed to do a certain job, and was performing these duties with excellent results. The type certificate number for the Warner powered 3C3-AT trainer was issued in May of 1929 and some 6 or more examples of this model were manufactured by Command-Aire, Inc. at Little Rock, Arkansas. R. B. Snowden Jr. was the president; C. M. Taylor Jr. was the V.P.; W. F. Moody was secretary-treasurer; Major John Carroll Cone, former Army Air Corps pilot, was the sales manager; Albert Voellmecke was the chief of engineering; and Wright "Ike" Vermilya was the chief pilot in charge of test and development. During this period of time, all sales and distribution for the Command-Aire was handled nation-wide by the Curtiss Flying Service, who also used the Command-Aire 3C3-T and the 3C3-AT occasionally in their training programs of the various flying schools they had scattered about the country.

Listed below are specifications and performance data for the Command-Aire biplane trainer model 3C3-AT as powered with the 110 h.p. Warner "Scarab" engine; length overall 24'8"; hite overall 8'4"; wing span upper and lower 31'6"; wing chord both 60"; wing area upper 169 sq.ft.; wing area lower 134 sq.ft.; total wing area 303 sq.ft.; airfoil "Aeromarine 2A"; wt. empty 1284; useful load 706; payload with 40 gal. fuel was 220 lbs.; gross wt. 1990 lbs.; max. speed 110; cruising speed 92; landing speed 35; climb 710 ft. first min. at sea level; service ceiling 13,000 ft.; gas cap. 40 gal.; oil cap. 5 gal.; cruising range 450 miles; price at the factory field in 1929 was $5500, lowered to $4165 in 1930.

The fuselage framework was built up of welded chrome-moly steel tubing, lightly faired to shape with fairing strips and fabric covered. Quite novel was the one-piece metal turtle-back that was easily removable for quick inspection and maintenance to components in the rear section of the fuselage. With the use of metal engine cowling, metal cockpit cowling, and the metal turtle-back, actually less than half of the fuselage was covered in fabric. The wing framework was built up of solid spruce spar beams and spruce and plywood built-up wing ribs; the completed framework was fabric covered. The ailerons which were on the lower panels only, were of a chrome-moly steel tube framework that

was fabric covered; the ailerons were of the "slotted hinge" type and were very effective at low air speeds, even down to the stall". Actuation of all controls was by push-pull tubes and bellcranks, no cables nor pulleys were used. The center-section bracing struts were of the normal "N" type, in combination with an extra strut that fastened from the upper wing to a fitting on the lower fuselage; this method of bracing eliminated the familiar cross-wires normally found in the center-section bay, and improved visibility straight ahead. The fabric covered tail-group was built up of welded chrome-moly steel tubing; the fin was ground adjustable and the horizontal stabilizer was adjustable in flight. The split-axle landing gear of 78 inch tread was built up of chrome-moly steel tubing in a streamlined section, and was of the normal 3-member type with two spools of rubber shock-cord to absorb the bumps; the tail-skid was of the spring-leaf type and wheels were 26x4. The fuel tank was mounted high in the fuselage just ahead of the front seat, and there was a direct reading fuel gauge that projected through the cockpit cowling. Instrumentation was kept at a minimum, and only the basic engine and flight instruments were installed. Wheel brakes, a metal propeller, inertia-type engine starter, and low-pressure "airwheels", were available as extra equipment. A later development of the Warner powered trainer was the model BS-14 that was on a Group 2 approval numbered 2-204. The next development in the Command-Aire biplane was the Curtiss "Challenger" powered model 5C3; see the chapter for A.T.C. # 184 in this volume.

Listed below are 3C3-AT entries that were gleaned from various records; this list may not be quite complete, but it does show the bulk of this model that were built.

NC-514; 3C3-AT (# W-53) Warner 110, converted 3C3-A.

NC-525E; 3C3-AT (# W-60) Warner 110.

C-962E; 3C3-AT (# W-107) Warner 110.

NC-971E; 3C3-AT (# W-109) Warner 110.

NC-975E; 3C3-AT (#W-113) Warner 110.

NC-976E; 3C3-AT (# W-114) Warner 110.

Fig. 168. The Laird "Speedwing" LC-R200 with a Wright J5 engine designed for the sportsman-pilot.

"Matty" Laird's airplanes have had a lasting and enviable reputation for being fast, sturdy, and a complete joy to own and to fly; adding further to this nation-wide reputation comes the Laird "Speedwing". The nimble "Speedwing", as the name would tend to imply, was a custom craft of outstanding qualities that was especially leveled at the discriminate sportsman-pilot who would want a swift and spirited airplane, an airplane he would be proud to own and to pamper, as if it were a thoroughbred horse. Strictly "sporting" is the word for the Laird "Speedwing", and it was every bit of that. Designed with speed to spare and a high cruising range, the LC-R was ideal for long cross-country jaunts, but it's sure-footed and delightfully responsive manuverability was it's outstanding quality, and the things possible in the air with this airplane were only limited by the whim and fortitude of the pilot. The "Speedwing" normally carried three in two open cockpits, but the front cockpit could be closed off with a metal cover; this metal panel which covered the gaping hole of the front cockpit, added several miles per hour to the top speed available. The model LC-R which was a custom-built airplane, was only built on order, usually to the customer's particular tastes and

fancies; built unhurried and with the greatest of care, it could be likened to a fine piece of precision machinery, but unlike a piece of impersonal machinery, the LC-R possessed and radiated a magnetic charm that one couldn't help but notice and enjoy. E. M. Laird advertised his airplanes as the "thoroughbreds of the air" and this could hardly be disputed, because the craft had proven time and again in years past that he had every justification to make this statement.

The Laird "Speedwing" was not entirely "designed" at one time to be as it is, but it would be better to say that it was more or less "developed", mainly the persuasion and efforts of Chas. "Speed" Holman and E. E. Ballough, two "Laird" pilots who had the habit of running rough-shod over all the competition in the major derbies and air-races. In 1927, two Wright J4 powered Laird "Commercial" were flown in the National Air Derby from New York to Spokane; "Speed" Holman romped into first place and E. E. Ballough came in 2cd only a half hour behind. In the 1928 National Air Derby from New York to Los Angeles, E. E. Ballough flew his same Laird, somewhat modified by now into an LC-R, to 2cd place only 19 minutes behind the winner. In a civilian free-for-all at the 1928 National Air Races held in Los Angeles, Ballough and this same LC-R averaged over 137 m.p.h. for the 75 mile course and came in first. In February of 1928, "Speed" Holman set a loop record of 1073 inside loops in the same Laird he used in 1927 to win the derby, but it had also been modified by now into a version approaching the LC-R; sometime later it was modified again and with this craft he performed two "outside loops", which was a feat that had been done only once before in a commercial type of aircraft. The Los Angeles to Cincinnati Air Derby in the fall of 1928, was won easily by "Speed" Holman; the Gardner Trophy Race of mid-1929, from St.Louis to Indianapolis and return, was a hotly contested race that was also won by Holman and his "Speedwing", averaging better than 157 m.p.h. for the 486 mile course. With the urge and determination to continually stay out on top, these two "Lairds" which were familiar to all air-minded folk by now, were being modified constantly and progressively by "Matty" Laird who was fiercely proud of these two airplanes, and from this background of developments stemmed the "Speedwing" version as described here. The type certificate number for the Laird "Speedwing" model LC-R200 as powered with the 220 h.p. Wright J5 engine, was issued in May of 1929 and some 3 or more of this model were manufactured by the E. M. Laird Airplane Co.

at Chicago, Illinois. Emil Matthew Laird was the president and chief of design; Lee Hammond was the V.P. in charge of sales; R. J. Hoffman and Charles Arens were engineers; and "Matty" Laird most always tested all new designs, with the assistance of Charlie Holman and Erv Ballough who were always ready, willing, able, and available.

Listed below are specifications and performance data for the Laird "Speedwing" model LC-R200 as powered with the Wright "Whirlwind" J5 engine of 220 h.p.; length overall 22'9"; hite overall 9'0"; wing span upper 28'0"; wing span lower 24'0"; wing chord upper 60"; wing chord lower 48"; wing area upper 121.3 sq.ft.; wing area lower 80.8 sq.ft.; total wing area 202 sq.ft.; airfoil "Laird"; wt. empty 1848; useful load 1066; payload with 76 gal. fuel was 390 lbs.; gross wt. 2914 lbs.; max. speed 145+; cruising speed 120+; landing speed 58; climb 1050 ft. first min. at sea level; service ceiling 16,000 ft.; gas cap. 76 gal.; oil cap. 6 gal. ;cruising range 650 miles; price at the factory was $10,600. The fuselage framework was built up of duralumin tubing with special steel clamps to make the joints, and braced in an X truss with heavy steel tie-rods; the framework was heavily faired to a deep streamlined shape and fabric covered. The large fuel tank was mounted high in the fuselage just ahead of the front cockpit with a direct-reading fuel gauge projecting through the cowling; a small

Fig. 169. The "Speedwing" design evolved from this J4-powered Laird Speedster, flown by "Speed" Holman.

Fig. 170.   Charles "Speed" Holman poses proudly by his famous Laird Speedster.

looking "I shaped" interplane struts were of laminated plywood and were shaved to a streamlined form. The fabric covered tail-group was built up of welded chrome-moly steel tubing; the fin was ground adjustable and the horizontal stabilizer was adjustable in flight. The robust split-axle landing gear was of the normal Laird type and used two spools of rubber shock-cord as shock absorbers; wheels were Bendix 30x5 and the tail-skid was of the spring-leaf type. Custom colors were available and the doped finish on the "Speedwing" was about the finest available. Wheel brakes, a metal propeller, navigation lights, and inertia-type engine starter were standard equipment. The next development in the Laird "Speedwing" was the LC-R300, see the chapter for A.T.C. # 176 in this volume.

Listed below are "Speedwing" LC-R200 entries that were gleaned from various records; this list may not be complete, but it does show the bulk of this model that were built.

X-7086; LC-RJ200 (# 166) Wright J4B, formerly # 149.

X-7087; LC-RJ200 (# 167) Wright J4B, formerly # 150.

C-7216; LC-R200 (# 168) Wright J5, later Wasp 450.

NC-634; LC-R200 (# 177) Wright J5, later as LC-R300.

NC-633; LC-R200 (# 178) Wright J5, later as LC-R300.

baggage compartment was in the turtle-back section of the fuselage, just in back of the pilot. The wing framework was built up of heavy-sectioned laminated spruce spar beams and spruce and plywood built-up wing ribs that were closely spaced together for added strength and to preserve the airfoil form across the span; the completed framework was fabric covered. There were four ailerons that were connected together in pairs by a streamlined push-pull strut; the distinctive

Fig. 171. An Ireland "Neptune" N2B with a 300 h.p. Wright J6 engine; note sweep-back of upper wing.

The seaplane, the flying boat, and the amphibian aircraft were just beginning to be appreciated again, and it's a wonder that this awakening had taken such a long time. After much struggle for a number of years to obtain airports that would be near enough to large metropolitan areas, the fact was clearly established that nearly every important city, or resort area in this country, was situated beside a large body of water, be it ocean, lake, or river. These large and convenient landing places were certainly too inviting to be ignored much longer, and a renewed interest was being shown in aircraft that could fly off water. Naturally, the most intelligent approach in this respect was the amphibious aircraft, which could fly off both land and water, and could pick and choose it's airport to the best advantage; bringing the passengers closer to the center of big cities, or country-homes on a lake, to resort areas, or perhaps fishing and hunting grounds, both on land or water, that often were only accessible by air. The amphibious aircraft actually saved the time, and offered the convenience and utility that made air-travel worthwhile.

Just such a versatile airplane was the Ireland "Neptune" model N2B; an amphibious craft which was a seaworthy flying-boat configuration with a retracting undercarriage that could be quickly extended for landings on an airport, or swung out of the way for landings on water. The N2B was a biplane arrangement that had it's powerplant placed high in a pusher-type nacelle that was located between the upper and lower wings, where it would be away from the water-spray, and be out of the way for those boarding or leaving the enclosed cabin. The cabin being in the forward part of the boat-type hull which seated five persons in good comfort, with panoramic vision for the pilot. For this type of craft, the pusher-type engine installation was very sensible and offered many advantages; it was safer for docking and anchoring, it's rearward location softened and muffled the engine noise, and allowed conversation in the air that need not be much over the normal tone. There was some sacrifice in performance to be expected in the amphibian aircraft because of the greater weight and increased drag, largely due to the increase in frontal area, but this sacrifice was a rather small price to pay for all of the extra benefits.

G. Summer Ireland was associated with Curtiss activities, both directly and indirectly for quite a time, and his first airplane designed for civil and commercial usage was the Ireland "Comet" of 1925. The "Comet" was a neat open cockpit biplane that carried three with the power of the Curtiss OX-5 engine, and was basically a Curtiss "Oriole" fuselage with a new landing gear and a new set of hi-lift wing panels. In 1926, Ireland introduced the all-new "Meteor", which was a rakish looking open cockpit biplane that carried 3 or 4 with the power of the Curtiss OX-5 engine, and also offered in combination with the Curtiss C-6 or the Wright "Whirlwind" J4. Ireland offered the amphibian as early as 1926, as an open cockpit flying-boat type with an Anzani engine, but the first example of the "Neptune" was not built until 1927. The prototype "Neptune" was actually quite similiar to the N2B, but was an open cockpit craft seating 4, and was powered with a Wright J5 engine; the upper wing panel was straight across and did not have the sweepback in the upper panel as characteristic of the model N2B. Several of the J5 powered N2 were built, but the performance in this combination was a bit soggy and not too sprightly, so Ireland decided to increase the power with the new Wright J6-300; the model N2B was the result. The type certificate number for the "Neptune" model N2B was issued in May of 1929 and some 15 or more of this model were built by the Ireland Aircraft, Inc. at Garden City, Long Island, New York. Bertram Work was the president; G. Sumner Ireland was V.P. in charge of design and sales; and D. J. Brimm Jr. was secretary, treasurer, and chief of engineering.

Listed below are specifications and performance data for the Ireland "Neptune" model N2B as powered with the 9 cyl. Wright J6 engine of 300 h.p.; length overall 31'; hite on wheels 12'2"; wing span upper 40'; wing span lower 34'; chord upper 72"; chord lower 66"; wing area upper 222 sq.ft.; wing area lower 154 sq.ft.; total wing area 376 sq.ft.; airfoil "Curtiss C-72"; wt. empty 2949; uesful load 1451; payload 730; gross wt. 4400 lbs.; max. speed 115; cruising speed 90; landing speed 45; climb 750 ft. first min. at sea level; service ceiling 11,000 ft.; gas cap. 85 gal.; oil cap. 5 gal.; cruising range 405 miles; price at the factory field was $22,500. The hull of the N2B was a framework built up of welded chromemoly steel tubing, that was surrounded with duralumin bulkheads and formers to establish the hull shape and provide a base for fastening the aluminum alloy outer skin. The cabin was in the forward portion of the hull and seated two in front, and a wide bench-type seat in back had ample room for three. The main fuel supply was in a tank that was located in the hull behind the cabin section; a gravity-feed auxiliary fuel tank was in the center-section panel of the upper wing. There was a tie-down shackle in the nose of the hull for mooring, and the anchor was stored in a compartment just in front of the cabin windshield. The wing panels were built up of solid spruce spar beams that were routed to an "I beam" section, with a combination of

Fig. 172. The "Neptune" N2B, with its ability to operate from land or water, offered worthwhile conveniences.

Fig. 173. An early "Neptune"; note Wright J5 engine and open cockpits.

wood and metal wing ribs; the leading edges were covered with dural sheet to preserve the airfoil form and the completed panels were fabric covered. The interplane struts were heavy gauge duralumin tubes in a streamlined section, and the interplane bracing was of streamlined steel tie-rods; metal framed and metal covered wing-tip floats were provided to keep the lower wing from digging into the water. The retractable landing gear was built up of chrome-moly steel tubing, and could be drawn up in about 35 seconds, and lowered in about 20 seconds; shock absorbers were air-oil struts, and the wheels were 30x5, with a 7 foot 10 inch tread. The fabric covered tail-group was a composite structure of duralumin beams and sheet steel ribs and formers; the fin was ground adjustable and the horizontal stabilizer was adjustable in flight. All movable control surfaces were aerodynamically balanced. Wheel brakes, metal propeller, engine starter, navigation and mooring lights, anchor, and a heaving line were stan-

dard equipment. The next development in the Ireland "Neptune" was the "Wasp" powered N2C which was on A.T.C. # 248.

Listed below are "Neptune" N2B entries that were gleaned from the Aeronautical Chamber of Commerce aircraft register; this list is not complete but it does represent the bulk of this model that were built.

C-9788; Neptune N2B (#   ) Wright J6, prototype.
NC-106H; Neptune N2B (# 32) Wright J6
NC-107H; Neptune N2B (# 33) Wright J6
NC-108H; Neptune N2B (# 34) Wright J6
NC-109H; Neptune N2B (# 35) Wright J6, not verified.
NC-110H; Neptune N2B (# 36) Wright J6, not verified.
NC-111H; Neptune N2B (# 37) Wright J6
NC-112H; Neptune N2B (# 38) Wright J6
NC-113H; Neptune N2B (# 39) Wright J6
NC-114H; Neptune N2B (# 40) Wright J6
NC-87K; Neptune N2B (# 42) Wright J6
NC-88K; Neptune N2B (# 43) Wright J6

Fig. 174. The International F-17-W "Sportsman" with a 220 h.p. Wright J5 engine.

The International "Sportsman" model F-17-W as pictured here, was an all-purpose airplane that was especially designed for the varied services that were performed by the average flying-service operator, plus the fact that it was also quite suitable to the requirements of the average sportsman-pilot. This model was a 3-4 place open cockpit biplane of typical "Fisk" configuration and was powered with the 9 cylinder Wright J5 engine of 220 h.p. As all of the "Internationals", and this one was surely typical, it had the characteristic octagonal shaped wooden fuselage structure that was such an unmistakeable trade-mark for these craft from many years back. The distinctive "International" airplanes were designed and developed by the modest Edwin M. Fisk, an honest man who plainly acknowledged that he got the idea for the "8 sided" fuselage configuration from that as used in dirigible construction. Edwin Fisk, one of the pioneers in aircraft development, had been designing and building airplanes since some time back in 1910; from then until 1929 he managed to design and develop at least

twenty different airplane types. None had ever achieved world-wide recognition, but "International" airplanes were known and talked about the country over, and loved by all who flew them.

Speaking with old-timers who have flown these airplanes, one could assume that the "International" was quite an impressive airplane in it's general behavior and probably considered a little better than average in all-round performance; you cannot dispute the love a pilot had for an airplane. The "International" was also known as a soft and smooth airplane, no doubt due to it's noise and vibration absorbing all-wood construction. The model F-17 in any combination was certainly no ravishing beauty, but it did impress one as a stalwart and capable airplane that would ask no favors. The International Aircraft Corp. seemed to have some difficulty in staying financed, and consequently they moved around a good bit. After their operation in Long Beach, Calif. for a year or so, they re-organized and moved to Ancor, Ohio which is just outside of Cincinnati. Not too successful with this operation either, they moved on to Jackson, Michigan where they quietly folded up sometime in 1930. This certainly should not be considered unusual or detrimental to this company's efforts, because many aircraft manufacturing concerns, some even of long standing, had to fold up in 1930 or shortly thereafter. The stock market crash and the resulting depression was grimly felt by one and all. The type certificate number for the model F-17-W as powered with the Wright J5 engine, was issued in May of 1929 and probably not more than 5 or 6 examples of this model were manufactured by the International Aircraft Corp. at Ancor, Ohio. Arthur H. Ewald was the president; Perry V. Ogden was V.P. and general manager; Edwin M. Fisk was chief of design and engineering; and H. A. Speer was sales manager and the chief pilot.

Listed below are specifications and performance data for the "Sportsman" model F-17-W as powered with the 220 h.p. Wright J5 engine; length overall 24′6″; hite overall 9′6″; wing span upper & lower 35′0″; wing chord both 60″; wing area upper 170 sq.ft.; wing area lower 155 sq.ft.; total wing area 325 sq.ft.; airfoil USA 27; wt. empty 1780; useful load 920; payload with 60 gal. fuel was 345 lbs.; payload with 40 gal. fuel was 465 lbs.; gross wt. 2700 lbs.; max. speed 130; cruising speed 110; landing speed 42; climb 1000 ft. first min. at sea level; service ceiling 15,000 ft.; gas cap. 60 gal.; oil cap. 7 gal.; cruising range at 12 gal. per hour was 500 miles; price at the factory field was $4350 less engine and propeller; with the price of the Wright J5 engine at approx. $5000, this would make the price around $9350.

The unusual fuselage framework was octagonal in section with 8 spruce longerons of varying thickness that extended from the engine firewall back to the tail-skid post; with octagonal shaped laminated bulkheads placed at frequent intervals, and the outer covering of plywood veneer which took up a good bit of the load and stresses, we might say this type of construction was semi-monocoque. A construction of rigidity and great strength that could be easily repaired with normal hand tools; steel and duralumin fittings for engine mount, wing attach, landing gear attach, etc., carried the main loads through the fuselage structure. The wing framework was built up of spruce and plywood box-type spar beams, with wing ribs of plywood webs that had circular cut-outs for lightness and were reinforced with spruce cap-strips; the leading edges of the panels were covered with plywood back to the front spar, and the completed framework was fabric covered. The main fuel tank was mounted high in the fuselage just

Fig. 175. The F-17-W was of all-wood construction, offering smooth easy-chair flying.

Fig. 176. This J5-powered International was Frank Clarke's Dole Race entry.

ahead of the front cockpit, and extra fuel was carried in a tank that was mounted in the center-section panel of the upper wing; there were four ailerons that were connected together in pairs by a stranded steel cable. The fixed surfaces of the tail-group were built up of wood spars and ribs, and the movable surfaces were built up of welded steel tubing; the leading edges of the vertical fin and horizontal stabilizer were reinforced with plywood, and all surfaces were fabric covered. The vertical fin was ground adjustable and the horizontal stabilizer was adjustable in flight. The landing gear was of the normal split-axle type with oleo-spring shock absor-

bers; wheels were 30x5 and brakes were standard equipment. A metal propeller, navigation lights, and inertia-type engine starter were available as optional equipment. The next development in the "International" biplane was the "Hisso" powered model F-17-H; see the chapter for A.T.C. # 155 in this volume.

Listed below are F-17-W entries that were gleaned from various records; there were more examples built than those listed, but registration records show only the following.
C-5655; F-17-W (# 46) Wright J5.
-7528; F-17-W (# 65) Wright J5.
NC-518K; F-17-W (# 76) Wright J5.

A.T.C. #155
(5-29)
INTERNATIONAL F-17-H

Fig. 177.   The International F-17-H had a V-8 Hispano-Suiza engine of World War I vintage.

The International "Sportsman" model F-17-H was a companion model to the F-17-W and was typical in most respects, except for the powerplant installation which in this case was the 8 cylinder vee-type "Hisso" (Hispano-Suiza) engine of 150-180 h.p. This might be looked upon as a rather old-fashioned installation for this particular time, but it is surprising how popular the "Hisso" became when OX-5 engine stocks were about gone. It has been said that the "International" was a better craft in many ways when in combination with a heavy water-cooled engine, and the performance of this model would tend to bear that out; it was quite sprightly with good manuverability in spite of it's apparent bulk, and shared all the other flight characteristics that made the "International" such a favorite, a favorite with those who had the opportunity to know this airplane. Beside the better-known installations such as the OX-5, the OXX-6, the Hisso A & E, and the Wright J5 engines, the model F-17 had also been powered with the 6 cylinder Curtiss K-6, the 9 cylinder Siemens-Halske, and

the 4 cylinder air-cooled Dayton "Bear"; all were combinations of particular interest and merit.

Though pretty much of a back-yard operation until 1927, "International" was formed at that time by 3 men and a lot of ideas; Edwin Fisk helped to nurture this beginning into quite an impressive operation in a year's time, but circumstances developing in the industry brought on fierce competition that made the going rather rough, perhaps more than his backers could stand. Typical of most any other engineer that was dedicated to his work, Ed Fisk didn't trouble himself with business-matters too much, and kept on with his development of modifications for existing types, or the designing of entirely new types. One of the most promising designs to come up was the model F-18 that was called the "Air Coach"; a 6 place cabin biplane that was powered with the Wright J5 engine. This craft was issued a Group 2 approval numbered 2-15, and at least 6 were built; a cargo version of this craft was called the "Mailman". The third craft in this series was powered with the

180 h.p. "Hisso" E engine. The Jackson, Michigan plant was announced as ready for production in July of 1929, and both the F-17-W and the F-17-H were scheduled; the model F-17 with OX-5 engine was also to be produced in an improved version that was issued a Group 2 approval numbered 2-100 for company serial numbers to # 100. Records of the Michigan operation are scare and it is hard to relate facts concerning the activity at this plant; though hangar-talk would indicate a limited production, that was curtailed by the final end. The type certificate number for the 3 place model F-17-H as powered with the 150-180 h.p. "Hisso" engine, was issued is May of 1929 and some 10 or more examples of this model were manufactured by the International Aircraft Co. which was now operating at Jackson, Michigan. T. P. Funk was the president; D. G. Morrison was V.P. and general manager; Edwin M. Fisk was V.P. in charge of engineering; and Chas. Hollerith was secretary and treasurer.

Listed below are specifications and performance data for the "Sportsman" model F-17-H as powered with the 150-180 h.p. "Hisso" engine; length overall 25'0"; hite overall 9'6"; wing span upper & lower 35'0"; wing chord both 60" wing area upper 170 sq.ft.; wing area lower 155 sq.ft.; total wing area 325 sq.ft.; airfoil USA 27; wt. empty 1755; useful load 951; payload with 60 gal. fuel was 386 lbs.; payload with 40 gal. fuel was 506 lbs.; gross wt. 2706 lbs.; max. speed 115 (120); cruising speed 97 (102); landing speed 44; climb 650 (720) ft. first min. at sea level; service ceiling 13,000 (14,000) ft.; figures in brackets are for 180 h.p. version; gas cap. 60 gal.; oil cap. 4-5 gal.; cruising range at 10-12 gal. per hour was 530-510 miles; price at the factory field less engine and propeller was $3500.

The fuselage framework was built up of spruce longerons and laminated bulkheads, covered with plywood veneer. The wing framework was built up of spruce and plywood box-type spar beams, with spruce and plywood wing ribs; the completed framework was fabric covered. The main fuel tank was mounted high in the fuselage just ahead of the front cockpit; extra fuel was carried in a gravity-feed tank that was mounted in the center-section panel of the upper wing. The interplane bracing struts were of heavy gauge chrome-moly steel tubes that were of round section; these tubes were encased in streamlined balsa-wood fairings that were covered in fabric. The engine coolant radiator was mounted in the leading edge of the centersection panel in the upper wing on some examples, and on some the radiator was underslung below the engine compartment, where it operated with a retracting mechanism for temperature control. The fabric covered tail-group had fixed surfaces of all-wood construction and the movable surfaces were of welded steel tube construction; the fin was ground adjustable and the horizontal stabilizer was adjustable in flight. The landing gear was of the normal split-axle type with oleo-spring shock absorbers on the later versions, and "rubber donut" rings in compression on the earlier versions. Wheel brakes, a metal propeller, navigation lights, and inertia-type engine starter were available as extra equipment.

Listed below are F-17-H entries that were

Fig. 178. The early Hisso-International had seating for four; note eight-sided fuselage.

Fig. 179. An early F-17 with 6 cylinder Curtiss K-6 engine.

gleaned from various records; this list may not be quite complete, but it does show the bulk of this model that were built.

C-1702; F-17-H (# 23) Hisso.
 -5756; F-17-H (# 35) Hisso.
 -7056; F-17-H (# 58) Hisso.
 -7057; F-17-H (# 59) Hisso.
 -7058; F-17-H (# 60) Hisso.
C-7937; F-17-H (# 66) Hisso.
C-7936; F-17-H (# 67) Hisso.

C-7935; F-17-H (# 68) Hisso.
C-7934; F-17-H (# 69) Hisso.
C-7933; F-17-H (# 71) Hisso.
NC-1675 serial # 22 had Siemens-Halske engine; C-1890 serial # 29 had Curtiss K-6 engine; X-5654 serial # 45 had Dayton "Bear" engine. There was a Group 2 approval numbered 2-57 (issued 6-29) for model F-17-H with serial # 34 through # 68.

Fig. 180. The Ford 5-AT-B had three "Wasp" engines; the "Tin Goose" was now a grand and gracious lady.

With scheduled air-travel on the increase, and the formation of several new air-lines in various parts of the country, stiff competition was coming into play, which led to demands and pleading for extra seating, better and more comfortable service, and all this hinged on faster and more frequent schedules. To comply with all this, Ford-Stout was obliged to develop the "Wasp" powered model 5-AT which was somewhat larger than previous types built, and a lot more powerful. A constant development and refinement of the "tri-motor" AT design was geared to anticipate demands as they were forseen, so it was really no great effort to be ready on this demand from lines already in operation, and those that were just forming. The new model 5-AT also had progressed through various changes from time to time, and had actually started out in 1928 as an experimental model 4-AT aircraft that was powered with three P & W "Wasp" engines of 450 h.p. each; three of this version were built and approved under a Group 2 approval numbered 2-32. The first example of this model went to Col. C. W. Deeds of Pratt & Whitney Aircraft Div., and was flown regularly from his "back lot" behind his home! The increased performance that was so anxiously sought was surely to be had in this particular model (5-AT-A) but it was a bit shy on seating, so the next step was

planned to increase the seating capacity by at least 2 or 3 more seats.

Out of this planning evolved the model 5-AT-B which was also powered with three "Wasp" engines of 450 h.p. each, and now had ample seating for 15-17 persons; with improvements in decor, besides increases in performance and utility, the lovable "Tin Goose" had blossomed into a grand and gracious lady that offered style and comforts to please the most demanding and most critical air-traveller of this day. Hurriedly pressed into service by N.A.T., Maddux Air Lines, T.A.T., Pan American Airways in Mexico, S.A.F.E., Stout Air Service, Colonial Air Transport, and some others, the 5-AT-B for a time at least, was the ultimate in air-travel and became the back-bone of scheduled air transportation. The type certificate number for the model 5-AT-B as powered with three P & W "Wasp" engines, was issued is June of 1929 and some 42 of this model were built by the Stout Metal Airplane Co., a division of Ford Motor Co. at Dearborn, Michigan. The earliest examples of the model 5-AT-B to come off the production line were put into service by National Air Transport (N.A.T.) and Transcontinental Air Transport (T.A.T.); these were high-performance 14 place airplanes that were powered with three "Wasp" engines, and were approved for service under

Fig. 181. The 5-AT-B was larger with greater span; it carried 15—17 passengers with increased performance.

Group 2 memo numbered 2-12 which was issued in February of 1929. From serial number 5-AT-4 and up, this approval covered some 8 or 9 airplanes of this type.

Listed below are specifications and performance data for the Ford-Stout "Tri-Motor" model 5-AT-B as powered with three "Wasp" engines of 420-450 h.p. each; length overall 49'10"; hite overall 12'8"; wing span 77'10"; wing chord at root 156"; wing chord at tip 92"; total wing area 835 sq.ft.; airfoil "Goettingen 386"; wt. empty 7576; useful load 5000-5674; payload 2720-3044; gross wt. 12,576-13,250 lbs.; weights varied with fuel load and cabin arrangement; max. speed 142; cruising speed 122; landing speed 62; climb 1200 ft. first min. at sea level; service ceiling 18,500 ft.; gas cap. 277-355 gal.; oil cap. 27-33 gal.; cruising range 540-610 miles; price at the factory field averaged at $55,000. The fuselage framework was built up of duralumin open channel sections that were riveted together in a bridge-like truss form, and was covered with corrugated "Alclad" sheet stock. The wing framework on the model 5-AT-B was typical to that as described in the chapter co-vering the model 4-AT-E, except that it spanned 3 feet and 10 inches wider. The outer wing panels were the same for all models, and any increases in the overall wing span were taken care of in the center-section panel, which was built integral with the fuselage structure. The tail-group was of a construction that was typical of all Ford "Tri-Motors" except that it was now much larger. Landing gear details, and other points not covered here, can be considered as similiar or typical to the model 4-AT-E; refer back to chapter for A.T.C. # 132 in this volume. Improvement in design and refinement of character had been constantly going on, despite the clamor for more production and more airplanes; as a consequence, improved models, or entirely different models, had been coming out one right after another. With 5-AT-B models still coming off the line, the improved model 5-AT-C was already making it's bows and anxious to take it's place among others serving in scheduled air transport. For a discussion of the "Wasp" powered model 5-AT-C, see chapter for A.T.C. # 165 in this volume.

Fig. 182. A Ford 5-AT-B at Ford Airport, the factory field in Dearborn, Michigan.

Fig. 183. A Boeing flying boat Model 204, with 450 h.p. engine. The 204 was developed from the B-1E.

It is interesting to note that most all of Boeing's early efforts in airplane design were examples of water-borne craft; no doubt this was greatly influenced by Boeing's location in Seattle, Washington where there is smooth water aplenty. First formed as the Pacific Aero Products in 1916, several float-mounted biplanes were produced in the first few years and the first boat-type model B-1 was introduced in 1919. Through several modifications and then a complete redesign, the B-1 series had evolved into the models B-1D and B-1E as described in U. S. CIVIL AIRCRAFT, Volume 1. These were two versatile aircraft that performed many a useful service, carrying air-mail, air-cargo, and occasional passengers in an area that abounded in lakes, rivers, and large bodies of water. With a note of regret it is to relate that gone from the scene now, almost entirely, is the romance and bountiful pleasures that were experienced with the "boat-type" airplane; a versatile craft that tends to combine the sport of managing a boat with the convenience and pleasures of flying. In certain parts of the country, areas that were dotted with lakes and coursed by

rivers, the flying-boat offered a multitude of sporting ventures and performed various duties that were almost impossible by any other means. A comfort to all aboard also, was that relaxed and reassured feeling that a nearby haven always existed for a hasty set-down in the event of trouble. Though very limited in numbers built, the Boeing flying-boats led a very active life and performed many a diversified service; such as transporting miners and supplies from seaboard areas to lakes in the interior, dusting pest-infested areas of forest with insecticides, spotting schools of fish for anxious fishermen, and hauling cannery supplies, to name just a few of it's every-day chores.

The newly developed Boeing model "204" was also a cabin biplane of the flying-boat type, and was quite typical to the previous models in most respects, except for some modifications that centered around an increased seating capacity. The model "204" had ample room and comfort for six, and was powered with the 9 cylinder Pratt & Whitney "Wasp" engine of 450 h.p.; a combination with plenty of muscle that translated itself into a very

Fig. 184. A 204 taxying to dock; bountiful pleasures were experienced with the boat type of airplane.

good performance for a craft of this type. The engine was mounted in a pusher-type nacelle that was placed between the wing panels; thus getting the whirling propeller away from the cabin section and minimizing any cause for danger, and removing most of the engine noise to a point further back. The arrangement of the cabin section in it's extreme forward position, was a sensible lay-out that provided an unlimited field of vision for the pilot and the passengers.

Registration records show five aircraft of the "204" type, but only two were actually completed at the factory; one example was delivered to Western Air Express where it served on the air-ferry line from the mainland of California to Catalina Island, a 22 mile jaunt

that was popular with joy-riders and visiting tourists. A second craft was assembled as a specially appointed personal air-taxi for the use of W. E. "Bill" Boeing; fitted with dual controls and a number of other deluxe features of appointments and equipment, records label this as the model "204-A". The third "204" was completed by a resident of Seattle who bought a complete set of parts from the factory; a sort of do-it-yourself project. The other two aircraft were never completed and it is likely that many components from these two unfinished aircraft are still in storage somewhere in the Seattle area as dust-covered reminders of the romantic past. Four of these flying-boats, an example of which is pictured here, were built by Boeing Aircraft of

Fig. 185. A Western Air Express 204 running a ferry service to Catalina Island.

Canada as the model C-204. The type certificate number for the Boeing Flying Boat model 204 as powered with the 450 h.p. "Wasp" engine, was issued in June of 1929 and it was manufactured by the Boeing Airplane Company at Seattle, Wash. Philip G. Johnson was the president; C. L. Egtvedt was V.P. and general manager; and C. N. Monteith was chief of engineering.

Listed below are specifications and performance data for the Boeing Flying Boat model 204 as powered with the 450 h.p. "Wasp" engine; length overall 32'6"; hite overall 12'2"; wing span upper and lower 39'8"; wing chord both 79"; wing area upper 248 sq.ft.; wing area lower 222 sq.ft.; total wing area 470 sq.ft.; airfoil "Boeing 103"; wt. empty 3371; useful load 1629; payload with 80 gal. fuel was 913 lbs.; gross wt. 5000 lbs.; max. speed 130; cruising speed 110; landing speed 60; climb 1000 ft. first min. at sea level; service ceiling 12,000 ft.; take-off run fully loaded was 750 ft. in 15 sec.; gas cap. 80 gal.; oil cap. 10 gal.; cruising range 400 miles; price at the factory was $20,000 and later reduced to $15,000 in the early part of 1931.

The hull framework was built up of oak, ash, and spruce wood members that were covered with two-ply mahogany planking; a sound-deadening construction that was relatively easy to maintain and repair. The cabin walls were sound-proofed and insulated with provisions for cabin heat and ventilation;

seats were upholstered comfortably in real leather and there was a large baggage compartment to the rear. The windows were of shatter-proof glass and could be rolled up and down; five hatches gave access to the cabin seating and the baggage compartment. A separate compartment was provided for storage of anchor and mooring lines. The wing framework was built up of solid spruce spars that were routed to an "I beam" section, with spruce and plywood truss-type wing ribs; the leading edges were covered with plywood back to the front spar and the completed framework was fabric covered. There were four ailerons that were connected together in pairs by a heavy stranded cable; the two fuel tanks were mounted in the root end of each upper wing panel for positive gravity feed. The fabric covered tail-group was built up in a combination of wood and steel tube construction; the vertical fin and the horizontal stabilizer was built up of spruce spars and ribs, and the rudder and elevators were of a welded chrome-moly steel tube framework. The fin was ground adjustable and the horizontal stabilizer was adjustable in flight; both movable surfaces of the tail-group were aerodynamically balanced. The wing-tip floats which kept the lower wings from digging into the water, were of a wood construction covered in mahogany veneer. Navigation lights, anchor and mooring lines, and inertia-type engine starter were standard equipment.

Fig. 186. A Boeing flying boat built especially for W.E. "Bill" Boeing as a 204-A, later sold to Percy Barnes.

Fig. 187. A Boeing flying boat built in Canada as a C-204; it performed useful service in the bush country.

The next Boeing development was the popular 40-B-4, a combination airmail and passenger carrying airplane that was powered with the P & W "Hornet" engine; for discussion of this model, see the chapter for A.T.C. # 183 in this volume.

Listed below are Boeing 204 entries that were registered with the Department of Commerce, Aero. Div. in 1929.

NC-874E; 204 (# 1076) Wasp 450.

NC-875E; 204 (# 1077) Wasp 450.
NC-876E; 204 (# 1078) Wasp 450.
NC-877E; 204 (# 1079) Wasp 450.
NC-878E; 204 (# 1080) Wasp 450.
Model 204 serial # 1076 was operated by Western Air Express; serial # 1077 was modified as private-craft for W. E. Boeing, designated 204-A; a third craft was built up from parts, using Boeing serial number and registration number already assigned.

Fig. 188. Sikorsky "Amphibion" S-38C with Pratt & Whitney "Wasp" engines; CF-ASO of Canadian Airways.

The Sikorsky "Amphibion" model S-38C was more or less typical to the S-38B as described previously, except that it carried 2 more passengers and carried less fuel load; the model S-38C with it's greater seating was mostly used in "coach service" on shorter hops, when more passengers would be carried on a more frequent schedule. The first example of the model S-38C was used for a time by several corporations, and then used by Northwest Airways on their route serving the cities of Minneapolis-St. Paul-Duluth. Four more of these craft, named the "Hawaii", "Maui", "Kauai", and "Molokai", were in scheduled service with the Inter-Island Airways of Hawaii. Starting in 1929, Inter-Island Airways initiated the first scheduled air-line routes in the Hawaiian Islands, operating in constant scheduled service throughout the islands for over 30 years without one fatality. The "Hawaii" (ship # 1) operated constantly from it's first flight in October of 1929 to February of 1938; the "Maui" (ship # 2) also operated in constant service from October of 1929, making the first scheduled air-mail flight between the Hawaiian Islands in 1934, and finally pulled out of service sometime in 1938; the "Kauai" (ship # 3) was put into active comission in January of 1930 and after a very exciting life period, was still on the active register as late as 1946; the "Molo-

kai" (ship # 4) was put into service during July of 1931 and led a busy life, performing many diversified services until sometime in 1947. Another example of the model S-38C was used by Canadian Airways on a 60 mile over-water hop from Vancouver to Victoria; this was the only S-38 registered in Canada. The accounts related above are meant to picture a busy craft that was working obediently on frequent schedules; all had active, useful lives of long duration which is but ample proof of the utility and inherent dependability that was engineered into these outstanding airplanes.

The Sikorsky "Amphibion" model S-38C was a slight modification of the basic layout; arranged for coach service, it had seating for 12, with a fuel load of 264 gallons, and was powered with two Pratt & Whitney "Wasp" engines of 420 h.p. each. Performance and flight characteristics were more or less typical, and the only variation evident was in the cabin seating arrangement. The landing gear was retractable and could be extended or retracted by hydraulic control in about 50 seconds; as a help in taxiing into position among close quarters, one wheel could be extended separately to cause drag and help bring the ship around. When retracted, the wheels lay flush under the lower wing; sponsons, which were like small floats, were attached to the outer ends of the lower wing to keep it from digging in during water take-offs or landings. Despite the maze of struts, braces, and booms, the S-38 was easy to inspect, service, and maintain; all struts were open on the ends for drainage and ventilation, the hull and the floats were easy to drain and ventilate, and the hull framework was of such rugged construction that very little seepage occured even in rough-water conditions. The sesqui-plane arrangement of the wings allowed for the structural rigidity of the biplane configuration, while still enjoying most of the advantages of monoplane efficiency, and keeping the greatest portion of the lifting surfaces well above the water. The engine nacelles were mounted high and well above the water, and the fuel and oil supply was in the upper wing; this to minimize fuel problems and keep annoying fumes out of the cabin area. Passenger seats were of wicker and easily removable to provide clear cargo area; the cabin windows were fixed but a sliding hatch in the roof of the cabin provided ample ventilation. Exit and entry to the cabin was through a hatch and down a ladder in the aft section of the hull; normal arrangment was four or five single seats down each side, but the cabin could be arranged in several different seating lay-outs. Some of the special "air yacht" versions were quite plush and contained many special conveniences. All in all, the S-38 "Amphibion" was an unforgettable airplane; the majesty of it's persona-

lity was quite intriguing and felt for many years. The type certificate number for the model S-38C as powered with two "Wasp" engines of 420 h.p. each, was issued in June of 1929 and some 8 or more examples of this model were manufactured by the Sikorsky Aviation Corp. at Bridgeport, Conn.; a division of the United Aircraft Corp. A. C. Dickinson was the president; W. A. Bary was V.P. and general manager; Igor I. Sikorsky was V.P. and chief of engineering; M. Gluhareff was assistant engineer; and Henry J. White was chief pilot and sales manager. The Curtiss Flying Service handled sales throughout the various stations they had in most parts of the country. The S-38 series were built into 1933, and some 20 or so were still in active service by 1939.

Listed below are specifications and performance data for the Sikorsky "Amphibion" model S-38C as powered with 2 "Wasp" engines of 420 h.p. each; length overall 40'3"; hite on wheels 13'10"; hite on water 10'2"; wing span upper 71'8"; wing span lower 36'0"; wing chord upper 100"; wing chord lower 59"; wing area upper 574 sq.ft.; wing area lower 146 sq.ft.; total wing area 720 sq.ft.; airfoil "Sikorsky GS-1"; wt. empty 6460; wt. empty does not include batteries and electric starters, add 96 lbs.; useful load 4020; payload with 264 gal. fuel was 2074 lbs.; gross wt. 10,480 lbs.; max. speed 125; cruising speed 110; landing speed 55; climb 880 ft. first min. at sea level; climb in 10 min. was 7250 ft.; service ceiling

Fig. 189. A Colonial-Western Airways S-38C taking on passengers.

Fig. 190.   An S-38C with American Airways; the unusual configuration was quite rugged and often stood up to 10—15 years of active service.

18,000 ft.; gas cap. 264 gal.; oil cap. 24 gal.; cruising range at 44 gal. per hour was 600 miles; basic price at the factory field was about $50,000 depending on equipment installed.

The hull framework was built up of oak and ash wooden frame members that were reinforced with "dural" metal plates and gussets at all joints; the outer hull covering was of heavy gauge "Alclad" sheet that was riveted together and fastened to the hull framework with wood screws, all seams were sealed with marine glue and fabric tape. The hull was arranged into 6 water-tight compartments for greater safety in case of hull puncture; there was a compartment in the extreme nose for stowing anchor and the mooring equipment. and an 82 cu. ft. compartment just back of that for up to 200 lbs. of baggage. The wing framework was built up of riveted and bolted duralumin girder-type spar beams, and riveted duralumin truss-type wing ribs; the

completed framework was fabric covered. Fuel tanks and oil tanks were mounted in the upper wing; the upper wing was of a center-section panel and two outer panels, and the lower wing was of L.H. and R.H. panels. The fabric covered tail-group was mounted on two booms that came aft from the upper wing and were braced to the hull by streamlined struts; the rudders were aerodynamically balanced. The split-axle landing gear of 116 inch tread was of the retractable type; wheels were 32x6 and brakes were available. Metal propellers, navigation lights, Eclipse inertia-type engine starters, fire extinguishers, anchor and mooring gear were standard equipment. The next Sikorsky "Amphibion" development was the single-engined S-39A which will be discussed in the chapter for A.T.C. # 340.

Listed below are all known Sikorsky S-38C entries that were gleaned from various records.
NC-199H; S-38C (# 214-3) 2 Wasp.
NC-111M; S-38C (# 214-7) 2 Wasp.
NC-112M; S-38C (#214-8) 2 Wasp.
NC-4V; S-38C (# 314-8) 2 Wasp.
NC-10V; S-38C (# 314-14) 2 Wasp.
NC-305N; S-38C (# 414-3) 2 Wasp.
NC-26V; S-38C (# 414-17) 2 Wasp.
NC-28V; S-38C (# 414-19) 2 Wasp.
Serial # 214-3 was with Northwest Airways; serial # 214-7, # 214-8, # 314-8, and # 414-3 were with Inter-Island Airways in Hawaii; serial 414-19 was with Canadian Airways as CF-ASO.

Fig. 191.   The S-38C was equally at home on water.

Fig. 192. The Curtiss "Thrush" with a "Challenger" engine; it was sometimes called the "Big Robin".

The boxy-looking "Thrush" was a direct development from the popular Curtiss "Robin" design, and was stretched out to a comfortable seating capacity for six. Following the configuration quite closely, the "Thrush" was a high wing cabin monoplane also of the typical "squarish" lines that proved to be quite "clean" in the aerodynamic sense, despite the feeling one would get from the slab-sided lines. Powered with the 6 cylinder Curtiss "Challenger" engine of 170-180 h.p., the big "Thrush" was an efficient design and a good load-carrier but tended to be somewhat underpowered with a full gross load, and performance fell somewhat below the average that was expected. Even with the engine rating at a maximum of 183 h.p. and operating with a gross load that was 100 lbs. less than the certificated allowance, the "Thrush" was flyable but far from being sprightly; it was just too much airplane for the power that was available. Development of this combination is rather mystifying, it may have been a misjudgement, or perhaps it was planned so to gain the strictest economy possible for a given payload, by way of maximum sacrifice in all-round performance.

The Curtiss organization which developed this design, built up three prototypes of the "Thrush" for extensive testing and planned to turn over the manufacture of this model to the Curtiss-Robertson Div. upon completion of the various testing programs. Included in the tests were various forms of N.A.C.A. type "low drag" engine cowlings with which Curtiss had hoped to gain extra performance in spite of the low available horsepower. The load-carrying abilities of the "Thrush" were certainly excellent and the flight characteristics were commendable, but performance in general was not adequate enough by comparison to the average aircraft of this type. The type certificate number for the 6 place Curtiss "Thrush" as powered with the 170 h.p. Curtiss "Challenger" engine, was issued in June of 1929 and 3 examples of this model were manufactured by the Curtiss Aeroplane & Motor Co., Inc. at Garden City, Long Island, N. Y. Glenn H. Curtiss was chairman of the board; C. M. Keys was the president; T. P. Wright was chief of engineering; and Charles S. "Casey" Jones was manager of sales and the chief pilot in charge of test and development.

Listed below are specifications and performance data for the Curtiss "Thrush" as powered with the 170 h.p. "Challenger" engine; length overall 32'4"; hite overall 9'3"; wing span 48'0"; wing chord 84"; total wing area 305 sq.ft.; airfoil "Curtiss C-72"; wt. empty 2160 (2232); useful load 1424 1468); payload with 60 gal. fuel was 850 (898) lbs.; payload with 110 gal. fuel was 554 (598) lbs.; gross wt. 3584 (3700) lbs.; the weights in brackets are as allowed in type certificate; the following performance figures are for

Fig. 193. A low-drag engine cowling was tested for gain in performance; the ship shown was the prototype.

craft with N.A.C.A. type low-drag engine cowling and 3600 lb. gross weight with 183 h.p. rating; max. speed 110; cruising speed 94; landing speed 50; stall speed 58; climb 465 ft. first min. at sea level; climb in 10 min. was 3900 ft.; service ceiling 10,100 ft.; performance figures for craft without N.A.C.A. cowling, and 3700 lb. gross load at 170 h.p. rating was proportionately lower; gas cap. 60 gal. normal, 110 gal. max.; oil cap. 5-9 gal.; cruising range with 60 gal. at 10 gal. per hour was 554 miles ;cruising range with 110 gal. at 10 gal. per hour was 1015 miles; cruise was based on 85% of maximum power; price at the factory field was approx. $10,000.

The fuselage framework was built up of duralumin tubing that was fastened together with duralumin wrap-around fittings and hollow duralumin rivets, points of greater stress were built up of welded chrome-moly steel tubing; the framework was faired to shape with fairing strips and fabric covered. The spacious cabin had ample room for six, and there were two large entrance doors on the right side with convenient steps for entry and exit. The semi-cantilever wing framework was

Fig. 194. The "Thrush" had ample room for six but lacked power; note special engine cowling tested.

built up of solid spruce spar beams with stamped-out "Alclad" wing ribs; the leading edge was covered with "Alclad" sheet and the completed framework was fabric covered. The two fuel tanks were mounted in the wing, one flanking each side of the fuselage. The wing was braced by parallel struts of chrome-moly steel tubing on either side; the wide-track landing gear of 116 inch tread was of the outrigger type and was tied into truss form with the front wing strut. Shock absorber struts were of the oleo-spring type, wheels were 30x5 and Bendix brakes were standard equipment; a steerable tail-wheel was provided for ease of ground manuvering. The fabric covered tail-group was built up of welded chrome-moly steel tubing; the fin was ground adjustable and the horizontal stabilizer was adjustable in flight. A Curtiss-Reed metal propeller was provided as standard equipment; navigation lights and engine starter were available as optional equipment. The Curtiss-Robertson "Thrush" with the Curtiss "Challenger" engine was slated for manufacture under A.T.C. # 160; for discussion of this model see the chapter for A.T.C. # 160 in this volume.

Listed below are Curtiss "Thrush" entries that were gleaned from various records; this list represents the total number of this model that were built.

C-7568; Thrush (# G-1) Challenger.
C-9787; Thrush (# G-2) Challenger.
C-9142; Thrush (# G-3) Challenger.

After extensive testing, these three craft were modified into the model J version that was powered with the 7 cylinder Wright J6 engine of 225 h.p.

Fig. 195. The Curtiss "Thrush" was to have been produced by Curtiss-Robertson but plans were cancelled.

The Curtiss-Robertson "Thrush" was more or less like a stretched-out and overgrown "Robin" that had seating for six; it was also of the typical "squarish" lines that had become very familiar and which oddly enough were quite "clean" in the aerodynamic sense, having been thoroughly proven in wind-tunnel tests. Powered by the 6 cylinder Curtiss "Challenger" engine of 170-180 h.p., the big "Thrush" tended to be somewhat underpowered with a full gross load, and performance fell short of that expected. Numerous tests had been performed on several different versions of the N.A.C.A. type "low drag" engine cowlings that diverted the air-flow around the engine cylinders into a smooth pattern, but apparently the tests had proven that low-drag "cowlings" were certainly no panacea to greater performance in a craft that definately needed more power, more than anything else.

The Curtiss organization at Long Island built three prototypes of the "Thrush" with Curtiss "Challenger" engines by May of 1929, and after preliminary tests which lasted through several months, were scheduled for some further testing by the Curtiss-Robertson Div. at St.Louis where they were to be manufactured. The Curtiss-Robertson plant had been progressively expanded until by now it contained units that totaled to more than 100,000 sq.ft. of shop area. Some 17 "Robins" were now coming off the line on an average per week, and factory space was already set aside to handle the manufacture of the "Thrush" in large quantity, but apparently further testing of the "Challenger" powered "Thrush" discouraged manufacture of this particular version. Consequently, all three of the prototypes were converted to the model J that was powered with the 7 cylinder Wright J6 engine of 225 h.p.; this new J combination was more in keeping with what the operators would be expecting from an airplane of this type. The type certificate number for the Curtiss-Robertson "Thrush" as powered with the 170 h.p. "Challenger" engine, was issued in June of 1929 and it was to be manufactured by the Curtiss-Robertson Div. at Anglum (St. Louis), Missouri. Due to the unsatisfactory reaction to this model, none were ever built in this version. Maj. Wm. B. Robertson was the president; C. M. Keys was V.P.; engineering was handled by the Curtiss corps of engineers under the direction of T. P. Wright; and Dale "Red" Jackson was the chief pilot in charge of test and development.

Listed below are specifications and performance data for the Curtiss-Robertson "Thrush" as powered with the 170-180 h.p.

Fig. 196. The squarish lines of the "Thrush" were quite clean in the aerodynamic sense; the configuration was proven in wind tunnel tests.

"Challenger" engine; length overall 32'4"; hite overall 9'3"; wing span 48'0"; wing chord 84": total wing area 305 sq.ft.; airfoil "Curtiss C-72"; wt. empty 2232; useful load 1468; payload with 60 gal. fuel was 898 lbs.; payload with 110 gal. fuel was 598 lbs.; gross wt. 3700 lbs.; the following performance figures are for a craft with N.A.C.A. type low-drag engine cowling and 3600 lb. gross weight with 183 h.p. rating; max. speed 110; cruising speed 94; landing speed 50; stall speed 58; climb 465 ft. first min. at sea level; climb in 10 min. was 3900 ft.; service ceiling 10,100 ft.; performance figures for a craft without N.A.C.A. "cowling" and 3700 lb. gross load at 170 h.p. rating was proportionately lower; gas cap. 60 gal. normal, 110 gal. max.; oil cap. 5 gal. normal, 9 gal. max.; cruising range with 60 gal. fuel at 10 gal. per hour was 554 miles; cruising range with 110 gal. fuel at 10 gal. per hour was 1015 miles; cruising speed was based on 85% of max. power; price at the factory field was tentatively set for $10,000.

The fuselage framework was built up of duralumin tubing that was fastened together in truss form with duralumin wrap-around fittings and hollow duralumin rivets, points of greater stress were built up of welded chrome-moly steel tubing; the framework was faired to shape with fairing strips and fabric

Fig. 197. The load-carrying ability of the "Thrush" was excellent but performance was below average.

covered. The spacious cabin had ample room for six, and there were two large entrance doors on the right side with two convenient steps for ease of entry and exit. The semi-cantilever wing framework was built up of solid spruce spar beams with stamped-out Alclad wing ribs; the leading edges were covered with a aluminum alloy sheet to preserve the airfoil form and the completed framework was fabric covered. The two fuel tanks were mounted in the wing for gravity feed, one flanking each side of the fuselage. The wing bracing struts were supported by a junction of auxiliary struts that prevented wing deflection during heavy aileron loads; all bracing struts were of chrome-moly steel tubing in streamlined section. The wide track split-axle landing gear of 116 inch tread was of the outrigger type that was tied into the front wing struts to make a rigid truss; shock absorbers were of the oleo-spring type, wheels were 30x5 and wheel brakes were standard equipment. A steerable tail-wheel was provided which made ground manuvering much easier for a ship of this weight and size. The fabric covered tail-group was built up of welded chrome-moly steel tubing; the vertical fin was ground adjustable and the horizontal stabilizer was adjustable in flight. A Curtiss-Reed forged metal propeller was standard equipment; navigation lights and engine starter were available as optional equipment. The next development in the "Thrush" series was the Wright J6 powered Curtiss model J that will be discussed in the chapter for A.T.C. # 236. The next Curtiss development is the "Challenger" powered "Fledgling Junior", a biplane trainer; see the chapter for A.T.C. # 182 in this volume.

Fig. 198. A Stinson "Junior" SM-2AB with a 220 h.p. Wright J5 engine; it was popular as a sportplane.

In reviewing the previous models of the Stinson "Junior" line, we follow the development from the first (A.T.C. # 48) 110 h.p. SM-2 (Warner "Scarab"), to the 165 h.p. (A.T.C. # 145) model SM-2AA (Wright 5) and see the trend of increasing horsepower; a trend that came about in answer to the constant clamor for a ship of this type with higher performance. It seems that this aircraft, with it's seating for four, was not looked at as a transport vehicle but more as a sport-craft, or fast air-taxi, that appealed to the business-man in a hurry, or the sportsman-pilot type of owner who demanded a high-performance airplane, regardless of cost or lack of economy, but would rather have it in the enclosed comfort of a plush cabin. So here we have the case of a design breaking away from original intent, and evolving into a definate pattern that was motivated by preferences shown by potential customers, or from owners of previous low-powered models. Eddie Stinson, actually happy with developments as they were forming, decided to go all the way to please, and gave them one with twice the horsepower; an earth-pawing vehicle of 220 h.p. that had punch enough and performance enough to please the most boisterous sportsman-pilot, or the most anxious business-man. This was the SM-2AB. The Stinson "Junior" model SM-2AB was powered with the 9 cylinder Wright J5 engine of 220 h.p. and possessed a spirit and perfor-

mance that had speed and climb to spare and manuverability that approached that of a sport-biplane.

To accomodate this extra power and the higher stresses it would impose, the SM-2AB was beefed-up here and there, but this didn't involve any great changes because the "Junior" structure already had strength with ample safety factor, as if in anticipation of the way things would go. The cabin interior was enlarged a bit for more elbow and leg room, the landing gear was lengthened for increased clearance and beefed-up for the added strain that would be generated by the higher performance, and the fuselage was lengthened and faired somewhat deeper; as a consequence, this "Junior" was a bit higher, longer, and bulkier. With customers actually waiting for airplanes to come off the line, the SM-2AB sold very well and some 32 or more of this model were built. The type certificate number for the model SM-2AB as powered with the Wright J5 engine, was issued in June of 1929 and amended in November of 1929; it was manufactured by the Stinson Aircraft Corp. at Wayne, Mich. Having outgrown their previous quarters at Northville, which were somewhat antiquated, Stinson moved operations into their new plant in Wayne, Mich, which was modern and spacious, and had an adjoining air-field that was developed into one of the finest in the area.

Listed below are specifications and perfor-

mance data for the Stinson "Junior" model SM-2AB as powered with the Wright J5 engine of 220 h.p.; length overall 28'4"; hite overall 8'4"; wing span 41'6"; wing chord 75"; total wing area 238 sq.ft.; airfoil "Clark Y"; wt. empty 2169; useful load 1060; payload with 70 gal. fuel was 470 lbs.; gross wt. 3229 lbs.; max. speed 130; cruising speed 112; landing speed 49; climb 840 ft. first min. at sea level; service ceiling 16,000 ft.; gas cap. 70 gal.; oil cap. 6 gal.; cruising range 600 miles; price at the factory field was $10,500. The fuselage framework on the SM-2AB was typical to previous models but beefed-up here and there for the added stresses generated by more power and higher performance; the fuselage was lengthened overall, and faired out a bit deeper to conform with the lines of the new powerplant installation. The wing framework was more or less the same, with heavy solid spruce spars that were routed to an "I beam" section, and fuel tanks were now of greater capacity to allow a fuel load of 70 gallons. The landing gear was lengthened for more ground clearance and beefed-up to take the added stresses of more weight and the greater loads that were generated by a higher landing speed and a quicker run down the runway on take-off. Wheel tread was the same at 110 inches, and wheels were heavy duty 30x5 or 32x6; tail-wheel cross-section was increased and the shock absorbing strut was of a heavier size. Doubling the power brought on the performance increases that were asked for, but also brought on greater strain to the structure which had to be modified somewhat to suit the needs. With the recent advent of the new Wright engine of the

J6 series, it was only logical to replace the Wright J5 with the new 7 cyl. J6 which developed 225 h.p. For a discussion of this new combination, see the chapter for A.T.C. # 194 in this volume which covers the model SM-2AC.

Listed below are SM-2AB entries that were gleaned from various records; this list may not be complete, but it does show the bulk of this model that were built.

NC-9699; SM-2AB (# 1047) Wright J5
NC-8423; SM-2AB (# 1048) Wright J5
NC-8442; SM-2AB (# 1050) Wright J5
NC-8427; SM-2AB (# 1053) Wright J5
NC-8434; SM-2AB (# 1056) Wright J5
NC-8438; SM-2AB (# 1059) Wright J5
          ; SM-2AB (# 1062) Wright J5, Grp. 2-267
NC-8443; SM-2AB (# 1064) Wright J5
NC-8444; SM-2AB (# 1065) Wright J5
NC-8474; SM-2AB (# 1073) Wright J5
NC-8480; SM-2AB (# 1074) Wright J5
 C-8473; SM-2AB (# 1078) Wright J5
NC-457H; SM-2AB (# 1079) Wright J5
 C-8476; SM-2AB (# 1080) Wright J5
 C-8479; SM-2AB (# 1082) Wright J5
 C-8483; SM-2AB (# 1083) Wright J5
NC-442H; SM-2AB (# 1084) Wright J5
 C-8482; SM-2AB (# 1085) Wright J5
NC-453H; SM-2AB (# 1086) Wright J5
NC-451H; SM-2AB (# 1087) Wright J5
NC-443H; SM-2AB (# 1088) Wright J5
NC-446H; SM-2AB (# 1092) Wright J5
NC-449H; SM-2AB (# 1094) Wright J5
NC-459H; SM-2AB (# 1096) Wright J5
NC-450H; SM-2AB (# 1097) Wright J5
NC-455H; SM-2AB (# 1098) Wright J5
NC-458H; SM-2AB (# 1100) Wright J5

Fig. 199. The clamor for performance caused increases in power; the "Junior" with the "Challenger" engine.

Fig. 200. The SM-2AB had performance to please the most - eager sportsman- pilot or businessman.

NC-464H; SM-2AB (# 1102) Wright J5
NC-477H; SM-2AB (# 1112) Wright J5
NC-479H; SM-2AB (# 1114) Wright J5
NC-469H; SM-2AB (# 1115) Wright J5

NC-490H; SM-2AB (# 1125) Wright J5
There were 14 unlisted aircraft, some may
have been SM-2AB, others were SM-2AA and
SM-2AC.

Fig. 201. The Fairchild KR-34C was a development of the Kreider-Reisner "Challenger" C-4-C.

The neat-looking Fairchild KR-34C was a later development of the popular Kreider-Reisner "Challenger" biplane, a craft that was typical to the best one could find in a general-purpose airplane. Designed specifically for the varied duties it would be asked to render by the average fixed-base operator, the "Challenger" biplane proved itself well in this capacity and managed to round-up quite a following in the two years it was being produced. So far it had been available with the OX-5, Comet, and Warner engines; the OX-5 combination was by far the most popular. Though not particularly an outstanding airplane in any sense, it was however, a compatible combination of many good qualities. Performance was well comparable to the overall average, it's utility and dependability in every-day service a paramount feature, and it's flight characteristics were best described as pleasant character and obedient nature.

The newly-developed Fairchild KR-34C which was actually a Kreider-Reisner "Challenger" C4C, was a 3 place open cockpit biplane of the general-purpose type, and was powered with the new 5 cylinder Wright J6 engine of 165 h.p. Rightfully proud of their new craft, two were entered by Fairchild in

the National Air Tour for 1929. Mrs. Keith Miller, a petite aviatrix of great prowess, turned in a stellar performance that brought her to 8th place, amid rounds of applause, in a very determined field of competition. A third craft was the official "press plane" for the 1929 tour. Seen only seldom in competition, the dutiful KR-34C was kept very busy working, and was becoming quite popular as a sport-craft, and as an air-taxi in the field of business-transport. The Fairchild Aircraft organization had their eye on the hustling and bustling Kreider-Reisner operation for some time, and in April of 1929 the Kreider-Reisner organization became a subsidiary of Fairchild by mutual agreement. The familiar "Challenger" name had soon been dropped and the K-R biplanes were now designated as the Fairchild "KR series". The type certificate number for the KR-34C as powered with the 165 h.p. Wright J6 engine, was issued in June of 1929 and some 60 or more examples of this model were manufactured by the Kreider-Reisner Aircraft Co., Inc. at Hagerstown, Maryland; now a division of the Fairchild Airplane Mfg. Co. Amos H. Kreider had been the president; treasurer; sales manager; and chief pilot; thereby keeping his finger on the pulse of K-R operations pretty well. In the re-organization, company officers were shifted around somewhat to read as follows: John S. Squires was the president; Louis E. Reisner was V.P.; F. E. Seiler was still handling the engineering chores, while taking on additional duties as production superintendent.

Listed below are specifications and performance data for the Fairchild KR-34C as powered with the 165 h.p. Wright J6 engine; length overall 23'2"; hite overall 9'3"; wing span upper 30'1"; wing span lower 28'9"; wing chord both 63"; wing area upper 154 sq.ft.; wing area lower 131 sq.ft.; total wing area 285 sq.ft.; airfoil "Aeromarine 2A Modified"; wt. empty 1524 (1457); useful load 844 (885); payload with 50 gal. fuel was 344 (385) lbs.; gross wt. 2368 (2342) lbs.; wts. in brackets are for earlier model C4C; max. speed 120; cruising speed 102; landing speed 45; climb 800 ft. first min. at sea level; service ceiling 14,100 ft.; gas cap. 50 gal.; oil cap. 4 gal.; cruising range at 9 gal. per hour was 510 miles; price at the factory field was $6575.

The fuselage framework was built up of welded chrome-moly steel tubing, faired to shape with wood fairing strips and fabric covered. There was an entrance door to the front cockpit, and a baggage compartment that was situated behind the rear cockpit and accessible through a door on the left side. The wing framework was built up of solid spruce spar beams that were routed-out for lightness, with spruce and plywood built-up

wing ribs; the completed framework was fabric covered. Ailerons were on the lower panels only, and were differentially operated for more positive control in the turns. The gravity-feed fuel tank was mounted high in the fuselage just ahead of the front cockpit, with a fuel-level gage on the dash-panel of the pilot's cockpit. The fabric covered tail-group was built up of welded chrome-moly steel tubing; the fin was ground adjustable and the horizontal stabilizer was adjustable in flight. The split-axle landing gear of 72 inch tread had shock absorber legs of rubber-rings in compression; wheels were 28x4 and Bendix brakes were standard equipment. The following was included as standard equipment; metal propeller; wiring for navigation lights; airspeed indicator; altimeter; oil pressure and temperature gage; compass; booster magneto; tachometer; dual controls; engine cover; cockpit covers; log book; first-aid kit; fire extinguisher; tie-down ropes; tool kit; and engine instruction book. The next development in the Kreider-Reisner biplane was the Curtiss "Challenger" powered C4D (KR-34B), which will be discussed in the chapter for A.T.C. # 208.

Listed below is a partial list of KR-34C entries that were gleaned from various records; a complete listing would be rather extensive, so we will submit about 20 or so.
NC-632E; KR-34C (# 260) Wright J6-5.
NC-695E; KR-34C (# 264) Wright J6-5.
NC-824E; KR-34C (# 265) Wright J6-5.
NC-890E; KR-34C (# 271) Wright J6-5.

Fig. 202. Mrs. Keith Miller flew a KR-34C to 8th place in 1929 Air Tour; Sherman Fairchild is at right.

Fig. 203.   The good performance and excellent flight characteristics of the KR-34C proved popular.

NC-329H; KR-34C (# 278) Wright J6-5.
NC-330H; KR-34C (# 279) Wright J6-5.
NC-338H; KR-34C (# 280) Wright J6-5.
NC-339H; KR-34C (# 281) Wright J6-5.
NC-340H; KR-34C (# 282) Wright J6-5.
NC-341H; KR-34C (# 283) Wright J6-5.
NC-342H; KR-34C (# 284) Wright J6-5.
NC-344H; KR-34C (# 291) Wright J6-5.
NC-345H; KR-34C (# 292) Wright J6-5.
NC-382H; KR-34C (# 293) Wright J6-5.
NC-383H; KR-34C (# 294) Wright J6-5.
NC-384H; KR-34C (#

NC-287K; KR-34C (# 326) Wright J6-5.
NC-261K; KR-34C (# 327) Wright J6-5.
NC-262K; KR-34C (# 328) Wright J6-5.
NC-289K; KR-34C (# 330) Wright J6-5.
All of the above listed aircraft were originally registered as "Challenger" model C4C, but were included in the re-designation to Fairchild KR-34C. The model KR-34C was later offered as a float-mounted seaplane on a Group 2 approval numbered 2-257 issued 8-13-30.

Fig. 204.   The Fairchild KR-34C was neat and well-mannered; Elinor Smith, the aviatrix, poses with this one.

Fig. 205. The Buhl "Sport Airsedan" CA-3D with a 300 h.p. Wright J6 engine.

The dashing Buhl "Sport Airsedan" was a 3 place enclosed sesqui-plane of high performance that was primarily leveled at the sportsman-pilot or the business man in a hurry; a discriminate owner who would want to own and enjoy a swift and manuverable airplane, but would much rather have it with the enclosed quiet and comfort of a cabin, and not have to be decked out in helmet and goggles. The brawny "Sport Airsedan" had nearly all the speed and verve of the high-powered open cockpit biplane, but with the added attraction of a roomy and comfortable cabin for three, and a payload high enough to allow ample baggage for all aboard. More or less a custom built sport model, the model CA-3D as powered with the 9 cylinder Wright J6 engine of 300 h.p., had many added custom features and was built to the individual tastes and dictates of the customer. The "Sport Airsedan" was predictable and flew well with a light touch, and with the power reserve of the Wright J6-9 engine, was an airplane of exceptional performance that was able to comply eagerly to the whims and fancies of any pilot.

Loren Mendell, pilot of the record-breaking "Angeleno", now had a model CA-3D "Sport Airsedan" that he called the "Angeleno Jr."; brimming with competitive spirit, he had flown this craft in the Oakland, Calif. to Cleveland, Ohio Air Derby that was run off in connection with the National Air Races for 1929 held in Cleveland. Pressed very hard by his competition, Mendell came in first as he flashed across the line some 3 minutes

ahead of second place. Mendell also flew in the Rim of Ohio Air Derby and came in a close third. Though the "Sport Airsedan" was in small number, they were extremely busy craft and were seen scurrying along on errands of all sorts, in all parts of the country. The type certificate number for the Buhl "Sport Airsedan" model CA-3D as powered with the 300 h.p. Wright J6 engine, was issued in June of 1929 and some 5 or more examples of this model were manufactured by the Buhl Aircraft Company at their plant in Marysville, Michigan. Like most of the aircraft manufacturers of this day, Buhl had a small engineering staff that never exceeded over 5 men, but Ettienne Dormoy who presided over the engineering staff like the "maestro" of a concert orchestra, was a brilliant man who kept the staff busy with a continuous flow of new ideas and new designs. Those ideas that materialized into projects, were designed, built, and flight tested in the matter of months; certainly a far cry from the tedious and lengthy procedures that are practiced in this day and age of aircraft manufacture!

Listed below are specifications and performance data for the Buhl "Sport Airsedan" model CA-3D as powered with the 300 h.p. Wright J6 series engine; length overall 25'0"; hite overall 8'2"; wing span upper 36'0"; wing span lower 20'10"; wing chord upper (constant) 72"; wing chord lower (tapered) 33" mean; wing area upper 190 sq.ft.; wing area lower 50 sq.ft.; total wing area 240 sq.ft.; wt. empty 2017; useful load 1183; payload with

90 gal. fuel was 430 lbs.; gross wt. 3200 lbs.; max. speed 150; cruising speed 125; landing speed 47; climb 1200 ft. first min. at sea level; service ceiling 20,000 ft.; gas cap. 90 gal.; oil cap. 5 gal.; cruising range at 15 gal. per hour was 700 miles; price at the factory field was $12,000.

The fuselage framework was built up of welded chrome-moly steel tubing, faired to shape with formers and fairing strips and covered in a combination of metal and fabric. The cabin portion of the fuselage was covered in metal panels, and the rear portion from the entry door back was covered in fabric; there was a large cabin entry door and a convenient step on each side. The cabin was sound-proofed and insulated and there were provisions for cabin heat and ventilation; a large baggage compartment was aft of the cabin and was accessible from inside or out. The wing framework was built up of spruce and plywood box-type spars for the upper panels and spars of solid spruce for the lower panels; wing ribs were built up of spruce and plywood in truss form and the completed framework was fabric covered. The two gravity-feed fuel tanks were mounted in the upper wing, one flanking each side of the fuselage. The ailerons in the upper wing panels were of the "balanced horn" type, but the "horn" was inset from the tip; all movable control surfaces were aerodynamically balanced. The fabric covered tail-group was built up of welded chrome-moly steel tubing; the fin was ground adjustable and the horizontal stabilizer was adjustable in flight. The landing gear was of the split-axle type in the normal 3-member arrangement, and used air-oil shock absorbing struts; wheels were 30x5. Wheel brakes, wiring for navigation lights,

and a metal propeller were standard equipment. It might be well to note that chrome-moly (molybdenum) steel tubing, which was now being used almost universally, had inherent properties quite attractive to the builders of aircraft; it was light and tough and was welded quite easily into major parts of the airframe. For example, the fuselage for the "Sport Airsedan" was built up of welded chrome-moly steel tubing and weighed under 200 pounds in the bare condition, or about one-tenth of the weight for the completed airplane. If this same fuselage were built up of even the highest grade of carbon steel, it would weigh nearly 400 pounds for a frame that would be equal in strength; one can readily see that this extra weight would be a serious loss to the useful load available. There was continuous development in various models of the "Airsedan" series for the next year or so, and several were modified into special-purpose airplanes mounting engines of higher horsepower. A light low-wing monoplane seating two side-by-side was developed, but never went beyond the prototype stage. Another monoplane developed was the high-performance "Airster" that was built as a 3 place airplane for the sportsman-pilot, or as a one-place craft for carrying airmail and air-cargo. The next Buhl development of any note was the popular "Bull Pup", which will be discussed in the chapter for A.T.C. # 405.

Listed below are Buhl CA-3D entries that were gleaned from various records; this list may lack one or two entries, but it does show the bulk of this model that were built.
NC-7448; CA-3D (# 35) Wright J6.
NC-9631; CA-3D (# 45) Wright J6.
NC-9635; CA-3D (# 49) Wright J6.
NC-8447; CA-3D (# 53) Wright J6.

Fig. 206. The "Sport Airsedan" was a clean design; note the sesquiplane arrangement.

Fig. 207. The CA-3D "Angelino Jr." was the sister ship to the earlier record-breaking "Angelino".

NC-8451; CA-3D (# 57) Wright J6.
Model CA-3D serial # 35 was originally a J5 powered CA-3C, and was modified to a CA-3D by installation of Wright J6-9 engine; models CA-3D serial # 45 and # 53 later as CA-3D Special 4 place craft on Group 2-72; model CA-3D serial # 57 later modified to CA-3E with Packard Diesel engine.

A.T.C. #164
(6-29)
FOKKER "STANDARD UNIVERSAL"

Fig. 208. The Fokker "Standard Universal" with the 300 h.p. Wright J6 engine offered increased performance.

Fokker airplanes certainly have figured prominently in the development of practical aviation during the early "twenties", doing well more than their share to awaken the world to the fact that the airplane was a good deal more than a machine of war, or a craft of sport. To gain the world's attention, the airplane necessarily had to set records or do the unusual, and Fokker airplanes were kept very busy setting new records and doing the unusual. Fokker airplanes had already flown across the North Pole, across the Atlantic Ocean, across part of the Pacific Ocean to Hawaii and clear across the vast Pacific to Australia, brought in air transportation that could surmount nature's many barriers in the "bush country" of Canada and Alaska, carried contract-mail on some of the earliest routes in the U.S.A. and Canada, established many of the early air-lines in this country, and proved to big-business that the airplane was the best available means of economical transportation. After all these years of pioneering, Fokker airplanes were now pretty much content to forsake the bally-hoo and get down to work, working quietly and unheralded at chores ranging through the whole gamut of things that were possible with an airplane.

Nearly 40 "Universals" were already scattered in many parts of the U.S. and Canada, at least 60 of the "Super Universal" had been built by now and had taken their places in varied services, and well over 50 of the majestic Fokker "Tri-Motor" had been criss-crossing the nation to accumulate over 25 million miles of experience on passenger-carrying routes. Now, added to this line-up of versatile air-transports was an improved version of the "Standard Universal". This new "Universal" was still quite typical in most all respects, but several improvements were now apparent; the long-suffering pilot was now comfortably enclosed in a weather-proof cabin, there were many improvements in minor detail, and a bit more power was added for increased performance. The new "Universal" was a high wing cabin monoplane seating 5 or 6 passengers plus the pilot, with room and comfort for all, and was powered with the 9 cylinder Wright J6 engine of 300 h.p. Performance was noticeably improved in all aspects over the earlier versions, besides retaining all the attractive qualities of all-round utility and dependability that made the "Universal" series such a popular choice. The spacious cabin had removable seats to provide 146 cubic feet of cargo area, with room available for 9 foot lengths. For use in the lake-dotted "bush country", the new

"Universal" was also available as a float-mounted seaplane. The type certificate number for the Fokker "Standard Universal" as powered with the 300 h.p. Wright J6 engine, was issued in June of 1929 and probably not more than 10 examples of this model were manufactured by the Fokker Aircraft Corp. at Teterboro Airport in Hasbrouck Heights, New Jersey. Harris M. "Pop" Hanshue was the president; Edw. V. "Eddie" Rickenbacker was V.P.; Wm. T. Whalen was V.P. and general manager; A. H. G. "Tony" Fokker was chief of engineering; and Alfred A. Gassner was the chief engineer. In late 1929 Fokker Aircraft became an affiliate of the General Motors Corp.; in May of 1930 there was another re-organization and Fokker Aircraft became the General Aviation Corp.

Listed below are specifications and performance data for the Fokker "Standard Universal" as powered with the 300 h.p. Wright J6 series engine; length overall 33'6"; hite overall 8'9"; wing span 47'9"; wing chord at root 96"; wing chord at tip 72"; total wing area 341 sq.ft.; airfoil "Fokker"; wt. empty 2482; useful load 1818; payload with 78 gal. fuel was 1140 lbs.; gross wt. 4300 lbs.; max. speed 130; cruising speed 110; landing speed 52; climb 830 ft. first min. at sea level; climb in 10 min. was 6800 ft.; service ceiling 14,000 ft.; gas cap. 78 gal.; oil cap. 5 gal.; cruising range at 15 gal. per hour was 500 miles; price at the factory field was $15,000 for landplane; $17,500 for seaplane; floats were mounted on landplane for $ 2,500 extra; landing gear complete with brakes and tail-skid could be bought separately for $1135; the following figures are for the float-mounted seaplane; wt. empty 2630; useful load 1670; payload with 78 gal. fuel was 990 lbs.; gross wt. 4300 lbs.;

max. speed 120; cruising speed 100; landing speed 50; climb 800 ft. first min. at sea level; climb in 10 min. was 6500 ft.; service ceiling 13,000 ft.; cruising range 450 miles; all other dimensions and data remained the same.

The fuselage framework was built up of welded chrome-moly & low-carbon steel tubing, lightly faired to shape and fabric covered. Seating could be arranged for 4, 5, or 6 passengers, and a baggage compartment of 30 cu. ft. was to the rear of the cabin. The semi-cantilever wing framework was built up with box-type spar beams of spruce caps and birch plywood sides, with plywood ribs and stringers that were covered with a birch plywood veneer. The wing was a one-piece structure that was bolted directly to the top of the fuselage. The two gravity-feed fuel tanks were mounted in the wing, one flanking each side of the fuselage. The cabin walls were soundproofed and insulated, with provisions for cabin heat and ventilation. The fabric covered tail-group was built up of welded chrome-moly steel tubing; the fin was ground adjustable and the horizontal stabilizer was adjustable in flight. The split-axle outrigger landing gear of 120 inch tread was tied into a truss with the front wing struts, shock absorbers were of rubber shock-cord; wheels were 32x6 and Sauzedde brakes were standard equipment. It is interesting to note that all control cables, fairleads, control horns, and pulleys, were exposed for ease of inspection and maintenance; this advantage for off-setting the small gain that would be realized through more aerodynamic cleanness. A metal propeller, wiring for navigation lights, Eclipse engine starter, and fire extinguisher were also standard equipment. The next development in the Fokker monoplane was the

Fig. 209. Later versions of the "Standard Universal" sported wheel pants and anti-drag engine cowlings.

Fig. 210. The "Standard Universal" served well in business flying; dependability made it a popular choice.

"parasol" transport model F-14 which will be discussed in the chapter for A.T.C. # 234.

Listed below are "Standard Universal" entries that were gleaned from various records; this list may not be complete, but it does show the bulk of this model that were registered.

NR-1776; Std. Univ. (# 421) J6-9-300.
NC-337N; Std. Univ. (# 437) J6-9-300.

NC-149H; Std. Univ. (# 440) J6-9-300.
NC-9774; Std. Univ. (# 441) J6-9-300.
 X-181H; Std. Univ. (#   ) J6-9-300.
NC-761Y; Std. Univ. (#   ) J6-9-300.
Serial # 421 originally was powered with Wright J5, converted with Wright J6-9-300 by Plane Speaker Corp.; X-181H was a special test-bed for Fokker Aircraft Corp.

Fig. 211.   The Ford Trimotor 5-AT-C was the ultimate in air travel and the backbone of scheduled air transport.

The Ford "Tri-Motor" model 5-AT-C can very well be considered as typical to the model 5-AT-B described previously, except perhaps for the following changes and improvements; only a few of these would actually meet the eye, but they were many to be sure. The fuselage was now lengthtened by nearly six inches, internal cabin dimensions were increased by some 60 cubic feet or more, and a total seating capacity now stood at 17 for the standard transport model. We say "standard transport" because the 5-AT-C was also offered in a 9-11 place deluxe "Club" model that was custom built as a winged "air yacht". Extra seating on the "transport" added up to a somewhat higher gross weight, but performance did not seem to suffer for it, in fact, the model 5-AT-C was so far the fastest of the Ford "Tri-Motor" series. Due in fact most likely, that the airplane was quite a bit "cleaner", especially in the engine nacelle area, which now quite often mounted the new "low drag" engine cowlings which were adding miles per hour. Added to this occasionally were streamlined "wheel pants" to improve the air-flow over the large wheels; these added still more speed. A normal top speed of 152 m.p.h. could easily be attained in this arrangement, and

speeds even up to 164 m.p.h. or more were recorded a time or two. The early 4-AT of a few years back, was a lumbering lady that did well to top over 100 m.p.h., but this latest creation was more sleek and "clean" and could cover 300 miles easily in 2½ hours. Scheduled air-lines could now offer a much faster service that was attracting more air-travellers, and the Ford "Tri-Motor" of which there were some 100 or more already, was fast becoming a familiar sight to people across the land; some 12 or more air-lines were already using the "Tin Goose" as standard equipment, and big business had long since learned to appreciate the value of owning the tri-motored Ford.

Production increases brought on several additions to the factory at Dearborn, and employees were being added constantly. From a working force of some 150 people at the end of 1927, this rose rapidly to a total of some 1200 by the end of 1928; further increases in production zoomed this figure still more and towards the end of 1929, there were some 1850 people employed in the building of Ford "Tri-Motors". Assembly line methods had been introduced to the manufacture of aircraft in this plant, and 3 "Fords" per week were now coming off the line. This was really quite an achievement when one stops to consider the amount of separate parts, components, and assembly work that would go into an airplane such as the "Wasp" powered 5-AT. Then too, much has already been said and written about the fast moving production lines at the Ford Motor Co. plants where they built the famous "Tin Lizzie", where one had hardly time to wipe his nose and take a chance on falling way behind in his production; this was not hardly the case in the "airplane plant" where more care and time was taken to produce products of greater excellence. Henry Ford would insist that the "Tri-Motor" be the best built, and the safest airplane that could be made; he actually carried this insistence out to some lengths. Somewhat unusual indeed was the fact that all Ford "Tri-Motor" customers were strongly urged to send their pilots, regardless of how much time and experience they already had, to the factory school where they could learn to fly the "Tin Goose", and do it properly; Henry Ford was not having any "accidents" that would be caused by poorly trained pilots, and therefore reflect back on his airplanes as being unsafe. The first model 5-AT-C was flown in May of 1929 and delivered to Maddux Air Lines in California, others were being delivered in succession and by the end of 1929, some 35 or so had been built. The type certificate number for the model 5-AT-C as powered with 3 "Wasp" engines was issued in June of 1929 and some 50 of this model were built altogether by the Stout Metal Airplane Co., a division of the Ford Motor Co. at Dearborn, Michigan. The adjoining factory airfield was a beautiful place called "Ford Airport" and was at that time one of the finest in the land, a popular stopover for transient aircraft that would often go out of their way to make the visit.

Listed below are specifications and performance data for the Ford "Tri-Motor" model 5-AT-C as powered with three P & W "Wasp" engines of 420-450 h.p. each; length overall 50'3"; hite overall 12'8"; wing span 77'10"; wing chord at root 156"; wing chord at tip 92"; total wing area 835 sq.ft.; airfoil "Goettingen 386"; wt. empty 7500; useful load 6000; payload 3720-3370; gross wt. 13,500 lbs.; empty weight and useful load will vary with fuel load and cabin arrangement; max. speed 142-152-164; cruising speed 122-130-140; max. speed and cruising speed will vary according to configuration, the highest speeds were attained with "low drag" cowlings over each engine and wheel pants; landing speed 65; climb 1050 ft. first min. at sea level; climb from sea level in 10 min. was 8000 ft.; service ceiling 18,500 ft.; gas cap. 277-355 gal.; oil cap. 27-33 gal.; cruising range 560-630 miles; price at the factory field for the standard transport was $55,000; the deluxe "Club" model averaged at $68,000.

The fuselage framework on this version was typical except for added length and a change in cabin arrangement, which included more seating, better heating and ventilation, improved lavatory facilities, and cabin lights for passenger conveniences. The deluxe "Club" model was more or less custom built, and cabin appointments would often include a divan, berths for sleeping, overstuffed chairs, typewriter and desk, radios and iceboxes, a steward's galley for dining enroute, and any number of other conveniences and luxuries for the discriminate air-traveller. The

Fig. 212. A 5-AT-C owned by Firestone; this ship led a very active life for twenty years.

Fig. 213. The deluxe version of the 5-AT-C had anti-drag cowlings and wheels pants for added performance.

wing framework on the model 5-AT-C was also typical of other models, but now contained mail compartments built into the root ends of the outer wing panels just outboard from the engine nacelles; these two compartments had a maximum capacity of 800 pounds. A 400 lb. baggage compartment was in the center-section panel of the wing, and another such compartment was later placed in the fuselage, aft of the main cabin. The engine nacelles were "cleaned up" to some extent, and the outboard engines were most often cowled-in with "low drag" cowlings of the "Townend ring" type which improved airflow around the engine cylinders; on some 5-AT-C versions, all 3 engines were cowled

in. The landing gear though typical, was improved in some details, and some versions of the model 5-AT-C sported streamlined wheel fairings, which were referred to as "wheel pants". Wheel brakes, metal propellers, and inertia-type engine starters were standard equipment. An unusual version of the 5-AT-C was the model mounted on huge twin floats as a seaplane, this was the 5-AT-CS and will be discussed in the chapter for A.T.C. # 296. The next Ford "Tri-Motor" development after the 5-AT-C as discussed here, was the model 6-AT as powered with three Wright J6-9 engines of 300 h.p. each; see chapter for A.T.C. # 173 in this volume.

Fig. 214. The Bourdon "Kitty Hawk" Model B-4 had a 100 h.p. Kinner engine.

The Bourdon "Kitty Hawk" was a trim three-place open cockpit biplane that was introduced early in 1928, for use as a general-purpose airplane of the type that was quite popular, and generally being used by most private-owners and the low budget fixed-base operator; this early version of the "Kitty Hawk", the model B-2, was described previously in the chapter for A.T.C. # 134. The model B-2 was powered with the seven cylinder Siemens-Halske engine of 105-113 h.p., and seven examples of this version were built, but delivery of engines from Germany was at times rather uncertain, so Bourdon decided to standardize with the five cylinder Kinner K5 engine of 90-100 h.p., which was manufactured here in the U.S.A. and was now becoming steadily available in good number. The new Bourdon "Kitty Hawk" model B-4, as powered with the Kinner engine, was quite typical in most respects; performance and flight characteristics were pretty much the same, and a slight reduction in weights was about the only difference between the two models.

Nancy Hopkins, a petite and very capable aviatrix, flew a Kinner-powered "Kitty Hawk" B-4 in the Women's Dixie Derby (Class B) from Washington, D.C. to Chicago and fi-nished in 4th place; this derby was in connection with the 1930 National Air Races held at Chicago, Illinois. She also captured two 2cd places and three 3rd places in the dead-stick landing contests held for women pilots. About a month later she was participating in the National Air Tour for 1930 and finished in 14th place against a line-up of many formidable entries. These results speak well for both Nancy Hopkins and the "Kitty Hawk" biplane. The type certificate number for the Bourdon "Kitty Hawk" model B-4 as powered with the Kinner K5 engine, was issued in June of 1929 for both landplane and seaplane versions, and some 17 or more of this model were built. The "Kitty Hawk" biplane was manufactured by the Bourdon Aircraft Corp. at Hillsgrove, Rhode Island; Allen P. Bourdon was the president, general manager, and spare-time chief pilot; C. L. Ayling was the V.P.; Roderick F. Makepeace was the secretary, treasurer, and purchasing agent; and John E. Summers was chief of design and engineering.

Listed below are specifications and performance data for the Bourdon "Kitty Hawk" model B-4 as powered with the 90-100 h.p. Kinner engine; length overall 23'0"; hite overall 8'6"; span upper and lower 28'0"; wing chord both 54"; area upper 123 sq.ft.; area lower 110 sq.ft.; total wing area 233 sq.ft.; airfoil U.S.A.-27; wt. empty 1107; useful load 768; payload 358; gross wt. 1875 lbs.; max. speed 110; cruising speed 92; landing speed 40; climb 660 ft. first min. at sea level; service ceiling 14,500 ft.; gas cap. 35 gal.; oil cap. 3.5 gal.; cruising range 475 miles; price at the factory field was $4800. The following figures are for the model B-4 as equipped with twin-float seaplane gear; wt. empty 1420; useful load 780; payload 370; gross wt. 2200 lbs.; max. speed 100; cruising speed 85; landing speed 47; climb 560 ft. first min. at sea level; service ceiling 12,000 ft.; cruising range 425 miles; all dimensions and other data remain the same.

The fuselage framework was built up of welded 1025 steel tubing, faired to shape with wood fairing strips and fabric covered. A front cockpit entrance door was provided on the left side, and a convenient step was provided for the pilot. The wing framework was built up of solid spruce spars that were routed to an "I beam" section; the wing ribs were built into a Warren truss form of spruce members and plywood gussets, the leading edges were covered with "dural" sheet to preserve the airfoil form and the completed framework was fabric covered. The upper wing was built in two panels that were joined

together at the center-line and fastened to a center-section cabane of inverted vee struts. Ailerons were on all four panels, and were connected together in pairs with a streamlined push-pull strut. The fabric covered tail-group was built up of welded 1025 steel tubing; the fin was ground adjustable and the horizontal stabilizer was adjustable in flight. The split-axle landing gear had two long telescopic legs that were of 7 foot 4 inch tread; shock absorbers were rubber "donut rings" in compression. Wheels were 26x4 and the tail-skid was of the springleaf type. The fuel tank was mounted high in the fuselage just ahead of the front cockpit; a baggage compartment was in the fuselage just behind the pilot's cockpit. Wheel brakes, metal propeller, inertia-type engine starter, and navigation lights were optional equipment. In 1930 the Bourdon Aircraft Corp. was merged into the Viking Flying Boat Company at New Haven, Conn. and manufacture of the "Kitty Hawk" biplane continued. The next development in the "Kitty Hawk" biplane was the model B-8 that was powered with the 125 h.p. Kinner engine; this model will be discussed in the chapter for A.T.C. # 392.

Listed below are "Kitty Hawk" model B-4 entries that were gleaned from various records; this list is not complete, but it does show the bulk of this model that were built.

Fig. 215. Nancy Hopkins flew this "Kitty Hawk" to 14th place in the 1930 National Air Tour.

Fig. 216. The "Kitty Hawk" B-4 was friendly and gentle; it operated well from grass airports.

Fig. 217. The B-4 was rather petite for a 3-place airplane; improved performance was a gain.

C-9460; B-4 (# 8) Kinner K5
    ; B-4 (# 9) Kinner K5
    ; B-4 (# 10) Kinner K5
NC-45M; B-4 (# 11) Kinner K5
NC-293M; B-4 (# 12) Kinner K5
NC-921M; B-4 (# 13) Kinner K5
NC-30V; B-4 (# 14) Kinner K5
NC-31V; B-4 (# 15) Kinner K5
NC-32V; B-4 (# 16) Kinner K5

NC-33V; B-4 (# 17) Kinner K5
NC-34V; B-4 (# 18) Kinner K5
NC-35V; B-4 (# 19) Kinner K5
NC-10460; B-4 (# 20) Kinner K5, also as
20-7.
        ; B-4 (# 21)
        ; B-4 (# 22)
    ; B-4 (# 23)
NC-10496; B-4 (# 24) Kinner K5

Fig. 218. The Great Lakes 2-T-1 biplane offered many the chance to play sportsman-pilot on a limited budget.

The cocky little swept-wing Great Lakes sport-trainer biplane was probably one of the most distinctive-looking and most easily recognizable airplanes of this time; a heart-warming combination of dash and performance that offered many the chance to play sportsman-pilot on a very modest budget. A little bit of genius, a stroke of luck, put them all together and we have a craft that has had a long span of exciting history and a popularity that hasn't waned a bit, even to this day. The prototype version of the 2-T-1 as shown here, is one that many may not know about or remember. Quite contrary to the general belief, the model 2-T-1 as it was first developed, was a straight-wing biplane and did not have the characteristic sweep-back in the upper wing panel, a feature of necessity that became such a distinguishing trade-mark for the sporty Great Lakes biplane.

The Great Lakes model 2-T-1, formally introduced to the public in prototype at the Detroit Air Show of March 1929, was a rakish and sporty biplane of rather diminutive proportions that seated two in tandem in open cockpits; it was powered with the four cylinder aircooled in-line "Cirrus" Mark 3 engine of 85 h.p. The configuration of this craft was quite average and in good harmony, but flight tests soon revealed that the aircraft, though flyable, was painfully tail-heavy and this was to pose a real problem; a problem because of the close-coupled design, and everything had to be just about where it was. Any relocation changes made to offset this tail-heaviness, would require almost a major re-design. Having already built two airplanes of this type and two already under construction, all with the same problems, this now called for some rather deep thinking. Discussion of the problem present, eventually brought on the realization that the only practical alternative was to put sweep-back in the wing panels; the upper wing panel was then swept-back to some 9 degrees on each side, the amount necessary to lengthen the nose-arm to offset the tail-heaviness, and there you have it, a most wonderful airplane was born. Many flying-folk are of the opinion that it was indeed a stroke of good luck and a very fortunate circumstance for all that the early Great Lakes biplane turned out to be tail-heavy, otherwise, it may have remained a "straight-wing" biplane and not acquired it's most attractive, most distinctive, and most lovable feature.

The Great Lakes Aircraft Corp. was formed late in 1928, and had taken over the

Fig. 219.   Charlie Meyers flying a Great Lakes 2-T-1 "Straight-Wing".

plant-site formerly used by the Glenn L. Martin Company at Cleveland, Ohio. Their first civil-commercial development was a craft called "Miss Great Lakes", an 8 place combination open and cabin transport biplane that was an adaption of the TG-1 (T4M-1) torpedo-bomber; the Great Lakes TG-1 was actually a modified or warmed-over version of a design inherited from the Martin Company. Two aircraft in this version were built and

sources closely associated with this project revealed the obvious, that it was pretty much a waste of time and good money. The 2-T-1 "Sport", already in the works, had hardly fared much better in the very beginning; the modification of this model into a swept-wing biplane was surely an example of good thinking, and a stroke of luck that really saved the day for the G.L.A.C.

The type certificate number for the Great

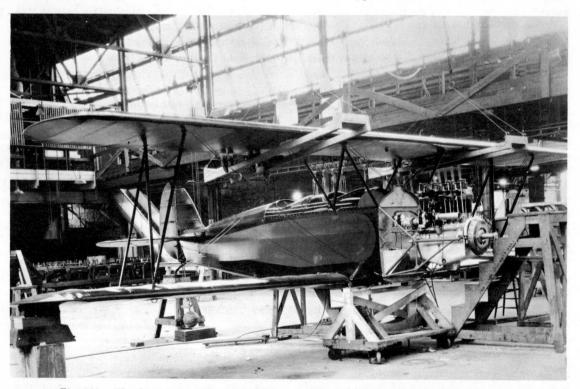

Fig. 220.   The Great Lakes "Straight-Wing" prototype being assembled; only four were built.

Fig. 221. A Great Lakes 2-T-1 showing the swept wing to improve balance; performance was sensational.

Lakes "Sport" model 2-T-1 (swept-wing) was issued in June of 1929 and it was manufactured by the Great Lakes Aircraft Corp. at Cleveland, Ohio. It has been difficult to determine just how many examples of the 2-T-1 were actually built under this certificate; some records estimate 40 or 50, and it is known that serial numbers in the fifties were already labeled as the 2-T-1A that came under A.T.C. # 228. However, serial # 7 to # 264 were eligible when compliance was made to certain tail-group conformances, likewise, all serial numbers eligible under A.T.C. # 167 were also eligible under A.T.C. # 228 when changed to conform; study of engineering memos would clarify this information. The G.L.A.C. had an impressive line-up of executive officers. Col. Benjamin F. Castle was the president; Charles F. Van Sicklen (formerly with Waco) was V.P. and sales manager; Capt. Holden C. Richardson was V.P. in charge of engineering; P. B. Rogers and Earl Stewart were project engineers; and Chas. W. Meyers was design-engineer, sales-engineer, and chief pilot in charge of test and development. Charlie Meyers, formerly design-engineer and test-pilot with Advance Aircraft Co., was largely responsible for the design and development of the "Waco 10" series, and the fabulous "Taperwing". He left Waco in 1928 to help form the new G.L.A.C., and it was Charlie Meyers who was largely responsible for the conception of the 2-T-1 "Sport" design and brought it on to it's swept-wing development. Though not a full-fledged engineer in the academic sense, Meyers however played a most important role in the design and development of the sport-

trainer series, and several other craft that were planned, drawing on experience and intuitive talent that had accumulated in him through participation in every facet of aviation imagineable since 1915.

The magnetic personality and the sprightly performance of the Great Lakes "Sport" was on everyone's lips and the news spread like wild-fire; the 2-T-1 was bought up as fast as they could be made and the company soon had a back-log of over 200 orders. Many fine airplanes have come out of this period, and the 2-T-1 was surely a shining example. Cocky and determined, the 2-T-1 could not be considered gentle, but it had no great vices and a good steady hand could transform it's playful nature into an experience of real pleasure. Because of it's nature it could not tolerate timidity, but it was easy to understand and it radiated with eagerness; a proper understanding between a pilot and this airplane usually transformed into mutual admiration and a lasting friendship. That the Great Lakes "Sport" was only too willing to do as bid, can be reflected back from the many outstanding records and achievements that were left behind during it's active life.

Listed below are specifications and performance data for the Great Lakes "Sport" model 2-T-1 as powered with the 85 h.p. "Cirrus" Mark 3 engine; length overall 20'4"; hite overall 7'10"; wing span upper and lower 26'8"; wing chord both 46"; wing area upper 97 sq.ft.; wing area lower 90.6 sq.ft.; total wing area 187.6 sq.ft.; airfoil straight-wing was M-6, swept-wing was M-12; wt. empty 1102; useful load 578; payload with 25 gal. fuel was 243 lbs.; gross wt. 1580 lbs.; max.

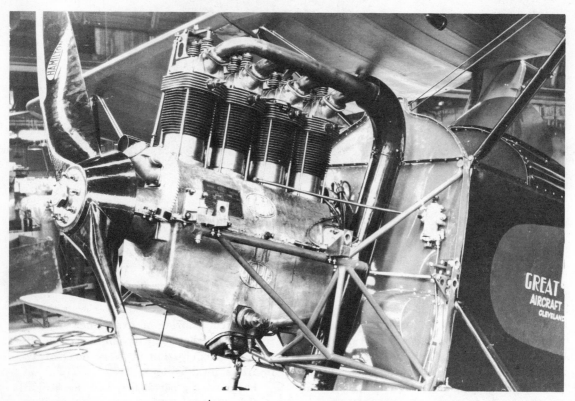

Fig. 222.   The 4 cylinder "Cirrus" Mark III of 85 h.p., shown mounted in a Great Lakes 2-T-1 trainer.

speed 115; cruising speed 97; landing speed 45; climb 780 ft. first min. at sea level; climb to 5,000 ft. was 8 min.; service ceiling 12,500 ft.; gas cap. 25 gal.; oil cap. 2 gal.; cruising range at 6 gal. per hour was 375 miles; price at the factory field was $4990, lowered to $3985; the American version of the 4 cyl. Cirrus engine sold for $1600.

The fuselage framework was built up of welded chrome-moly steel tubing, faired to shape with metal formers and metal fairing strips and fabric covered. A hinged deck in the rear portion of the front cockpit cowling swung out to make room for easy access to the forward cockpit; bucket-type seats were provided that had wells for a parachute pack. The 2-T-1 was designed for sport-flying and it was certainly excepted that it would be stunted on occasion, the occupants would then normally wear parachutes. The wing framework was built up of solid spruce spar beams that were routed-out to an "I beam" section, and the wing ribs were of stamped-out 17 ST duralumin sheet; the leading edge of the panels were covered with "dural" sheet to preserve the airfoil form and the completed framework was fabric covered. The gravity-feed fuel tank was mounted in the center-section panel of the upper wing, and the

ailerons of the Friese offset-hinge type were mounted in the lower panels. The ailerons were operated by metal push-pull tubes and bellcranks, the rudder and the elevators were operated by braided cables; there were adjustable limit-stops on all controls. All of the interplane struts were of heavy gauge chrome-moly steel tubing in a streamlined section, and the interplane bracing was of heavy gauge streamlined steel wire. The fabric covered tail-group was built up of riveted duralumin ribs and formers; the fin was fixed and the horizontal stabilizer was adjustable in flight. The split-axle landing gear of 70 inch tread was of the rugged out-rigger type and used oleo-spring shock absorbers; wheels were 24x4 and brakes were available. The prototype version had the rubber donut-ring shock absorbers, but they were too bouncy, so the change was made to "oleo". The upper wing of the straight-wing prototype was built up in one piece, but the swept-wing version was built up with a center-section panel and two outer panels; the lower wing was set back with 25 inches of stagger and had 2 deg. of dihedral. The "Cirrus" was a 4 cyl. aircooled in-line engine of British design and manufacture, that was now being built under license here in the U.S. by American Cirrus

Engines; the Mark 3 engine was rated 85 h.p. at 1900 r.p.m., and 95 h.p. at 2100 r.p.m. The next development in the Great Lakes sport-trainer series was the improved model 2-T-1A, which will be discussed in the chapter for A.T.C. # 228. Extensive tests were continually being conducted on this sport-trainer series; it was flown on skis, on twin-floats as a seaplane, with a winter canopy enclosure over the cockpits, with anti-stall "wing slots" in the upper wing, with Cirrus and Menasco inverted engines, as a one-place racing craft, and even tested with the 5 cyl. Kinner K5 engine.

Listed below is a partial list of 2-T-1 entries that were gleaned from various records; a complete listing would be rather extensive, so we submit about 20 or so.

-577; 2-T-1 (# 3) Cirrus.
-620; 2-T-1 (# 4) Cirrus.
-3020; 2-T-1 (# 5) Cirrus.
-9413; 2-T-1 (# 6) Cirrus.
C-9461; 2-T-1 (# 7) Cirrus.

C-9462; 2-T-1 (# 8) Cirrus.
NC-9463; 2-T-1 (# 9) Cirrus.
NC-9464; 2-T-1 (# 10) Cirrus.
NC-376H; 2-T-1 (# 11) Cirrus.
NC-377H; 2-T-1 (# 12) Cirrus.
NC-378H; 2-T-1 (# 13) Cirrus.
NC-379H; 2-T-1 (# 14) Cirrus.
NC-380H; 2-T-1 (# 15) Cirrus.
NC-838H; 2-T-1 (# 16) Cirrus.
NC-381H; 2-T-1 (# 17) Cirrus.
NC-839H; 2-T-1 (# 18) Cirrus.
NC-840H; 2-T-1 (# 19) Cirrus.
NC-841H; 2-T-1 (# 20) Cirrus.
NC-842H; 2-T-1 (# 21) Cirrus.
NC-843H; 2-T-1 (# 22) Cirrus.
NC-844H; 2-T-1 (# 23) Cirrus.
NC-845H; 2-T-1 (# 24) Cirrus.
NC-846H; 2-T-1 (# 25) Cirrus.

The first two aircraft manufactured were G.L.A.C. transports # 1 and # 2; sport-trainer series start with # 3; NC-200K thru NC-214K were # 26 thru # 40; NC-535K thru NC-544K were # 41 thru # 50.

Fig. 223. A Waco BS-165 (BSO) with a 5 cylinder Wright J6 engine. Shown here is the prototype.

Waco biplanes were so popular and so numerous, they were almost to the point of being taken for granted. New models of the basic "Ten" were often introduced with complete acceptance but very little to-do from the flying populace, as if to say, here is another new model of the Waco biplane and it's just bound to be a good one; they always were. The Waco biplane, which was certainly a perennial favorite the country over, had already been presented in at least 15 different models and some 1900 or more of these aircraft had been manufactured to this time. An addition to this great line of utility aircraft was the new model "165" (BS-165), a compatible combination of a tried and true airplane with the power of a new engine that was showing every possibility of being a great one.

The Waco model BS-165 was a 3 place open cockpit biplane of the general-purpose type which was long the mainstay of the average flying-service operator, and was powered with the new 5 cylinder Wright J6 engine of 165 h.p. Typical in most all respects, the new model "165" was of the familiar "Waco 10" configuration from the engine firewall back, with some improvements in de-

tail, and shared all the other inherent characteristics that made the Waco biplane such a favorite the country over. This new combination was one of efficiency and good manuverability, and it's short-field performance with a good-sized payload aboard, was well in keeping with established Waco tradition. Though leveled at a limited field of prospective buyers, the model "165" was built in good number; the OX-5 powered "Ten" was still being built by the hundreds, there were many orders coming in for the fabulous "Taperwing", and of course the "Whirlwind-Waco" and the "Hisso-Waco" were still great favorites and were also produced in pretty good quantity.

Designation of Waco models had now been hitched to engine horsepower, so the well-loved "Waco 10" with the OX-5 engine became the Waco 90, the Siemens-Halske powered Waco was the model "125", the Hisso powered models were the "150" and "180", the "J5 Waco" was the model "220", and so on. In a designation code used sometime later, these became the GXE, the DSO, the ASO, and the ATO, with the BS-165 as the BSO; this designation system is a study in itself, which we will try to discuss in some length later on. About June of 1929, the Advance Aircraft Co. changed it's name and was now known as the Waco Aircraft Co. to better identify the firm with it's popular product; there had also been some changes in executive personnel. The type certificate number for the model BS-165 (BSO) as powered with the 5 cylinder Wright J6 series engine, was issued in June of 1929 and some 35 or more examples of this model were manufactured by the Waco Aircraft Co. at Troy, Ohio. Clayton J. Bruckner was the president and general manager; Lee N. Brutus was the V.P.; R. E. Lees had replaced Chas. Van Sicklen as sales manager, who was now with the Great Lakes Aircraft Corp.; Russell F. Hardy had replaced E. E. Green as the chief of engineering; and Freddie Lund was the chief pilot in charge of test and development, in place of Charlie Meyers who had also gone to Great Lakes Aircraft Corp.

Listed below are specifications and performance data for the Waco biplane model BS-165 as powered with the 165 h.p. Wright J6 engine; length overall 23'6"; hite overall 9'1"; wing span upper 30'7"; wing span lower 29'3"; wing chord both 62.5"; wing area upper 155 sq.ft.; wing area lower 133 sq.ft.; total wing area 288 sq.ft.; airfoil "Aeromarine Modified"; wt. empty 1529; useful load 956 (1001); payload with 65 gal. fuel was 345 (390) lbs.; gross wt. 2485 (2530) lbs.; figures in brackets were for later version; max. speed 120; cruising speed 100; landing speed 44; climb 750 ft. first min. at sea level; service ceiling 18,000 ft.; gas cap. 65 gal.; oil cap. 7 gal.; cruising range at 10 gal. per hour was approx. 600 miles; price at the factory field was $6370; the following figures are for a special-purpose 1 place version of the BS-165; wt. empty 1563 lbs.; useful load 1037; payload with 65 gal. fuel was 426 lbs.; gross wt. 2600 lbs.; performance figures were affacted very little, and all the other dimensions and data remained the same.

The fuselage framework was built up of welded chrome-moly steel tubing, faired to shape with wood fairing strips and fabric covered. The cockpits were roomy and well upholstered, with provisions for dual controls; there was a baggage compartment in the turtle-back section of the fuselage just behind the rear cockpit. The wing framework was built up of solid spruce spar beams and spruce and plywood built-up wing ribs; the completed framework was fabric covered. There were four ailerons that were connected together in pairs by a streamlined push-pull strut; the ailerons were operated differentially for better control. The landing gear of 78 inch tread was of the familiar Waco split-axle type, using two long oleo-spring telescopic legs; later versions as shown were fitted with the out-

Fig. 224. The performance of the Waco BS-165 was ideal for general purpose service.

Fig. 225. The later version of the BS-165 (BSO) had improved landing gear and other refinements.

rigger type using "Aerol" shock-absorbing struts. Wheels were 30x5 and Bendix brakes were standard equipment. The fabric covered tail-group was built up of welded chrome-moly steel tubing; the vertical fin was ground adjustable and the horizontal stabilizer was adjustable in flight. Wings wired for navigation lights, and a wooden propeller was standard equipment, but a metal propeller was available for $275 extra; an inertia-type engine starter was also optional at extra cost. The next development in the Waco biplane was the popular "straight-wing" model CS-225 (CSO) as powered with the 7 cylinder Wright J6 engine of 225 h.p.; this model will be discussed in the chapter for A.T.C. # 240.

Listed below are BS-165 entries that were gleaned from various records; this list may not be quite complete, but it does show the bulk of this model that were built.

C-8543; BS-165 (# A-68) Wright 5.
C-931H; BS-165 (# A-143) Wright 5.
C-944H; BS-165 (# A-144) Wright 5.
C-940H; BS-165 (# A-145) Wright 5.
NC-750K; BS-165 (# A-147) Wright 5.
NC-952H; BS-165 (# A-148) Wright 5.
C-959H; BS-165 (# A-149) Wright 5.
C-958H; BS-165 (# A-150) Wright 5.

NC-752K; BS-165 (# A-153) Wright 5.
NC-751K; BS-165 (# A-157) Wright 5.
NC-765K; BS-165 (# A-158) Wright 5.
NC-261M; BS-165 (# A-159) Wright 5.
NC-767K; BS-165 (# A-161) Wright 5.
NC-768K; BS-165 (# A-162) Wright 5.
NC-769K; BS-165 (# A-163) Wright 5.
NC-770K; BS-165 (# A-164) Wright 5.
NC-784K; BS-165 (# 3001) Wright 5.
NC-785K; BS-165 (# 3002) Wright 5.
NC-786K; BS-165 (# 3003) Wright 5.
NC-267M; BS-165 (# 3004) Wright 5.
NC-203M; BS-165 (# 3005) Wright 5.
NC-861M; BS-165 (# 3006) Wright 5.
NC-603N; BS-165 (# 3041) Wright 5.
NC-608N; BS-165 (# 3070) Wright 5.
NC-618N; BS-165 (# 3110) Wright 5.
NC-614N; BS-165 (# 3131) Wright 5.
NC-617N; BS-165 (# 3132) Wright 5.
NC-623N; BS-165 (# 3144) Wright 5.
NC-625N; BS-165 (# 3145) Wright 5.

Serial numbers for the following BS-165 aircraft are unknown, C-2671, C-900H, NC-609N, and NC-770N; NC-788K is also listed as serial # A-153 which is probably in error; NC-952H serial # A-148 is also listed as a model CSO; C-8543 serial # A-68 was the prototype model BS-165.

The superb outline of the "Vega" wooden monococque fuselage was created in simplicity by joining two half-shells that were formed in a simple mold called a "concrete tub", and every Vega fuselage emerged from the very same tub. The differences in the various models were created after the fuselage halves left the "tub", were joined together and it was then the windows, doors, engine and it's mounting were decided upon. The same basic fuselage was also used for the parasol-wing "Air Express", the low-wing mounted "Explorer", the "Sirius", and so on; the deciding factor of what model would finally emerge from the factory door was the placement of the wing, the powerplant that was chosen, and the arrangement of the seating capacity. This versatility in the manufacture of the "wooden Lockheeds" was a very interesting feature of design and certainly a practical arrangement for fabrication of the numerous models; by the same token, the quite simple procedure in modification of one certain model into a model of another type, makes for some bewilderment when trying to pin-point the exact model designation of a certain aircraft. Let us say that a certain Vega with a certain registration number, started out in it's life as a Wright J5 powered Vega 1; it could have been easily modified by the installation of a Wright J6-9-300 engine into the Vega 2, or perhaps a "Wasp" engine was installed and it could have become a Vega 5; a Vega 5 could have been yet further modified into a model 5-B or 5-C. This is not to say that this sort of modification development was detrimental in any way, not in the least, but it is rather perplexing some 30 years later to the photo-collector or the aviation historian. It reminds one of the antics of a pretty lady that decides to change her hair-do and make-up, then comes up with an almost unrecognizable configuration, to all but the expert eye.

The Lockheed "Vega" listed for approval on A.T.C. # 169 is described as a 6 place version that was powered with a 450 h.p. Pratt & Whitney "Wasp" engine; as there are no known Vega model designation numbers to fit this description, we can only assume that an extra seat was added to the basic 5 passenger lay-out to come up with this particular 6 place version, or perhaps one seat was left out from the arrangement of the early basic 7 place version. Some Vegas were reported to have sort of a fold-up buddy seat installed to allow seating for 5 in the basic cabin arrangement which normally only had four seats, and this modification of seating may very well apply to this approval, but we cannot be sure. It is also fact that some Vegas were so highly modified for special purposes that they did not conform to any of the basic model designations, although they may have started out as a Vega 5, or whatever the case may be. In general, the most readily accepted tally for the amount of Lockheed "Vega" that were built is 130; of this number some 28 started out as the Vega 1, about 8 or so as the Vega 2 and several variants from this as the Vega 2-D, some 20 or 30 as the Vega 5, about 40 as the Vega 5-B, and so on; in narrowing the field down to the amount of probables that might come under A.T.C. # 169, we are left with only one Vega that would fit the provisions in this approval. This particular Vega was a 6 place craft that operated for Pan American-Grace Airways; all other registered Vegas were listed as 5 place or 7 place models. As there seem to be no other Vegas that fit the description of this approval, we can be led into assuming that this modification into a 6 place airplane, is a seating arrangement that could apply to any "Wasp" powered "Vega"; one that could be arranged to conform to these provisions as set down by this particular approval specification.

Listed below are specifications and performance data for the 6 place Lockheed "Vega" as powered with the 450 h.p. Pratt & Whitney "Wasp" engine; length overall 27'8"; hite overall 8'6"; wing span 41'0"; wing chord at root 102"; wing chord at tip 63"; M.A.C. 80"; total wing area 275 sq.ft.; airfoil at root Clark Y-18; airfoil at tip Clark Y-9.5; wt. empty 2465; useful load 1568; payload with 100 gal. fuel was 720 lbs.; gross wt. 4033 lbs.; max. speed 170 (185); cruising speed 140 (155); figures in brackets are for craft fitted with

N.A.C.A. type low-drag engine cowling; landing speed 55; climb 1350 ft. first min. at sea level; service ceiling 19,750 ft.; gas cap. 100 gal. max.; oil cap. 10 gal.; cruising range at 20 gal. per hour was 630 (690) miles; basic price at the factory field ranged from $20,000.

Construction details, typical for all "wooden Lockheeds", have been discussed quite thoroughly in the chapter for A.T.C. # 102 and # 140; the following details are also typical to all Lockheeds as built during this period. The tail-cone which faired off the aft end of the fuselage was a cone of stamped-out sheet metal; the striking "wheel pants" or wheel fairings were 6 feet long and added a noticeable increase to the top speed, but were later being discarded because of various problems during average operation. The interior cabin walls were lined in a heavily padded pig-skin, which gave the cabin interior an elegant sense of richness, with good wea-

ring qualities; several of the cabin windows had sliding panels for ventilation. The newly-developed N.A.C.A. type low-drag engine cowling was often mounted on the Vegas of this period, and it's use added up to 18 m.p.h. to the maximum speed; the use of "wheel pants" added another 6 m.p.h., therefore, many of the Vegas were approaching speeds of 200 m.p.h. The later famous Wiley Post, a flyer's flyer of top ability, was chief test-pilot for Lockheed at this time and remained at this job until June of 1930. The next development in the "Vega" series was the 7 place model 5-B, which will be discussed in the chapter for A.T.C. # 227.

Listed below is the only example of a "Wasp" powered Lockheed "Vega" that was registered as a 6 place airplane, but several others may have been modified to conform to this arrangement.

NC-9424; Vega 5 (# 78) Wasp 450.

Fig. 226. The Kreutzer "Air Coach" Model K-3 had three LeBlond 90 engines; this light trimotor seated six.

In a planned effort to sell the utility of air-travel by private-owned airplane, to the business man or the private owner, several aircraft manufacturers thought it wisest to stress safety and the reliability of operation above all else; an assurance of reliability would be most appealing to those interested in air travel but yet concerned with the chance of engine failure. By far the easiest way to insure the sense of safety and reliability in an airplane, was the multi-engined configuration; with multiple engines there was always present that sense of comfort and assurance to know that power for continued flight was still available, even in the case of failure to one powerplant. Past examples of this type of aircraft were the large "twins" and "tri-motors" which had been with us for some years, and now the trend was beginning to swing over to the light multi-engined craft; among the first of these in this new concept was the Kreutzer tri-motored "Air Coach".

The good-looking Kreutzer "Air Coach" was a "baby tri-motor" that had ample room and comfort for six, in a spacious and well appointed cabin, with a promise for an extra margin of safety and reliability provided by three engines. The model K-3 which will be discussed here, mounted three 7 cylinder Le-Blond engines of 90 h.p. each; this provided a total of 270 horsepower which would be sufficient for the job intended, a reserve of power that translated itself into good performance with ample power still available (180 h.p.) for continued flight in the case of failure to one engine. The Kreutzer tri-motor flew well with one engine dead during the carefully planned simulated tests, but how it would act under actual emergency conditions is not known. Well suited for feeder-line service over rough country, a Kreutzer model K-3 was put into service by the Apache Air Lines out of Globe, Arizona; an area of treacherous terrain where an extra margin of safety would surely be appreciated by all aboard.

The prototype of the Kreutzer "Air Coach" series, as shown here, was introduced in the latter part of 1928 as a four place aircraft that was powered with 3 Velie engines of 55 h.p. each; tests were quite satisfactory but a planned seating for six pointed to the necessity of increasing the available power. The next version was considerably improved in many ways and was powered with one Le-Blond 90 in the nose and two LeBlond 60 in the outer nacelles, for a total of 220 h.p.; this was the model K-2 as discussed in the next chapter. The model K-3 with three LeBlond 90 engines was almost a parallel development and proved to be somewhat more suitable. The

Fig. 227. The Kreutzer "Air Coach" prototype had three Velie engines; this one is being christened here by Racquel Torres with a bottle of champagne.

type certificate number for the model K-3 as powered with 3 LeBlond engines of 90 h.p. each, was issued in June of 1929 and at least three examples of this model were manufactured by the Joseph Kreutzer Corp. at their plant in Venice, Calif. Joseph Kreutzer was president and general manager; D. B. Shidler was V.P.; Otto Konstanzer was secretary-treasurer; A. F. White was sales manager; Albin K. Peterson was chief of design and engineering; and Henry H. Ogden of round-the-world fame was the chief pilot in charge of test and development.

Listed below are specifications and performance data for the Kreutzer "Air Coach" model K-3 as powered with three LeBlond 90 engines; length overall 33'6"; hite overall 9'6"; wing span 48'10"; wing chord 84"; total wing area 315 sq.ft.; airfoil "Goettingen 398"; wt. empty 2846; useful load 1654; payload with 85 gal. fuel was 900 lbs.; gross wt. 4500 lbs.; max. speed 120; cruising speed 103; landing speed 46; climb 750 ft. first min. at sea level; service ceiling 15,000 ft.; gas cap. 85 gal.; oil cap. 9 gal.; cruising range at 15 gal. per hour was 550 miles; price at the factory field was $17,950.

The fuselage framework was built up of welded chrome-moly steel tubing, faired to shape with duralumin formers and fairing strips, and fabric covered. The cabin was spacious and well appointed with provisions for heating and ventilation; a dead-air space in the cabin walls provided sound-proofing and insulation. The forward section of the cabin was fitted with shatter-proof glass, and dual controls were provided; the baggage compartment was to the rear of the cabin section, with a separate locker for the pilot. The wing framework, in two halves, was built up of box-type spar beams of laminated plywood webs and heavy spruce cap-strips, with spruce and plywood truss-type wing ribs; the completed framework was fabric covered. The fuel supply was carried in two gravity-feed tanks that were mounted in the root end of each wing half; an oil tank of three gallon capacity was mounted in each nacelle. A steel tube truss arrangement of the wing brace struts and the landing gear struts, provided a junction for the engine nacelles that were mounted under the wing; the landing gear of 162 inch tread was of the split-axle outrigger type and used "Aerol" shock absorbing struts. Wheels were 30x5 and Bendix brakes were standard equipment; a steerable tail-wheel was a great help in ground manuvering. The fabric covered tail-group was built up of welded chrome-moly steel tubing; the vertical fin was ground adjustable and the horizontal stabilizer was adjustable in flight. Navigation lights, metal propellers, and engine starters were optional equipment. Another development in the "Air Coach" series was the model K-2; see the chapter for A.T.C. # 171 in this volume.

Listed below are Kreutzer model K-3 entries that were gleaned from various records; this list may not be complete, but it does show the bulk of this model that were built

NC-9354; Kreutzer K-3 (# 104) 3 LeBlond 90.

NC-9494; Kreutzer K-3 (# 105) 3 LeBlond 90.

NC-211M; Kreutzer K-3 (# 109) 3 LeBlond 90.

Model K-3 serial # 104 was converted from a model K-2; identity of serial # 107 is unknown, it may have been a model K-3.

Fig. 228. The Kreutzer "Air Coach" K-2 had a LeBlond 90 in the nose and two LeBlond 60's in the wings; it offered luxury and dependability.

The Kreutzer model K-2 was a companion model to the model K-3, and was typical in most all respects except for the engine installations, which in this case was one 7 cylinder LeBlond 90 in the nose and two 5 cylinder LeBlond 60 in the wing nacelles for a total of 220 h.p. Planned as an economy model of the "baby tri-motor", the model K-2 was somewhat cheaper in first cost, and somewhat cheaper to operate but performance suffered by about 50 horsepower. The performance with 3 engines in operation, was still good and quite adequate for the service intended, but performance with one engine dead, be it nose-engine or wing-engine, would be quite marginal and would present some unexpected problems. More than likely it is this power percentage loss that convinced operators into modification of the lower-powered K-2 version into the higher-powered K-3 series, which stood a better chance in this respect with more available horsepower. A version was planned to mount 3 Warner "Scarab" engines, which would have provided a total of 330 h.p.; performance of this aircraft would have been outstanding by comparison, and the power left available, even with one engine gone, would have been more than adequate for normal operation to a haven of refuge. As far as is known, this Warner powered version was never built, but

a later version of the "Air Coach" mounted 3 Kinner engines as the model K-5.

The Kreutzer "Air Coach" series were designed and engineered by one Albin K. Peterson, a man of great talent who had an unmistakeable flair for proportion with lines of graceful flow; evidence of this surely shows up in the "Air Coach" design, and also in the configuration of the huge tri-engined "Albatross" which Peterson designed and engineered for the Zenith Aircraft Co. Albin K. Peterson had also been the chief engineer at Zenith Aircraft where he designed the trim-looking model Z-6 cabin biplanes; the similiarity of technique in lines and proportion is plainly evident. The "Albatross" design surely had influence in the Kreutzer "Air Coach" configuration, and is plainly evident in the "Emsco" series that were introduced a year or two later. It is remarkable how the inner personality of a man and his feeling for flow and proportion, can be recognized in his work like a trade-mark. Henry H. Ogden who was technical advisor and chief pilot in charge of test and development for the Kreutzer Corp., was also a man of great talent; probably well remembered as one of the Army's round-the-world flyers that circumnavigated the globe by airplane in 1924. An extremely able pilot with good engineering sense, Ogden supervised and conducted early tests on the

"Air Coach" series, but chose to leave the organization shortly after for plans of his own; plans which culminated into development of the tri-engined Ogden "Osprey".

The Kreutzer "Air Coach" model K-2 was a high wing cabin monoplane of the "baby tri-motor" type that was designed to offer a margin of safety and more reliability of operation, in an aircraft that would be suitable for the business man as a light executive-transport, as an air-liner for small loads on frequent schedule, or the private-owner who had an urge to travel by air in company with a number of friends. Designed with ample room and comfort for six, the cabin was richly and tastefully appointed with provisions for heating and ventilation. There were two large entrance doors at the rear of the cabin section, and a large baggage compartment was accessible from either inside or the outside; there was a separate locker for the pilot. The model K-2 delivered good performance for the power invested, with good handling and good flight characteristics; opinions would suggest that economy of operation in place of adequate power reserve, was on the wrong side of the ledger for this particular model. The Kreutzer "Air Coach" model K-2 as powered with one LeBlond 90 and two LeBlond 60, was issued a type certificate number in July of 1929 and at least 3 examples of this model were manufactured by the Joseph Kreutzer Corp. with executive offices in Los Angeles, Cal.; the plant site was in Venice, Calif. Corporation officers were listed in the previous chapter.

Listed below are specifications and performance data for the Kreutzer "Air Coach"

model K-2 as powered with one LeBlond 90 and two LeBlond 60 engines; length overall 33′6″; hite overall 9′6″; wing span 48′10″; wing chord 84″; total wing area 315 sq.ft.; airfoil "Goettingen 398"; wt. empty 2697; useful load 1748; payload with 85 gal. fuel was 1000 lbs.; gross wt. 4445 lbs.; max. speed 115; cruising speed 97; landing speed 45; climb 670 ft. first min. at sea level; service ceiling 13,000 ft.; gas cap. 85 gal.; oil cap. 9 gal.; cruising range at 13 gal. per hour was 575 miles; price at the factory field was $15,800; the following figures are for the model K-2 as fitted with electric engine starting equipment; wt. empty 2807; useful load 1693; payload with 85 gal. fuel was 948 lbs.; gross wt. 4500 lbs.; performance was affected very little by the additional weight.

The fuselage framework was built up of welded chrome-moly steel tubing, faired to an oval cross-section by duralumin former bulkheads and numerous duralumin fairing strips that carried out the smooth flows from nose to tail; the completed framework was fabric covered. The wing framework was built up of box-type spar beams of laminated plywood webs and heavy spruce cap-strips, with spruce and plywood truss-type wing ribs; the leading edge of the wing was covered with plywood back to the front spar to preserve the entering form of the airfoil, and the completed framework was fabric covered. Two long narrow-chord ailerons were of all-wood construction similiar to the wing; the fuel tanks were mounted in the root ends of each wing half, which was joined together and bolted to the top of the fuselage. All wing brace struts, engine nacelle struts, and landing gear struts

Fig. 229. The trimotor K-2 was designed to offer increased reliability of operation.

Fig. 230. The trimotor Kreutzer was designed by Albin K. Peterson; at right is Henry Ogden, chief test pilot.

were of heavy gauge chrome-moly steel tubing in round section; all struts were encased by balsa-wood fairings into a streamlined section. Wheels were 30x5 and Bendix brakes were standard equipment. Navigation lights, metal propellers, and electric engine starters were optional equipment. The next development in the Kreutzer "Air Coach" series was the Kinner powered model K-5, which will be discussed in the chapter for A.T.C. # 223.

Listed below are Kreutzer model K-2 entries that were gleaned from various records; this list represents the total number that were built.

X-71E; Kreutzer K-1 (# 101) 3 Velie.

NC-612; Kreutzer K-2 (# 102) LeBlond 90 & 60.

NC-9493; Kreutzer K-2 (# 103) LeBlond 90 & 60.

NC-9354; Kreutzer K-2 (#104) LeBlond 90 & 60.

NC-847H; Kreutzer K-2 (# 106) LeBlond 90 & 60.

Model K-1 serial # 101 was prototype; serial # 104 was later converted into a model K-3; identity of serial # 107 is unknown, it most likely was a model K-3.

Fig. 231. A Bach "Air Yacht" 3-CT-8 with a "Hornet" in the nose and two 5 cylinder Wright J6's in the wings.

The Bach tri-motored "Air Yacht" was making a great showing and provoked more than the average interest in aviation circles everywhere, especially on our western coast; it's all-round high performance and it's adaptability to constant rugged service was soon established and became a matter of discussion among the men who fly. Interest in the Bach "Air Yacht" was increasing steadily from many angles, and plans for it's manufacture were geared accordingly; by May of 1929, a continuous plant expansion and additions of factory tools and equipment brought about a potential of production capacity that was estimated to be at least five completed airplanes per month. A nation-wide dealer organization was soon under way and contracts were received for the total output at the plant for the year of 1929; this was indeed a rosy future to look forward to. Paralleling these developments, were plans for various new models to fit the calculated needs, and practically a constant improvement in each successive craft as it came off the line. Promotion of the "Air Yacht" was adequate and it's thundering appearance at various air-meets was always a source of interest to many. In a display of speed and remarkable manuverability, Waldo Waterman and Wm. "Billy" Brock came in 1-2 in the speed event for multi-motored craft over a 5 mile closed course of ten laps duration, during the Na-

tional Air Races for 1929 held at Cleveland, Ohio. The winner clocking an average of better than 136 m.p.h., and polishing the pylons with the agility of a sprightly sport-plane.

The latest development in the Bach "Air Yacht" series was the model 3-CT-8, which was typical to the earlier versions in a great extent, except for several minor improvements here and there and the addition of a bit more power. The model 3-CT-8 had ample room with elegant comfort for 8 passengers, plus pilot and co-pilot, and it was powered with one 525 h.p. Pratt & Whitney "Hornet" engine in the nose of the fuselage and two 165 h.p. Wright J6 (5 cyl.) engines in the wing nacelles; a combined total of 855 h.p. which translated into an earth-pawing performance. Pickwick Airways, an affiliate of the Pickwick Stage bus system, had several of the earlier 3-CT-6 version already in scheduled operation and soon added a few of the 3-CT-8 to their growing fleet and expanding service. Had it not been for the tragic business-crash of late 1929, and the crippling "depression" of early 1930, the rosy future of the Bach "Air Yacht" may have well blossomed into an operation of substantial proportion, but the misfortune had taken it's toll and the scars left behind were hard to heal. The type certificate number for the Bach "Air Yacht" model 3-CT-8 was issued in July of 1929 and at least 4 examples of this model

were manufactured by the Bach Aircraft Co.,
Inc. at the L. A. Metropolitan Airport in Van
Nuys, Calif. George J. Bury was the president for a time, followed by B. L. Graves;
L. Morton Bach was V.P. in charge of engineering; C. W. Faucett was secretary-treasurer; and Waldo D. Waterman was the chief
pilot in charge of test and development.

Listed below are specifications and performance data for the Bach "Air Yacht" model
3-CT-8 as powered with one 525 h.p. "Hornet" engine and two 165 h.p. Wright J6 engines; length overall 36'10"; hite overall 9'9";
wing span 58'5"; wing chord at root 132";
wing chord at tip 96"; total wing area 512
sq.ft.; airfoil "Clark Y"; wing tapered in
planform only; wt. empty 4785; useful load
3195; payload with 200 gal. fuel was 1685
lbs.; gross wt. 7980 lbs.; max. speed 157;
cruising speed 133; landing speed 60; climb
1180 ft. first min. at sea level; service ceiling
18,500 ft.; gas cap. 200 gal.; oil cap. 18 gal.;
cruising range at 42 gal. per hour was 590
miles; price at the factory field was $39,500.

The fuselage framework was built up of
six wooden longeron members that were bolted together with steel plates and fittings, the
framework was cleverly arranged so that the
cabin area was clear of braces, and there were
no obstructions in the doors or windows; the
completed framework was covered with plywood veneer and an outer covering of fabric
for surface smoothness and added strength.
The cabin was fitted with windows of shatter-
proof glass and arranged with 8 passenger

*Fig. 232. The cabin interior of the 3-CT-8 offered
speedy transport for eight.*

seats; the pilot and co-pilot were seated in a
separate section up front with access to the
main cabin, and the baggage compartment of
50 cu. ft. capacity was to the rear of the cabin
with access from the inside or the outside.
The semi-cantilever wing framework was
built up of spruce and plywood box-type
spar beams, with plywood wing ribs that were
routed for lightness and reinforced with

Fig. 233. The Bach 3-CT-8 featured high performance and rugged structure; the fuselage was all-wood.

Fig. 234. The Bach 3-CT-8; note rugged truss bracing.

spruce diagonals and cap-strips; the leading edges were covered with plywood and the completed framework was fabric covered. The ailerons were of a wood framework built around a box-type spar, with fabric covering; all movable controls were operated by double cables running through micarta pulleys. The wing was braced to the fuselage by two parallel struts of heavy chrome-moly tubes that were faired to an Eiffel 376 airfoil section, for stability and added lift; the engine nacelles were built into a truss framework with the undercarriage and the wing brace struts. The fuel tanks were mounted in the root ends of each wing half, and the fuel to the engines flowed through a selector valve in the pilot's cabin. The plywood covered tail-group was built up in a wood framework; the vertical fin was built integral to the fuselage aft section and the horizontal stabilizer was adjustable in flight. The split-axle landing gear of 18 foot tread was fastened from the fuselage to the engine nacelles and used rubber-hydraulic shock absorbing struts; a dual tail-wheel was

provided, main wheels were 36x8 and Bendix brakes were standard equipment. An interesting accessory was the "fenders" that were mounted over each wheel to deflect debris from the spinning wheels away from the whirling propellers. Metal propellers, inertia-type engine starters, and navigation lights were standard equipment. The next new development in the Bach "Air Yacht" series was the model 3-CT-9 which will be discussed in the chapter for A.T.C. # 271.

Listed below are Bach 3-CT-8 entries that were gleaned from various records; this list may not be quite complete, but it does show the bulk of this model that were built.

C-8069; 3-CT-8 (# 11) Hornet & 2 Wright 5.

NC-245K; 3-CT-8 (#   ) Hornet & 2 Wright 5.

NC-53M; 3-CT-8 (# 16) Hornet 2 Wright 5.

NC-54M; 3-CT-8 (# 17) Hornet & 2 Wright 5.

Fig. 235. A Ford Trimotor 6-AT-1 on floats; this hybrid craft led an interesting life in Canadian service.

The "Tri-Motor" model 6-AT-A was one of the non-standard versions in the Ford-Stout series; basically planned and calculated to handle the maximum capacity and payload with the utmost in operating economy. Being of almost special-purpose category, only three of this version were built, but they led a very active and rather interesting life. The first example of this model, test-flown in May of 1929, was purchased and delivered to the R.C.A.F. for government-sponsored tests and experiments in "forest dusting"; operating in the timber areas from the lakes and rivers of Canada as a float-mounted seaplane. Flying as low as 20 feet above the tree-tops, the 6-ATS carried up to 1400 lbs. of arsenate dust and could lay a swath of pest extermination some 250 feet wide. The experiments of "dusting timber" had proven quite successful, and were carried on with this same craft for several more years. The second example of this model was first built up as a 6-AT-A, and later in the year was modified into the rare 7-AT; as the model 7-AT it was flown by Myron Zeller to 3rd place in the National Air Tour for 1929, and flown to first place by Harry Russell in the National Air Tour for 1930. The third example of this model was delivered to the Colonial-Western Airways as a 6-AT-A where it served on their scheduled routes until May of 1930; this craft

was then modified into the more powerful 5-AT-C version and put back into operation by the Colonial Air Transport. The original 6-ATS as delivered in Canada, then remained as the only example of this model; sold by the R.C.A.F. in 1937, it operated in air transport service for several more years.

The tri-motored model 6-AT was actually a hybrid aircraft with the larger wing span and larger cabin interior of the 5-AT-C, with the more economical power installation of the model 4-AT-E. This resulted in a large capacity transport that could operate with good efficiency in the strictest economy. Powered with three Wright J6 engines of 300 h.p. each, the model 6-AT was a combination hampered by some loss in performance, but the loss in performance was a fair exchange for the increased payload and the economy gained. The first example of the 6-AT that was delivered to the R.C.A.F., was one of the few Ford "Tri-Motors" that had operated on wheels, skis, or floats, and the first large airplane to be used successfully as a "duster". The model 6-AT-A normally had seating for fifteen, and with the cabin seats removed had a volume capacity of over 500 cubic feet to carry better than 2800 lbs. of payload. Performance of the model 6-AT-A was more or less comparable to the J5 powered 4-AT-B, but was quite adequate for the job intended. The type

*Fig. 236. The 6-ATS afloat on one of Canada's numerous bodies of water, which made easily accessible
landing areas for ease of operation.*

certificate number for the "Tri-Motor" model 6-AT-A as powered with 3 Wright J6
engines of 300 h.p. each, was issued in July
of 1929 and only 3 examples of this model
were manufactured by the Stout Metal Airplane Co., a division of the Ford Motor Co.
at Dearborn, Michigan.

Listed below are specifications and performance data for the Ford "Tri-Motor" model
6-AT as powered with 3 Wright J6 engines of
300 h.p. each; length overall 50'6"; hite overall 14'1"; wing span 77'10"; wing chord at
root 156"; wing chord at tip 89"; total wing
area 835 sq.ft.; airfoil "Goettingen 386"; the
following figures are for the model 6-ATS
seaplane; wt. empty 8250; useful load 4250;
payload with 231 gal. fuel was 2344 lbs.; gross
wt. 12,500 lbs.; max. speed 120; cruising
speed 100; landing speed 62; climb 650 ft.
first min. at sea level; climb in 10 min. was
5,000 ft.; service ceiling 11,000 ft.; gas cap.
231 gal.; oil cap. 24 gal.; cruising range at 45
gal. per hour was 475 miles; price complete
with twin-float gear was $92,650; the follo-
wing figures are for the 6-AT-A landplane
with 231 gal. fuel; wt. empty 7009; useful
load 4721; payload 2800 lbs.; gross wt. 11,730
lbs.; max. speed 130; cruising speed 110; landing speed 60; climb 800 ft. first min. at sea
level; service ceiling 13,000 ft.; cruising range
550 miles; the following figures are for the
6-AT-A landplane with 277 gal. fuel; wt.
empty 7048; useful load 5096; payload 2900
lbs.; gross wt. 12,144 lbs.; max. speed 130;
cruising speed 108; landing speed 62; climb

720 ft. first min. at sea level; climb in 10 min.
was 5500 ft.; service ceiling 12,500 ft.; price
at the factory was close to $50,000.

The fuselage framework was built up of
riveted duralumin channels into bulkheads
and stringers that were covered with corrugated "Alclad" skin; the seats were removable to provide up to 529 cubic feet of cargo
capacity, plus a baggage compartment capacity of 30 cubic feet. The huge cantilever wing
framework was built up with spar beams of
riveted "dural" channel and open-section
stock, and covered with corrugated "Alclad"
skin; the fuel tanks were mounted in the wing.
The tail-group was also built up with riveted
duralumin channel and open-section stock,
and covered with corrugated "Alclad" skin;
the vertical fin was ground adjustable and the
horizontal stabilizer was adjustable in flight.
The split-axle landing gear of 172 inch tread
had telescopic legs that were fastened to each
engine nacelle, and shock absorbers were
rubber donut-rings in compression; wheels
were 36x8 and a steerable tail-wheel and individual wheel brakes were standard equipment. Wiring for navigation lights, metal propellers, and inertia-type engine starters were
also standard equipment. The next development in the Ford "Tri-Motor" series was the
rare model 7-AT, which will be discussed in
the chapter for A.T.C. # 246.

Listed below are 6-AT entries that represent the total production of this model.

G-CYWZ; 6-ATS (# 6-ATS-1) 3 Wright
J6-9-300.

Fig. 237. The original Ford "Flivver" with 3 cylinder Anzani engine, at one time planned for quantity production. The little craft offers a great contrast to the familiar trimotor Ford transport.

-8485; 6-AT-A (# 6-AT-2) 3 Wright J6-9-300.
NC-8486; 6-AT-A (# 6-AT-3) 3 Wright J6-9-300.
Serial # 6-ATS-1 was originally approved as 2 place seaplane on a Group 2 approval numbered 2-80 issued 6-14-29, and reissued 11-8-29 as a 12 place seaplane; serial # 6-AT-2 was modified to a 7-AT on A.T.C. # 246; serial # 6-AT-3 was modified to a 5-AT-C Special. Complete histories of the Ford 6-AT, and all other models as well, can be found in the "FORD STORY" by Wm. T. Larkins; it is highly recommended for a study of this series.

Fig. 238. The Barling NB-3 with LeBlond 60 engine, manufactured by Nicholas-Beazley.

The Nicholas-Beazley "Barling NB-3" was an unusual little light monoplane that was born of many revolutionary ideas; the fact that it could carry 3 people with good performance on 60 horsepower was well worthy of some applause in itself. The unusual NB-3 incorporated many new principles in it's construction, and it's overall design was a refreshing approach to the never-ending problem of marketing a useful light airplane for personal use, one that would be both economical and efficient in it's service. That the NB-3 delivered a performance far beyond the average with this amount of horsepower, is very well proven by it's record. During the 1929 National Air Races held at Cleveland, "Barney" Zimmerley flew an NB-3 to first place in the efficiency race for light airplanes; earlier in the year he climbed an NB-3 to 20,862 feet to set an altitude record for light airplanes, later boosting this record to 25,100 feet. On a flight from Brownsville, Texas to Winnipeg in Canada, Zimmerley established a non-stop distance record for light airplanes by flying 1650 miles. The 1650 mile trip, at a cost of about $25, was accomplished in 16 hours of flight which calculates to an average speed of about 103 m.p.h.; a feat well deserving and well worthy of praise for a craft of this type. Designed with a sturdy construction

that would render care-free service for many years, with an economy of operation that would be hard to beat per seat-mile, the Barling NB-3 posed as an attractive buy for the private-owner or as a general-purpose airplane for the small fixed-base operator.

Walter H. Barling who designed the "Barling NB-3", was one of our foremost aeronautical engineers who pioneered many interesting concepts in the art of aircraft design and construction. He will probably be best remembered for his design of the "Barling Bomber", a huge and ugly brute that was designed for the Air Service. Though designed well, to specifications set forth by the Army, it was sometimes called the greatest aerial monument to misplaced ingenuity and misspent money; a big and somewhat ugly brute that was adjudged just about useless as a practical war-machine. The well-known "Lawson Air-Liner" of 1925 was another design by Walter Barling that had great potential, but came out at the wrong time. Barling joined Nicholas-Beazley in May of 1927 with plans to develop the NB-3; development of this revolutionary craft was necessarily rather slow because of the many new concepts in it's construction that had to be fabricated and proven, but a prototype was finally introduced in the latter part of 1928. First powered in tests with a 60 h.p. Anzani engine, it fulfilled the expectations as promised and bore unmistakeable signs of a great future; production was soon scheduled with the 5 cylinder LeBlond 60.

The Nicholas-Beazley Airplane Co. was formed by Russell Nicholas and Howard Beazley back in 1921 to handle sales of supplies and accessories for aircraft; in just a few years they became one of the world's largest aviation supply-houses where one could get everything from a cotter-pin or a pair of goggles, to a brand-new OX-5 engine or a complete and ready to fly airplane. Selling war-surplus airplanes to aspiring aviators or former war-heroes, became a big business in this country right after the first world war and Nicholas-Beazley sold hundreds of the "Standard" biplanes that were offered ready to fly for about $750. They also developed and marketed an "NB-Standard" that was modified a good deal into a cleaner airplane that delivered a better performance, for about $850. Marketing of the Barling NB-3 began in 1929 and it started off rather slow because it was an entirely new concept, and it took some time to surmount the barriers of long-established fallacies, and it took some time in convincing the flying and buying public into accepting this new form.

The Barling NB-3 was a 3 place open cockpit monoplane of the low-winged type, with an internally braced cantilever wing; an all-metal wing of thick section that was strong but yet light, with an efficient airfoil section of Barling design. The main section of the wing panel was straight without any dihedral angle, but the outer tip-panels of 5½ foot length were swept-up at a 5 degree dihedral angle which provided excellent lateral stability. Powered with the 5 cylinder LeBlond 60

engine of 65 h.p. maximum, the NB-3 delivered an outstanding performance in this combination, and it's flight characteristics were best described as obedient and gentle. Being of good aerodynamic proportion and inherently stable, it was gentle in the stall and described as spin-proof, but it would have been much better to say that it would "spin", but with almost instant recovery. Short-field performance of the NB-3 with gross load was excellent and one could say that it was a good example of the ideal airplane for the average private-owner. The type certificate number for the Barling NB-3 as powered with the 65 h.p. LeBlond engine, was issued in July of 1929 and some 50 or more examples of this model were manufactured by the Nicholas-Beazley Airplane Co. at Marshall, Missouri. Russell Nicholas was the president; Chas. M. Buckner was V.P.; Howard Beazley was secretary-treasurer; Hayes Walter was sales manager; Walter H. Barling was chief of design and engineering; and Claude Sterling "Barney" Zimmerley was the chief pilot in charge of test and development.

Listed below are specifications and performance data for the Barling NB-3 as powered with the 65 h.p. LeBlond engine; length overall 21'11"; hite overall 6'10"; wing span 32'9"; wing chord 62"; total wing area 159 sq.ft.; airfoil "Barling 90-A"; wt. empty 744; useful load 629; payload with 18 gal. fuel was 340 lbs.; gross wt. 1373 lbs.; max. speed 100; cruising speed 85; landing speed 37; climb 600 ft. first min. at sea level; climb to 12,500 ft. was 45 min.; service ceiling 12,500 ft.; gas cap. 18 gal.; oil cap. 1.5 gal.; cruising range at 4.5 gal. per hour was 300 miles; price at the factory field was $3600.

The fuselage framework was built up of welded chrome-moly steel tubing and stream-

Fig. 239.   The record-breaking NB-3 flew 1600 miles for $25.00; note extra fuel tanks in the forward fuselage.

Fig. 240. *The Barling NB-3 carried three on 60 h.p.; the craft incorporated many outstanding design advancements.*

lined to shape with aluminum alloy fairing structures that faired the fuselage into the wing attach junction, and also formed the cockpit cowling and the turtle-back section; the completed framework was fabric covered. The fuel tank was mounted high in the fuselage just ahead of the front cockpit, and was actually shaped to form the cockpit cowling as it faired back from the engine. The all-metal wing framework was built up of box-type spar beams that were formed from stamped-out U sections of aluminum alloy sheet that were riveted together into beams; the wing ribs were riveted together of aluminum alloy sheet stampings. The leading edge of the wing was covered with aluminum alloy sheet to preserve the airfoil form and the completed framework was fabric covered. The Barling wing structure was an innovation at this particular time and weighed only 1.06 lbs. per sq. ft., which was about half of the weight for other metal-wing designs of comparable strength. Another useful feature was the ability to procure prefabricated sections of the wing to replace damaged areas. The fabric covered tail-group was built up of welded chrome-moly steel tubing; the vertical fin and the horizontal stabilizer were ground adjustable for engine torque and variations in weight. The split-axle landing gear of 57 inch tread was built up of welded chrome-moly steel tubing in streamlined section and used rubber shock-cord to absorb the bumps; wheels were 24x3 or 24x4 and no brakes were provided. The tail-skid was of the leaf-spring

type, with a removable hardened shoe. The next development in the Barling NB-3 series was the Velie powered NB-3V which will be discussed in the chapter for A.T.C. # 230.

Listed below are Barling NB-3 entries that were gleaned from various records; a somplete listing would be rather extensive, so we submit a partial listing of 20 aircraft or so.

-355; NB-3 (# 1) Anzani 60.
-356; NB-3 (# 2) Anzani 60.
-357; NB-3 (# 3) Anzani 60.
-358; NB-3 (# 4) Superior Radial.
-359; NB-3 (# 5) Superior Radial.
NR-546; NB-3 (# 6) LeBlond 60.
-547; NB-3 (# 7) LeBlond 60.
NR-548; NB-3 (# 8) LeBlond 60.
; NB-3 (# 9) LeBlond 60.
-2833; NB-3 (# 10) LeBlond 60.
; NB-3 (# 11)
-2834; NB-3 (# 12) LeBlond 60.
NC-9311; NB-3 (# 13) LeBlond 60.
-9312; NB-3 (# 14) LeBlond 60.
-9313; NB-3 (# 15) LeBlond 60.
-9314; NB-3 (# 16) LeBlond 60.
-9315; NB-3 (# 17) LeBlond 60.
C-9316; NB-3 (# 18) LeBlond 60.
NC-9317; NB-3 (# 19) LeBlond 60.
NR-9318; NB-3 (# 20) LeBlond 60.
C-9319; NB-3 (# 21) LeBlond 60.
C-9320; NB-3 (# 22) LeBlond 60.

Serial # 1, # 2, and # 3 later powered with LeBlond 60; serial # 1 had Warner Jr. as NR-355; serial # 6 later had Genet 80; serial # 16 later had Pobjoy engine; serial # 18 later as 2 pl. craft on Group 2-157.

Fig. 241. The Travel Air Model SA-6000-A was a twin-float version of the A-6000-A.

In answer to the pleas received for a high performance work-horse transport, a transport that could operate efficiently and profitably under the exacting conditions prevailing in the bush-country of Canada and our Pacific-northwest, Travel Air was obliged to develop the model SA-6000-A. The model SA-6000-A was more or less an A-6000-A land-operated monoplane that was mounted on Edo JF-1 twin-float seaplane gear, and the seats were made easily removable for hauling cargo; there was about 8 feet of clear floor space length for bulky loads in a total capacity for about 90 cubic feet. The model A-6000-A as described in a previous chapter, was a bear-cat for performance and the SA-6000-A was hampered only very little by the bulky float installation; the agility of this large craft, whether in the air or on the water, was quite surprising to say the least. An 800 pound payload was carried easily, and with the performance possible, this craft was indeed a sound investment for "work in the bush".

The Travel Air model SA-6000-A was a large high wing cabin monoplane with ample room and comfort for six, with quickly removable seats to provide over 20 sq. ft. of clear floor space for bulky or odd-shaped cargo loads. There was a large door to the rear of the cabin section for loading the cargo, and there was a door up forward to allow entry and exit for the pilot. The cabin walls were sound-proofed and heavily insulated to ward off extremes in temperature, and cabin heat was provided from the engine's exhaust-manifold; these features were designed for and greatly appreciated by the pilots on cold wintry days. Though demand for a versatile craft of this type was strong, the demand was never too great, so as a consequence only a small number of this model were built. The type certificate number for the Travel Air monoplane model SA-6000-A as powered with the Pratt & Whitney "Wasp" engine of 450 h.p. and mounted on Edo floats, was issued in July of 1929 and at least 2 or more examples of this model were manufactured by the Travel Air Company at Wichita, Kansas.

Listed below are specifications and perfor-

Fig. 242. Performance of the SA-6000-A was ideal for bush country service; one is shown on a Minnesota lake.

mance data for the Travel Air monoplane model SA-6000-A as powered with the 450 h.p. "Wasp" engine; length overall (including floats) 33'9"; hite overall (in water) 10'6"; wing span 54'5"; wing chord 84"; total wing area 340 sq.ft.; airfoil "Clark Y-15"; wt. empty (wt. empty includes 91 lb. of equipment) 3676; useful load 1824; payload with 130 gal. fuel was 800 lbs.; gross wt. 5500 lbs.; max. speed 130; cruising speed 108; landing speed 65; stall speed 65; climb 800 ft. first min. at sea level; service ceiling 16,000 ft.; gas cap. 130 gal.; oil cap. 9 gal.; cruising range at 25 gal. per hour was 540 miles; price at the factory field was around $20,000.

The fuselage framework was built up of welded chrome-moly steel tubing, faired to shape with wood formers and fairing strips, then fabric covered. A 28 cubic foot baggage compartment was to the rear of the cabin section, with provisions for stowing seaplane gear. The semi-cantilever wing framework was built up of spruce and plywood box-type spar beams, with spruce and plywood Warren-truss wing ribs; the completed framework was fabric covered. The gravity-feed fuel tanks were mounted in the wing with one flanking each side of the fuselage; fuel-level gauges were visible in the cabin. The wing bracing struts were heavy gauge chrome-moly steel tubes that were encased in streamlined balsa-wood fairings; ailerons were of the offset-hinge Friese type. The fabric covered tail-group was built up of welded chrome-moly steel tubing and sheet steel formers; the fin was ground adjustable and the horizontal stabilizer was adjustable in flight. The undercarriage for the floats was built up of welded chrome-moly steel tubing in streamlined section. A metal propeller and inertia-type engine starter were provided as standard equipment. The next Travel Air development was the model E-4000 biplane, see the chapter for A.T.C. # 188 in this volume.

Listed below are the only two SA-6000-A entries that could be found in available records.

NC-8111; SA-6000-A (# 883) Wasp 450.
CF-AFK; SA-6000-A (#   ) Wasp 450.

Fig. 243. The Laird "Speedwing" LC-R300 with a 300 h.p. Wright J6 engine; the craft was custom-built with great care.

The Laird model LC-R300 was a progressive development of the "Speedwing" series, a version that was typical to the earlier Wright J5 powered LC-R200 in most every respect, except for the change in engine installations which in this case was the 9 cylinder Wright J6 of 300 h.p. Basically, the LC-R300 as it first came out, was only a modification of existing LC-R "Speedwing" examples; in fact, two of the LC-R200 craft were converted to the LC-R300 type, merely by the change of engines and other slight modifications necessary for this combination. The development of the "Speedwing" series actually stems back to the two "Laird-Specials" that were flown in contest by Chas. "Speed" Holman and E. E. Ballough since 1927; entered in every national competitive event for the past 3 seasons, these two craft were continuously being modified and speeded-up to defy the efforts of competitors, and the record hung up by these two sensational airplanes is quite remarkable. Desires voiced from several sportsman-pilots on the possibility of owning and flying comparable aircraft, led to the development of the Laird "Speed-

wing" series; the "Speedwing" was first introduced as the Wright J5 powered LC-R200, and now it was being offered as the improved model LC-R300.

The Laird "Speedwing" model LC-R300 was a 3 place open cockpit biplane of high performance that was leveled at the sportsman-pilot, a breed of man who would appreciate and enjoy an airplane that had the heart of a thoroughbred, with a performance and muscle to match. The "Speedwing" biplane normally carried three, but the front cockpit could be covered off with a metal panel when not in use, to gain several miles per hour in available cruising speed. With the ability to carry a fair-sized payload and the fuel capacity for a high cruising range, the LC-R300 was ideal for long cross-country jaunts and was usually fitted with the very best in navigational and flight instruments. Being an airplane of almost special-purpose category, the "Speedwing" biplane was custom-built to customer's order and consequently, only a small number were built. Those few that were built, were a shining beacon for Laird performance and quality; with E. M. Laird, quality

Fig. 244.   A later version of the LC-R300 as modified for air show work.

and performance had become a well-earned tradition. The model LC-R300 "Speedwing" biplane remained in development into the "thirties" and the latest versions were considerably improved with N.A.C.A. type "low drag" engine cowlings and much better streamlining throughout as shown; no doubt borrowing from valuable experience gained with the famous Laird "Solution" and "Super Solution" racing biplanes. The type certificate number for the "Speedwing" model LC-R300 as powered with the 300 h.p. Wright J6, was issued in July of 1929 and some 6 or more examples of this model were manufactured by the E. M. Laird Airplane Co. at Chicago, Illinois. The Laird airplane factory was just a short way from the airfield, and manufactured assemblies were trucked to the field for final assembly and flight test. Emil Matthew Laird was the president; Lee Hammond was V.P. and sales manager; with these engineers presiding over the engineering department at different times in the following order; Raoul J. Hoffman, Charles Arens, and R. L. Heinrich.

Listed below are specifications and performance data for the Laird "Speedwing" model LC-R300 as powered with the 300 h.p. Wright J6 engine; length overall 22'7"; hite overall 9'0"; wing span upper 28'0"; wing span lower 24'0"; wing chord upper 60"; wing chord lower 48"; wing area upper 121 sq.ft.; wing area lower 81 sq.ft.; total wing area 202 sq. ft.; airfoil "Laird"; wt. empty 1922; useful load 1088; payload with 76 gal. fuel was 390 lbs.; gross wt. 3010 lbs.; max. speed 155; cruising speed 128; landing speed 60; climb 1200 ft. first min. at sea level; service ceiling 17,000 ft.; gas cap. 76 gal.; oil cap. 10 gal.; cruising range at 15 gal. per hour was 575 miles; price at the factory field was $11,500. The performance figures for the later version with N.A.C.A. low-drag cowling were comparable except for the following; max. speed 175; cruising speed 140; landing speed 65.

The fuselage framework was built up of duralumin tubing with special steel clamps to make the joints, and braced in an X truss with heavy steel tie-rods; the framework was heavily faired to a deep streamlined shape and fabric covered. The later model with the N.A.C.A. cowling was covered in metal panels to the rear of the pilot's cockpit. The large fuel tank was mounted high in the fuselage just ahead of the front cockpit, with a direct-reading fuel gauge projecting through the cowling; a baggage compartment of some 50 lb. capacity was in the turtle-back section of the fuselage, in back of the rear cockpit. On the later models, another compartment was added just below. The wing framework was built up of heavy-sectioned laminated spruce spar beams, with spruce and plywood built-up wing ribs that were closely spaced together for added strength and to preserve the airfoil form across the span; the leading edges were covered and the completed framework was fabric covered. There were four ailerons

Fig. 245. A modified version of the LC-R300; craft shown originally was a Wright J5-powered Model LC-R200.

that were connected together in pairs by a streamlined push-pull strut; the distinctive looking "I shaped" wing struts were of a laminated plywood sandwich that was shaved and shaped to a streamlined form. The fabric covered tail-group was built up of welded chrome-moly steel tubing, except for the horizontal stabilizer which was built up of wood spars and ribs on the earliest models; the fin was ground adjustable and the horizontal stabilizer was adjustable in flight. The split-axle landing gear of 84 inch tread was of the normal type as used by Laird, and used two spools of rubber shock-cord to absorb the bumps; the later models used oleo-spring shock absorbing struts. The tail-skid was of the spring-leaf type; wheels were 30x5 and Bendix brakes were standard equipment. Custom colors were available and the quality of the finish on the "Speedwing" was about the finest to be seen anywhere. A metal propeller, navigation lights, and inertia-type engine starter were standard equipment. The next development in the Laird "Commercial" was the 300 h.p. LC–B300 which will be discussed in the chapter for A.T.C. # 353, and the next development in the Laird "Speedwing" was the "Wasp Jr." powered LC-RW300 which will be discussed in the chapter for A.T.C. # 377.

Listed below are Laird LC-R300 entries that were gleaned from various records; this list is not complete, but it does show the bulk of this model that were built.

NC-634; LC-R300 (# 177) J6-9-300.
NC-633; LC-R300 (# 178) J6-9-300.
NC-9419; LC-R300 (# 179) J6-9-300.
NC-734K; LC-R300 (# 181) J6-9-300.
NC-56K; LC-R300 (# 182) J6-9-300.
NC-14803; LC-R300 (#   ) J6-9-300.

Fig. 246. A version of the LC-R300 as built in the early 1930's; design improvements added speed.

Fig. 247. *A Cunningham-Hall PT-6 with a 300 h.p. Wright J6 engine; shown is the prototype.*

The James Cunningham & Sons, Co. founded in 1838, was long famous for high quality horse-drawn carriages and coaches, and also famous for fine custom-made automobiles. Upon study of the trend, they decided it was perfectly logical and quite reasonable for them to branch out into the manufacture of the latest mode of transportation, the airplane. Formed as a subsidiary unit, with operating space in the factory building of the James Cunningham & Sons, Co., the Cunningham-Hall Aircraft Corp. launched an enthusiastic program to manufacture the Cunningham-Hall PT-6. The PT-6 (also CHPT-6) designed as a personal transport, was a fairly large and rather buxom cabin biplane of handsome and generous proportions, with comfortable seating for six; though labelled as an "all-metal airplane", it was not in the true sense. The PT-6 was of all-metal framework but fabric covering was used for most of the airframe; construction data at the end of this chapter will explain this in more detail. Powered with the new 9 cyl. Wright "Whirlwind" J6 series engine of 300 h.p., this craft had a sprightly performance with an efficient utility that offered payload, comfort, and spaciousness. The passenger's compartment in the main cabin section seated four, and was quite large to avoid cramped quar-

ters; entrance and exit was gained through a large door at the rear. The pilot's compartment up forward, was separate from the main cabin section but access was possible through a bulkhead door; this forward compartment, raised slightly above the main cabin section, seated the pilot and one passenger with separate entry or exit possible by a side-panel door on each side. Visibility in the main cabin was quite good and the pilot also, in his elevated position, had exceptional visibility in most all directions.

The Cunningham-Hall PT-6 (CHPT-6) was designed and developed by Randolph F. Hall, who had been active as an aircraft designer, engineer, and stress analyst with various aircraft companies since 1915; his broad experience and practical good sense was well reflected in the handsome lines of the PT-6. The first PT-6, as shown here, was introduced in the early part of 1929; it was X-461E with serial # 2961. The type certificate number for the Cunningham-Hall PT-6 as powered with the Wright J6-9-300 was issued in July of 1929 as a dual-controlled airplane, and re-issued in September of 1929 with single controls. The PT-6 (Personal Transport-6place) was manufactured by the Cunningham-Hall Aircraft Corp. at Rochester, New York and about 6 of this type were built in all. F. E. Cunningham was the president; Randolph F. Hall was V.P. and chief engineer; Paul D. Wilson was chief pilot, and Wm. T. Thomas, one of the pioneers in American aviation, was consulting engineer.

Though very well designed with a bag full of attractive features, the PT-6 got off to a rather bad start; a second airplane (NC-692 W, serial # 2962) was not built until 1930. The Guggenheim Safe Airplane Contest of 1929, prompted entry in the tests with the special-built Cunningham-Hall model X which was an unusual "inverted sesqui-plane" with the largest wing on the bottom. This 2 place craft was powered with a Walter (Czechoslovakian) engine and though performance was excellent, it did not survive the preliminary judging. A model PT-6F was built as a freighter in the latter "thirties", and though only 6 of this type were built in all, there is still at least one that has been serving usefully as a hard-working "bush-plane" in our northwest country.

Listed below are specifications and performance data for the Wright J6 powered Cunningham-Hall model PT-6; length overall 29'8"; hite overall 9'7"; span upper 41'8"; span lower 33'8"; chord upper 78"; chord lower 54"; wing area upper 239 sq.ft.; wing area lower 131 sq.ft.; total wing area 370 sq.ft.; airfoil "Clark Y"; wt. empty 2670; useful load 1680; payload 910; gross wt. 4350 lbs.; max. speed 136; cruising speed 115; landing speed 45; climb 900 ft. first min. at sea level; service ceiling 17,500 ft.; gas cap. 90 gal.; oil cap. 6 gal.; normal range at 15 gal. per hour

was 6 hours or 690 miles; price at the factory was $13,900. The fuselage framework was built up of welded chrome-moly steel tubing, and quite unusual is the fact that the longeron tubes were squared at all of the joint intersections to eliminate the curved bevel ends that would be necessary with the joining of round tubes; the structure was then faired to shape and the cabin portion was covered with corrugated "dural" sheet. The balance of the fuselage aft, was fabric covered. The cabin walls were sound-proofed and insulated against temperature changes, noise, and vibration, with thick blankets of Balsam-wool; a large baggage compartment was located to the rear of the main cabin section. The upper wing framework was built up of two spar-beams that were fabricated of chrome-moly steel tubing in girder form, bolted together; the spars in the lower wing panels were single, large diameter, chrome-moly steel tubes. The wing ribs for both panels were built up of riveted duralumin tubes in truss-type form and the leading edges of all panels were covered with dural sheet to preserve airfoil form; all completed panels were then fabric covered. Two fuel tanks of 45 gallon capacity each, were placed in the root end of each upper wing panel. Ailerons of construction similiar to the wings, were of the Freise offset hinge type and were in the upper wing panels only. The fabric covered tail-group was built up of welded steel tubing frame members, with sheet steel formers and ribs; the fin was ground adjustable and the horizontal stabilizer was adjustable in flight. The rudder was aerodyna-

Fig. 248. The PT-6 had a rugged metal structure; its large capacity was ideal for air freight service.

Fig. 249. The PT-6F was designed to haul freight in the bush country.

mically balanced, but it was somewhat unusual in that it used the offset-hinge type that was similiar to that used on the ailerons; this type of control surface balance was used quite frequently, later on, but was still quite rare at this time. The split-axle landing gear was of the outrigger type and was fastened at one point to the spar of the lower wing; shock absorbers were "Aerol" struts, wheels were 32x6, and Bendix brakes were standard equipment. A normal type, shock-cord sprung tail-skid was used, but a tail-wheel was available as optional equipment. A metal propeller, inertia-type engine starter, exhaust collector-ring, and wings wired for lights were also standard equipment.

The last version of this model, also built under this A.T.C. number, was the PT-6F of the latter "thirties", that was powered with a 9 cylinder Wright engine of 330-365 h.p.; this version was specially arranged for hauling air-freight but the cabin could be easily arranged to carry four passengers. The configuration and construction of the model PT-6F were typical except for the low-drag cowling that was mounted over the engine; large low-pressure tires were used on the main gear and a swiveling tail-wheel was used in place of the tail-skid. Cabin interior of 156 cubic foot capacity was lined with corrugated aluminum alloy sheets, and had a large opening in the ceiling for lowering freight into the cabin compartment. A large 56 inch door in the side of the fuselage, allowed entry for bulky articles, Listed below are the changes in weights and performance of the PT-6F (Freighter) to compare with the earlier PT-6; dimensions were more of less the same. Weight empty 2875; useful load 1675; payload 915; gross weight 4550 lbs.; max. speed 150; cruising speed 130; landing speed 55; climb 1100 ft. first min. at sea level; usable ceiling 20,000 feet.

Fig. 250. The Alliance "Argo" with a Hess "Warrior" engine; the performance of this ship was exceptional.

The local airfield was all a-buzz and everyone was watching the skies; someone had heard that an "Argo" demonstrator was coming in. Before long, two specks above the far horizon materialized into full-sized airplanes, and we were treated to a demonstration of jubilant flying that was a joy to behold. Flashing across our 'field, the two gyrated into low-level loops, snap-rolls, and hammerhead-stalls; we were given just about the full treatment, and then they wound up the shebang with several hair-raising "buzz-jobs". To say that everyone was thoroughly impressed with these two airplanes and their devil-may-care display, would be putting it very mildly. One may say that flying such as this borders on the foolhardy, but that's the kind of airplane the "Argo" was; not foolhardy mind you, but spirited, sure-footed, and extremely capable, transmitting this to the pilot which made the average fellow feel extra-special when flying in one of these delightful craft. A sensitive and eager airplane that was rugged enough to withstand the abuse of the amateur, as well as the high stresses imposed upon it by the veteran pilot who would often vent his exuberance in the form of "stunts".

Pictured here, we can see that the "Argo" was of rather conventional configuration, and was put together with a nice balance of trim and functional lines. It was a 2 place open cockpit sport-trainer biplane seating two in tandem, and it was powered with the new 7 cylinder Hess "Warrior" engine. The Hess "Warrior" of 115-125 h.p. soon proved itself to be a mighty fine powerplant; a number of manufacturers had slated the "Warrior" for some of their new models, but the entire output, barring a few, was used for Alliance built aircraft. The type certificate number for the Alliance "Argo" was issued in July of 1929 and it was manufactured by the Alliance Aircraft Corp. in Alliance, Ohio. William E. Trump was the president; Aubrey W. Hess was V.P., general manager, and chief of design; Adrian T. Hess (brother) was in charge of production; Gordon T. Waite was chief engineer for aircraft development; John E. Everett was chief engineer for engine development; F. A. Giles and Edward Leedy were test pilots.

Long interested in aviation, the Hess brothers, Aubrey and Adrian, organized the Hess Aircraft Co. in Wyandotte, Michigan during the early part of 1926 to manufacture the Hess "Bluebird"; the good-looking "Bluebird" was a 3 place open cockpit biplane that was offered with the Curtiss OX-5 engine,

Fig. 251. The Alliance "Argo" prototype at the 1929 Detroit Air Show.

but variants of this model were offered with the "Hisso" A and E (Hispano-Suiza) engines and also with the Wright "Whirlwind". Only a small number of the "Bluebird" type were built, and the company ceased manufacture sometime in 1927. Reorganizing later as the Alliance Aircraft Corp. of Alliance, O., a factory building was under construction during April of 1928. Their first airplane (X-6876, serial # A-1) was introduced about mid-1928 and was first called the "Bluebird Sport"; shortly after, it was designated officially as the "Argo". The prototype "Argo" was shown at the Detroit Air Show for 1929, and soon after, additional aircraft were coming off the production line. The "Argo" made a big hit and was well received in flying-circles all over the country, but the oncoming "depression" and the shaky period afterwards, was hard-felt at Alliance Aircraft; in all, some 20 of the "Argo" sport-trainer were built and of this number possibly 2 or 3 are still in existance. It is a down-right pity that fate and the circumstances didn't allow this fine airplane to make a better showing in the scheme of things; a showing of which it was surely capable. Some years later, the "Alliance" plant was revived again for aircraft manufacture; it was put to use by the honor-able C. G. Taylor for building the little "Taylorcraft" monoplane.

Listed below are specifications and performance data for the Alliance "Argo" as powered with the Hess "Warrior" engine; length overall 20'4"; hite overall 7'9"; span upper 28'8"; span lower 26'; chord both 48"; wing area upper 110.5 sq.-ft.; wing area lower 92.5 sq.ft.; total wing area 203 sq.ft.; airfoil "Clark Y"; wt. empty 1085; useful load 586; payload 221; gross wt. 1671 lbs.; (these wts. for serial # 102 and up, with 8 gal. reserve fuel tank); max. speed 120; cruising speed 102; landing speed 44; climb 1050 ft. first min. at sea level; service ceiling 16,000 ft.; gas cap. 26 gal.; oil cap. 4 gal.; cruising range 350 miles; price at the factory was $4500. The fuselage framework was built up of welded chrome-moly steel tubing, faired to shape with fairing strips and fabric covered. The wing framework was built up of solid spruce spars that were routed to an "I beam" section, and spruce and plywood built-up ribs; completed framework was fabric covered. There were four narrow-chord ailerons that were connected in pairs by a push-pull strut; the fuel supply was carried in the center-section panel of the upper wing. The landing gear on the prototype was of the split-axle type using two long oleo-

Fig. 252. The "Bluebird" was a Hess design of 1926-27; it was a 3-place craft with the OX-5 or Hisso engine.

spring legs, but this was modified and subsequent aircraft had the normal tripod gear with 2 spools of shock-cord to absorb the bumps; tread was 72″ and wheels were 28x4. The fabric covered tail-group was also built up of welded chrome-moly steel tubing; the fin was ground adjustable and the horizontal stabilizer was adjustable in flight. Wheel brakes, metal propeller, and engine starter were available as optional equipment. The Hess "Warrior" was also designed, developed, and manufactured by the Alliance Aircraft Corp., it was a 7 cylinder radial air-cooled engine of many advanced features. Weight was 295 lbs. without prop hub or starter, and it developed a maximum of 125 h.p. at 1850 r.p.m.; price at the factory was $2250.

Listed below are "Argo" entries that were gleaned from the Aeronautical Chamber of Commerce aircraft register; this list is not complete but does show the bulk of the "Argo" that were built.

X-6876; Argo (# A-1) "Warrior", prototype a/c.
NC-3601; Argo (# 101) "Warrior".
NC-594K; Argo (# 102) "Warrior".
NC-9399; Argo (# 103) "Warrior".
NC-595K; Argo (# 104) "Warrior", also as NC-595.

NC-596K; Argo (# 106) "Warrior", no listing on # 105.
NC-1M; Argo (# 107) "Warrior".
NC-2M; Argo (# 108) "Warrior", no listing on # 109.
NC-5M; Argo (# 110) "Warrior", no listing on # 111.
NC-6M; Argo (# 112) "Warrior".
NC-7M; Argo (# 113) "Warrior".
NC-8M; Argo (# 114) "Warrior", no listing on # 115.
NC-79N; Argo (# 116) "Warrior".
NC-80N; Argo (# 117) "Warrior".
NC-81N; Argo (# 118) "Warrior".
NC-82N; Argo (# 119) "Warrior".
NC-83N; Argo (# 120) "Warrior".

Fig. 253. The Alliance "Argo" was one of the few airplanes that could withstand abnormal stresses of the outside loop; the craft shown here performed two in a row.

Fig. 254. A Parks P-1 with Curtiss OX-5 engine; early Parks airplanes were classroom projects.

Parks Air College had long been noted for it's excellent flying school, one of the largest in the country; a flying school that had trained thousands of pilots. Parks Air College also had a mechanics course, an extensive course for future airplane mechanics where they were taught the art of airplane repair and manufacture to it's fullest extent; to the extent that an actual airplane was built in the classrooms. Actual construction of the airplane was followed through, from the "pile of steel tubing" stage, through all the welding, all the woodwork, the covering and doping, up to the time that the craft was ready to fly. Several of the airplanes that were actually built by students of the school in the course of their training, airplanes that were built along the designs and specifications of one manufacturer or another, had actually been flown in test and then put into service on the flight line of the school for pilot training. From the fact related above and from the similiarity of it's general configuration it is then perfectly logical to speculate or assume that the "Parks" airplane might have started out as a "student built" project that was patterned and built along Kreider-Reisner "Challenger" lines and specifications. It is certainly easy to see that the KR "Challenger" biplane and the "Parks" biplane were look-alikes.

The Parks model P-1 was a 3 place open cockpit biplane of the familiar general-purpose type that had flourished with great success for over five years now, and it was also powered with the trusty 8 cylinder Curtiss OX-5 engine of 90 h.p. Performance and flight characteristics of this airplane were typical, and it was a good average for an airplane of this type. Parks Air College, founded in December of 1927, had long used the "Travel Air" biplane as it's standard equipment for all phases of it's flight training course, but was actually planning to replace all this equipment with "Parks" built airplanes; a few of the Parks biplanes did find their way into school service, but most of the Travel Air biplanes remained. The Parks airplane was launched as an ambitious project and besides the model P-1, the design was developed into the P-2 series which were identical airframes but were powered with radial air-cooled engines of medium horsepower; there was also the model P-3 which was a sport-type 4 place high wing cabin monoplane that was powered with the Wright J5 engine, and the proposed model P-4 which was to have been a high wing cabin monoplane of the light transport type with seating for six, and powered with the 300 h.p. Wright J6 engine. Had circumstances been more favorable, there is no doubt these two Parks monoplanes would have been developed more fully for manufacture in quantity. The type certificate number for the Parks model P-1 as powered with the Curtiss OX-5 engine, was issued in July of 1929 and an amendment to the certificate was issued in November of 1929; some 45 or more examples of this model were manufactured by the Parks Aircraft Corp. at E. St. Louis, Illinois; a subsidiary of Parks Air Lines, Inc. which owned Parks Airport and Parks Air College, which later became a division of the Detroit Aircraft Corp. Oliver L. Parks was the V.P. and general manager.

Listed below are specifications and performance data for the Parks model P-1 as powered with the 90 h.p. Curtiss OX-5 engine; length overall 24'1"; hite overall 9'1"; wing span upper 30'1"; wing span lower 28'9"; wing chord both 63.5"; wing area upper 156 sq.ft.; wing area lower 134 sq.ft.; total wing area 290 sq.ft.; airfoil "Aeromarine Modified"; wt. empty 1331; useful load 747; payload with 32 gal. fuel was 355 lbs.; gross wt. 2078 lbs.; max. speed 100; cruising speed 85; landing speed 37; climb 540 ft. first min. at sea level; service ceiling 12,000 ft.; gas cap. 32 gal.; oil cap. 4 gal.; cruising range 340 miles; price at the factory field was $3165 with new OX-5 engine, when available; also sold less engine and propeller.

The fuselage framework was built up of welded chrome-moly steel tubing in truss form, faired to shape with wood fairing strips and fabric covered. The fuel tank was mounted high in the fuselage, just ahead of the front cockpit; the front cockpit had a door on the left side for ease of entry and exit. The engine coolant radiator was mounted below the engine, just ahead of the landing gear; the radiator was shuttered for temperature control. The wing framework was built up of solid spruce spar beams and spruce and plywood built-up wing ribs; the completed framework was fabric covered. There were four ailerons that were connected together in pairs by a streamlined push-pull strut; interplane bracing was of streamlined chrome-moly steel tube struts and streamlined steel tie-rods. The fabric covered tail-group was built up of welded chrome-moly steel tubing; the fin was ground adjustable and the horizontal stabilizer was adjustable in flight. The robust landing gear was of the simple straight-axle type and used two spools of rubber shock-cord to snub the bumps; the tail-skid was a shock-cord sprung chrome-moly steel tube with a hardened shoe that was removed for replacement as it wore down. The tail-skid on any airplane was somewhat of a nuisance, but it served nicely as a drag or a brake on sod surfaces, for airplanes that were not equipped with wheel brakes. Ground manuvering an airplane without brakes, is a long forgotten art that called for a generous blast from the engine and the proper amount of rudder at the right time, or else! The next development in the Parks biplane was the Axelson powered model P-2; see the chapter for A.T.C. # 200 in this volume.

Listed below are Parks P-1 entries that were gleaned from various records; this list is not complete, but it does show the first production run of this model.

Fig. 255. The Parks P-1 was a sturdy craft used on the flight line of Parks Air College.

Fig. 256. A typical 3-place open cockpit biplane of the period, the Parks P-1 had an amiable nature and able performance.

C-10008; Parks P-1 (# 212) OX-5, first as KR-31.

NC-7993; Parks P-1 (# 224) OX-5, first as KR-31.

C-7877; Parks P-1 (# 228) OX-5, first as KR-31.

C-7996; Parks P-1 (# 233) OX-5, first as KR-31.

NC-11H; Parks P-1 (# 1951) OX-5, Parks manufacture.

          ; Parks P-1 (# 1961) OX-5, no record.

NC-91H; Parks P-1 (# 1962) OX-5, Parks manufacture.

NC-89H; Parks P-1 (# 1963) OX-5, Parks manufacture.

NC-90H; Parks P-1 (# 1964) X-5, Parks OX-5 manufacture.

NC-331K; Parks P-1 (# 1975) X-5, Parks OX-5 manufacture.

NC-349K; Parks P-1 (# 1976) X-5, Parks OX-5 manufacture.

NC-350K; Parks P-1 (# 1977) X-5, Parks OX-5 manufacture.

NC-351K; Parks P-1 (# 1978) X-5, Parks OX-5 manufacture.

NC-352K; Parks P-1 (# 1979) X-5, Parks OX-5 manufacture.

NC-353K; Parks P-1 (# 197-10) X-5, Parks OX-5 manufacture.

NC-361K; Parks P-1 (# 197-11) X-5, Parks OX-5 manufacture.

NC-363K; Parks P-1 (# 197-12) X-5, Parks OX-5 manufacture.

NC-362K; Parks P-1 (# 197-13) X-5, Parks OX-5 manufacture.

NC-376K; Parks P-1 (# 197-14) X-5, Parks OX-5 manufacture.

NC-377K; Parks P-1 (# 197-15) X-5, Parks OX-5 manufacture.

NC-912K; Parks P-1 (# 197-16) X-5, Parks OX-5 manufacture.

The next block of airplanes were with serial # 198-1 thru # 198-16; and block after that were serial # 199-1 thru # 199-10, as far as the records show. Of this total number, some 20 or more were still active some 10 years later.

Fig. 257. The Timm "Collegiate" trainer with a Kinner K-5; the prototype was tested with an Anzani engine.

The "parasol" monoplane with it's wing suspended high above the fuselage, was a pleasant configuration and an efficient aerodynamic arrangement that possessed inherent "pendulum stability" and usually afforded a very good range of vision; good vision was very desirable while flying close to home in heavy round-the-airport traffic, and good stability was certainly appreciated by the student engaged in pilot training. Thus, these two important requisites were developed into the "Collegiate", which as the name implies, was developed as a primary-trainer type of craft. The bulky Timm "Collegiate" was a two place open cockpit parasol monoplane, with seating for two in tandem, and was powered with the 5 cylinder Kinner K5 engine of 90-100 h.p. Built robust and rugged to withstand the abuse under the inexperienced hands of fledgling pilots, it was docile and placid by nature and quite easy to fly. The type certificate number for the Kinner powered "Collegiate" was issued in July of 1929 and only one of this model was manufactured by the Timm Aircraft Corp. at Glendale, California; 40 craftsmen of all sorts were employed and production capacity was estimated at two airplanes per week. Four more of the "Collegiate" were built, but each of these was powered with a different engine; in combination with powerplants such as the Comet 165, the Challenger 170, the Continental A-70 of 165 h.p.; and the MacClatchie "Panther"

of 150 h.p. A coverage of activities in the new plant, reported ten airplanes under construction, but these do not show up in any reports. Otto W. Timm was the president, and chief of design and engineering; brother Wally P. Timm was V.P.; and Jack Gardener was secretary, manager of sales, and in charge of purchasing.

Otto W. Timm, a true pioneer and one of our "Early Birds", built his first airplane way back in 1910, at 17 years of age. After racing souped-up automobiles at county fairs the two years previous, he decided auto-racing was too dangerous and would try his luck in building and flying "aeroplanes". His first airplane never flew because of side-tracking circumstances, but he taught himself to fly in a Curtiss "Pusher" while acting as mechanic for an exhibition flyer. Graduating into exhibition flying, Timm made good money but spent most of it in improving his "aeroplanes" or designing and building new types. His first tractor-type biplane was designed and built in 1913; a stick and wire craft with the engine in front. In addition to constant experimentation in aircraft, Timm also spent much time and money in developing aircraft engines; one of the first successful "radial type" engines was developed by Timm and he used it in one of his own airplanes. Uncle Sam called him up in 1917, and Timm became a senior instructor at Rockwell Field. During the latter days of the first World War, he had

Fig. 258.   The parasol-wing design of the "Collegiate" trainer was stable and offered excellent visibility.

designed a two place training craft and the engine to fly it, too; the government showed considerable interest in it, but the war's end cancelled further developments in the project. After the war he continued flying and constructed several more airplanes; one of which was a "racer" with a Curtiss OX-5 engine. In the year 1922, Timm joined Ray Page of Nebraska Aircraft (Lincoln-Standard) as their chief engineer. Nebraska Aircraft, which later became Lincoln-Standard Aircraft, had Timm redesigning the "Standard" biplane of World War 1 into 5 place transports (LS-5) and 3 place sport-craft; the "Lincoln-Standard" was a familiar sight around the country throughout the "twenties". While at Lincoln, he gave "Lindy" (Charles A. Lindbergh) his first airplane ride and some instruction. Feeling that his usefulness at Lincoln had run it's course, Timm left in 1925 and free-lanced as a consulting engineer to various individuals and concerns; he also supervised the construction of several custom built airplanes. His fling at free-lance engineering was followed by the organization of his own company at Glendale, Calif. to build aircraft of varying designs; among one of his first was the "Collegiate" monoplane. The prototype "Collegiate" was first powered and tested with an Anzani engine, but this was replaced with the Kinner K5 for certification.

Listed below are specifications and performance data for the Timm "Collegiate" as powered with the Kinner engine of 90-100 h.p.; length overall 24'7"; hite overall 8'5"; wing span 35'; wing chord 84"; total wing area 236 sq.ft.; wt. empty 1309; useful load 643; payload with 40 gal. fuel was 183; gross wt. 1952 lbs.; max. speed 108; cruising speed 92; lan-

ding speed 35; climb 640 ft. first min. at sea level; service ceiling 16,000 ft.; gas cap. 40-50 gal.; oil cap. 6 gal.; cruising range 540-630 miles; price at the factory field was $5500. The fuselage framework was built up of welded chrome-moly steel tubing, heavily faired to shape with wood fairing strips and fabric covered. The wing framework was built up of spruce spar planks that were laminated into 5 plys and routed to an "I beam" section, with spruce and plywood truss-type wing ribs; the completed panels were fabric covered. The wing, in three sections, was perched atop a center-section cabane and was braced to the fuselage with two chrome-moly steel tubes on each side. The fuel supply was carried in a tank that was built into the center-section panel of the wing; ailerons were of very narrow chord and spanned nearly the full length of each wing panel. The fabric covered tail-group was built up of welded chrome-moly steel tubing; the fin was ground adjustable and the horizontal stabilizer was adjustable in flight. The control mechanism was provided with well-placed inspection plates that were quickly removable for pre-flight inspection or maintenance. The landing gear of 90 inch tread, was of the long leg split-axle type and used air-oil shock absorbers; wheels were 28x4. Wheel brakes, metal propeller, and inertia-type engine starter were offered as optional equipment.

Listed below are "Collegiate" entries that were gleaned from the Aeronautical Chamber of Commerce aircraft register, and various other reports; from all indications, this was the extent of production in this series.

C-337; Collegiate K-100 (# 101) Kinner K5, later had Comet 165 on Grp. 2-209 approval.

Fig. 259. The Timm "Collegiate"; an early version powered by a 10 cylinder Anzani engine.

C-887E; Collegiate C-165 (# 102) Comet 165, Grp. 2-209.

C-888E; Collegiate C-170 (# 103) Curtiss Challenger 170, Grp. 2-202.

X-16E; Collegiate TC-165 (# 104) Continental 165, Grp. 2-265.

X-279V; Collegiate M-150 (# 105) MacClatchie Panther, Grp. 2-239.

A.T.C. #181
(7-29)
LINCOLN-PAGE PT

Fig. 260. The Lincoln-Page PT (Page Trainer) with OX-5 engine was a modification of the popular LP-3.

Evidently spurred on by Swallow's development of the TP trainer and it's resulting success, plus the steadily gaining success of the 2 place Command-Aire trainer, Lincoln-Page foresaw the possibilities for a craft of this type and modified the standard LP-3 somewhat into a two place training craft; a craft designed specifically for the use of flying schools in their primary flight training programs. Using the basic configuration of the LP-3 as a pattern, actually quite a bit of modification was performed, though it may not seem too apparent at first. The major change was in a narrowing down of the fuselage to fit the tandem seating, and an increase in overall length to de-sensitize the bucking and pitching movements that would develop with the nervous hand of a fledgling-student on the stick. The wing cellule was changed a bit to one of lesser area, using a good stable airfoil; the center-section cabane was about the same as that used on the later versions of the LP-3 type. The instruments, what little there were, were grouped on the front cockpit cowling so as to be seen from either cockpit; Bloxham safety-sticks were used to enable the instructor to disengage the student's control-stick in the case of a freeze-up on the controls during a moment of panic. Back in the old "Jenny days", a konk on the head

with a fire-extinguisher bottle accomplished the same purpose, but this was frowned upon and considered a rather primitive method at this particular time. Well balanced aerodynamically, the Lincoln PT was stable, easy to fly, and was quite forgiving in nature; in other words, the PT was not too fussy and would tend to overlook a good bit of pilot error.

The Lincoln-Page PT (Page Trainer) was a 2 place open cockpit biplane seating two in tandem, and was powered with the 8 cylinder Curtiss OX-5 engine of 90 h.p., or the OXX-6 of 100 h.p. The PT was kept bare of the unnecessary frills but not to the point of slab-sided severeness, and therefore presented a neat and trim appearance that would be of some interest to the average private-owner as an economical sport-plane; especially attractive was the price of $1985 less engine and propeller. The PT achieved early success and was employed as the standard training equipment with several large flying schools in the country; the service record logged by this craft was quite commendable. Being easily adaptable to modification, the PT trainer was soon planned to mount various engine installations up to 110 h.p.; among these were the Kinner K5, the Warner "Scarab", and the Brownback 90. Though not fully acrobatic in

Fig. 261. The two-place PT was specially designed for flight training and low-cost sport flying.

the true sense, the PT was quite versatile and able to perform most of the basic acrobatic manuvers used in advanced pilot training, without strain or too much coaxing. The type certificate number for the Lincoln PT as powered with the Curtiss OX-5 engine, was issued in July of 1929 and some 28 or more examples of this model were manufactured by the Lincoln Aircraft Co. at Lincoln, Nebraska. Victor Roos was president and general manager; Ray Page was V.P. in charge of sales and purchasing; A. H. Saxon was chief engineer for a time, and so was Ensil Chambers. The term "Lincoln-Page" was dropped in favor of just plain "Lincoln", but the flying-folk the country over were familiar with Lincoln-Page for many years and the name was not easily forgotten.

Listed below are specifications and performance data for the Lincoln PT trainer as powered with the 90 h.p. Curtiss OX-5 engine; length overall 26'2"; hite overall 8'10"; wing span upper 32'3"; wing span lower 31'9"; wing chord both 58"; wing area upper 150 sq.ft.; wing area lower 140 sq.ft.; total wing area 290 sq.ft.; airfoil "Goettingen 436"; wt. empty 1428; useful load 540; payload with 29 gal. fuel was 170 lbs.; gross wt. 1968 lbs.; max. speed 100; cruising speed 84; landing speed 37; climb 600 ft. first min. at sea level; climb in 10 min. was 4200 ft.; service ceiling 11,500 ft.; gas cap. 29 gal.; oil cap. 4 gal.; cruising range at 8 gal. per hour was 300

miles; price at the factory field with OX-5 engine was $2565, later raised to $2900; price less engine and propeller was $1985.

The fuselage framework was built up of welded chrome-moly steel tubing in a modified Warren truss, faired to shape with wood fairing strips and fabric covered. Dual controls were provided and seats were of the bucket type with wells to accomodate a parachute pack. The wing framework was built up of solid spruce spar beams, with wing ribs built up of bass-wood; the completed framework was fabric covered. There were four ailerons that were connected together in pairs by a streamlined push-pull strut. The fuel tank was mounted high in the fuselage just ahead of the front cockpit, with a direct-reading fuel gage that projected through the cowling; the engine radiator was underslung below the engine compartment and was shuttered for temperature control. The fabric covered tail-group was built up of welded chrome-moly steel tubing; the fin was ground adjustable and the horizontal stabilizer was adjustable in flight. The landing gear of 76 inch tread was of the normal split-axle type, using two spools of rubber shock-cord to absorb the bumps; wheels were 26x4 and no brakes were provided. If the Lincoln PT was bought less engine, the customer shipped his own engine and propeller for installation at the factory. The next development in the Lincoln PT trainer was the Kinner powered model PT-K; this model will be discussed in the chapter for A.T.C. # 279.

Listed below are Lincoln PT entries that were gleaned from various records; this list may not be quite complete, but it does show the bulk of this model that were built.

-486E;	PT (#) OX-5.
NC-617;	PT (# 302) OX-5.
NC-618;	PT (# 303) OX-5.
NC-2482;	PT (# 304) OX-5.
C-9414;	PT (# 305) OX-5.
C-9415;	PT (# 306) OX-5.
NC-789K;	PT (# 307) OX-5.
NC-790K;	PT (# 308) OX-5.
;	PT (# 309) OX-5.
NC-531M;	PT (# 310) OX-5.
NC-532M;	PT (# 311) OX-5.
-533M;	PT (# 312) OX-5.
NC-314V;	PT (# 313) OX-5.
NC-298V;	PT (# 314) OX-5.
NC-78N;	PT (# 315) OX-5.
NC-56N;	PT (# 316) OX-5.
NC-57N;	PT (# 317) OX-5.
NC-140N;	PT (# 318) OX-5.
NC-141N;	PT (# 319) OX-5.
NC-142N;	PT (# 320) OX-5.
NC-184N;	PT (# 321) OX-5.
NC-185N;	PT (# 322) OX-5.
NC-186N;	PT (# 323) OX-5.
;	PT (# 324) OX-5.
;	PT (# 325) OX-5.
NC-33W;	PT (# 326) OX-5.
NC-34W;	PT (# 327) OX-5.
NC-35W;	PT (# 328) OX-5.

Fig. 262. The Curtiss "Fledgling Jr." with a 6 cylinder "Challenger" engine, a lesser-known version of the trainer.

The lovable and almost comical Curtiss "Fledgling" is certainly an airplane that most everybody knows, a motherly craft that was training thousands of student pilots all over the country for years at the famous Curtiss Flying Service schools. But as shown here, we have a "Fledgling" that hardly anybody knows or perhaps has ever seen; this is the "Fledgling Jr.", almost an orphan in the Curtiss line-up of well known airplanes. In contrast to the unmistakeable configuration of the standard "Fledgling" trainer with it's acres of wing area and a literal forest of wing struts, the "Fledgling Jr." was a trim and efficient looking single-bay biplane of more average and more normal proportions. The fact that it was so much more normal than the standard "Fledgling", is what makes the "Fledgling Jr." such an oddity.

The "Fledgling Jr." was a 2 place open cockpit biplane seating two in tandem, that was developed by Curtiss engineers as a craft to train student pilots; it bore typical resemblance to the standard "Fledgling" in most all respects except for less wing area which was done up in the standard single-bay wing cellule form, instead of the double-bay layout that hadn't been used on a light airplane for quite a few years. The Curtiss "Fledgling Jr."

was powered with the 6 cylinder air-cooled Curtiss "Challenger" engine of 170-180 h.p., and it's performance figures which were based on actual tests, support the guess that it might have been quite a good airplane. Among the many features plainly visible, good vision is evident by the excessive stagger between the upper and the lower wing panels, with a trailing edge cut-out for vision upward. It is also evident that everything was designed for quick inspection and easy maintenance, in order to keep the "shop time" down to the barest minimum. Perhaps the performance was not as expected, or perhaps it was just not suitable to the job of training student pilots; whatever the case was, it kept this craft from being produced in quantity, and as far as can be determined, the example shown here was the only one ever built. The type certificate number for the Curtiss "Fledgling Jr." as powered with the 170 h.p. "Challenger" engine, was issued in July of 1929 and it was manufactured by the Curtiss Aeroplane & Motor Co. at Garden City, Long Island, New York.

Listed below are specifications and performance data for the Curtiss "Fledgling Jr." as powered with the 170-180 h.p. Curtiss "Challenger" engine; length overall 27'10";

Fig. 263. The Curtiss "Fledgling Jr." was a rare type; note single-bay wing cellule.

hite overall 10'4"; wing span upper and lower 31'6"; wing chord both 60"; total wing area 289 sq.ft.; airfoil "Curtiss C-72"; wt. empty 1921; useful load 671; payload with 40 gal. fuel was 210 lbs.; gross wt. 2592 lbs.; the following performance figures were based on 180 h.p. rating; max. speed 107; cruising speed 91; landing speed 50; climb 575 ft. first min. at sea level; climb to 5,000 ft. was 11 min.; climb to 10,000 ft. was 33 min.; service ceiling 10,600 ft.; gas cap. 40 gal.; oil cap 3.5 gal.; cruising range at 10 gal. per hour was 358 miles; cruise was based on 85% of max. power; load factor of safety at high-incidence was 8.25 to 1.

The fuselage framework was built up of welded chrome-moly steel tubing, faired to shape with wood fairing strips and fabric covered. The seats were adjustable and were shaped with wells for a parachute pack; dual controls were of stick and rudder pedals. The wing framework was built up of solid spruce spar beams with spruce and birch plywood built-up wing ribs; the leading edge of the upper wing was covered with aluminum alloy sheet and the leading edge of the lower wing was covered with plywood back to the front spar, then the completed framework was fabric covered. There were four ailerons of welded steel tube construction that were con-

Fig. 264. The "Lark" was an early Curtiss design which led to the development of the "Fledgling" series.

Fig. 267.   The 40-B-4 was a stout-hearted pioneer with earned a respected niche in the annals of air transport.

tilation, and his seat was adjustable for vision and comfort. The first successful plane-to-ground radio communication by voice had been accomplished by now, and the pilots of the 40-B-4 could be kept informed of the weather ahead and other necessary flight data. Being of a rather placid nature and obedient to the varying needs or commands, the 40-B-4 made a tough job much easier and therefore

was a great favorite among the pilots. The model 40-B-4 was built into 1931 and some 38 examples were reported by factory records; the last 20 examples of this model were completely equipped with radio, had a steerable tail-wheel, and were fitted with the "Townend ring" type of low-drag engine cowling. The type certificate number for the model 40-B-4 as powered with the 525 h.p.

Fig. 268.   A Boeing 40-B-4 taking off on a night mail run.

Pratt & Whitney "Hornet" engine, was issued in July of 1929 as a 5 place combination passenger and mail carrier, or as a 1 place mailplane; it was manufactured by the Boeing Airplane Co. at Seattle, Wash., a division of the United Aircraft Transport Corp. Philip G. Johnson was the president; C. L. Egtvedt was V.P. and general manager; Erik H. Nelson was sales manager; C. N. Monteith was chief of engineering, a department that was by now composed of some 90 engineers and draftsmen; Eddie Allen, Les Tower, and Eddie Campbell were pilots in charge of testing and development.

Listed below are specifications and performance data for the Boeing model 40-B-4 as powered with the 525 h.p. "Hornet" engine; length overall 33'3"; hite overall 11'9"; wing span upper & lower 44'3"; wing chord both 79"; wing area upper 285 sq.ft.; wing area lower 260 sq.ft.; total wing area 545 sq.ft.; airfoil "Boeing 103"; wt. empty 3809 (3709); useful load 2271 (2371); payload with 140 gal. fuel was 1160 (1260) lbs.; gross wt. 6080 lbs.; weights in brackets are for 1 pl. mailplane; max. speed 137; cruising speed 115; landing speed 57; stall speed 57; climb 800 ft. first min. at sea level; climb in 10 min. was 6320 ft.; service ceiling 15,100 ft.; gas cap. 140 gal.; oil cap. 12 gal.; cruising range at 25 gal. per hour was 575 miles; price at the factory field was $24,500., reduced to $22,500 in 1931.

The fuselage framework was built up of welded chrome-moly steel tubing, faired to shape with wood fairing strips; the forward portion of the fuselage was covered with aluminum alloy panels and the rear portion was covered in fabric. Entrance to the sound-proofed and insulated cabin was by two doors on the left side; the seats were comfortably upholstered and were slightly staggered for more shoulder room. The pilot's cockpit was raised slightly above the cabin roof line for better vision, and the seat was adjustable for comfort. A 25 cu. ft. compartment for mail or cargo was just back of the engine firewall, and two smaller compartments fore and aft of the cabin were for passenger's baggage. The wing framework was built up of solid spruce spar beams that were routed-out for lightness, with spruce and plywood built-up wing ribs; the leading edges were covered with plywood and the completed framework was fabric covered. There were four ailerons that were connected together in pairs by stranded steel cable; retractable landing lights were built into the lower wing. The fuel supply was carried in one tank that was mounted in the fuselage, and two tanks that were mounted in the root ends of the upper wing; fuel flow was by gravity from the wing tanks and the fuel from the fuselage tank was pumped into the upper tanks by engine-driven fuel pump or a hand-operated wobble pump. The fabric covered

Fig. 269.   One of several 40-B-4 types which were modified from the 40-C version.

Fig. 270. A Varney Air Lines 40-B-4 preparing to load at the terminal; the latest buildings were a far cry from the "operations shacks" of a few years earlier.

tail-group was built up of welded chrome-moly steel tubing; the fin was ground adjustable and the horizontal stabilizer was adjustable in flight. The split-axle landing gear of 88 inch tread was of the normal Boeing cross-axle type, with Boeing oleo-spring shock absorbers; wheels were 36x8 and brakes were standard equipment. A metal propeller; Bendix brakes, parachute flares, engine-driven generator, battery, fire extinguishers, shatterproof glass, electric inertia-type engine starter, oil cooler, fuel pumps, bonding and shielding for radio, and navigation lights, were standard equipment. A full complement of night-flying equipment was available. The next Boeing development was the tri-motored 80-A that will be discussed in the chapter for A.T.C. # 206.

Listed below is a partial listing of Boeing 40-B-4 entries that were gleaned from various records.

NC-5389; 40-B-4 (# 1042) Hornet.

C-5390; 40-B-4 (# 1044) Hornet.
C-178E; 40-B-4 (# 1096) Hornet.
C-179E; 40-B-4 (# 1097) Hornet.
C-180E; 40-B-4 (# 1098) Hornet.
NC-740K; 40-B-4 (# 1147) Hornet.
NC-741K; 40-B-4 (# 1148) Hornet.
NC-742K; 40-B-4 (# 1149) Hornet.
NC-743K; 40-B-4 (# 1150) Hornet.
NC-830M; 40-B-4 (# 1155) Hornet.
NC-831M; 40-B-4 (# 1156) Hornet.
NC-832M; 40-B-4 (# 1157) Hornet.
NC-833M; 40-B-4 (# 1158) Hornet.
NC-834M; 40-B-4 (# 1159) Hornet.
NC-842M; 40-B-4 (# 1168) Hornet.
NC-843M; 40-B-4 (# 1169) Hornet.
NC-10340; 40-B-4 (# 1421) Hornet.
NC-10341; 40-B-4 (# 1422) Hornet.

Serial # 1042, # 1044, # 1096, # 1097, # 1098, were originally Wasp powered model 40-C (A.T.C. # 54) in service with Pacific Air Transport that were later modified to 40-B-4 specifications.

Fig. 271. A Command-Aire 5C3 with a Curtiss "Challenger" engine; performance was excellent.

The Command-Aire biplane in the model 5C3 version was probably the firm's best and most popular airplane, that is, of the models that were powered with radial air-cooled engines. The OX-5 powered Command-Aire had the edge on them all in number, but that was mainly because of it's low first-cost and fairly economical operation. The model 5C3 was also a three place open cockpit biplane of typical Command-Aire configuration, and was powered with the 6 cylinder Curtiss "Challenger" engine of 170-185 h.p.; a combination which provided ample power reserve for a sprightly performance. The 5C3 had a terrific short-field performance and the postage-stamp sized plots it could use for it's "airport" was amazing; a feat that would have to be seen to be fully appreciated. Though fairly fast and quite manuverable, the 5C3 had excellent slow-speed characteristics and complete control of the airplane was possible at about 40 m.p.h. The inherent stability and good low-speed control is what prompted the Curtiss Flying Service to choose the 5C3 "duster" for dusting cotton in the southern states; very essential characteristics when working in close quarters at practically a stalled attitude near the ground. Quite often

when the terrain would permit, 3 or 4 "dusters" would fly past in staggered formation and dust a swath nearly a half mile wide; Curtiss Flying Service had at least 16 of the Command-Aire 5C3 "duster version" in service, and treated over 200,000 acres of cotton in 1929.

Major John Carroll Cone, a former Army Air Corps pilot, who was the sales manager at Command-Aire, flew a "Challenger" powered 5C3 to 7th place in the 1929 National Air Tour in hot competition with a very impressive list of contestants. A month previous to that, he had flown in the Miami to Cleveland Air Derby and finished second in the Class C category, right behind a "Challenger" powered Rearwin "Ken Royce". Major Cone and the 5C3 also entered in several closed-course speed races while at the National Air Races that were held at Cleveland, but the Command-Aire was never touted for speed as one of it's main attributes, and Cone had to be content with a position about midway in the field each time. The capable Ruth Nichols tried her hand with the 5C3 in the Womens Australian Pursuit Race, but was forced out of competition for some reason or another; her praise for it's handling was high.

With a very useful utility and a very capable performance, in combination with a reasonable price-tag, the model 5C3 was certainly a good investment for the private-owner, the fixed-base operator, or the business man in a hurry who had need for a versatile air-taxi. The popularity of the "Challenger" powered 5C3 was laid on a good foundation, so it is no great wonder that at least 20 of these craft were still in active service some 10 years later. The type certificate number for the model 5C3 as powered with the Curtiss "Challenger" engine, was issued in July of 1929 for both a landplane and a seaplane version; some 35 or more examples of this model were manufactured by Command-Aire, Inc. at Little Rock, Arkansas.

Listed below are specifications and performance data for the Command-Aire biplane model 5C3 as powered with the 170 h.p. Curtiss "Challenger" engine; length overall 24'5"; hite overall 8'4"; wing span upper and lower 31'6"; wing chord both 60"; wing area upper 169 sq.ft.; wing area lower 134 sq.ft.; total wing area 303 sq.ft.; airfoil "Aeromarine 2A"; wt. empty 1482; useful load 883; payload with 55 gal. fuel was 340 lbs.; gross wt. 2365 lbs.; max. speed 123; cruising speed 103; landing speed 37; climb 850 ft. first min. at sea level; service ceiling 14,000 ft.; gas cap. 55 gal.; oil cap. 5 gal.; cruising range 500 miles; price at the factory field was $6325, and later reduced to $5950 in early 1930. The following figures are for the seaplane version of the 5C3 as mounted on Edo DeLuxe twin-float gear; wt. empty 1664; useful load 883;

payload with 55 gal. fuel was 340 lbs.; gross wt. 2547 lbs.; max. speed 115; cruising speed 98; landing speed 42; climb 770 ft. first min. at sea level; service ceiling 13,000 ft.; price at the factory was $6950; all other dimensions and data remained the same.

The fuselage framework was built up of welded chrome-moly steel tubing, lightly faired to shape with fairing strips and fabric covered. On the "duster" version, the front cockpit which normally seated two, was converted to a "dust bin" with a capacity for 500 pounds of dust that was used to treat the cotton for extermination of the trouble-some boll weevil. The fuel tank on the standard model was mounted high in the fuselage just ahead of the front cockpit, but the "duster" version utilized all the available fuselage capacity ahead of the pilot's cockpit for a dust bin, so the two gravity-feed fuel tanks were mounted in the root ends of each upper wing panel. The wing framework was built up of solid spruce spar beams with spruce and plywood built-up wing ribs; the completed framework was fabric covered. The "Aeromarine 2A" airfoil and the abundance of effective wing area in a good aerodynamic arrangement was the combination that gave the Command-Aire the ability to operate at such low air-speeds, and the "slotted hinge" ailerons plus the aerodynamic effectiveness of the other control surfaces permitted positive control at near-hovering speeds, and almost down through the "stall". The "Challenger" powered 5C3 was fitted a little more deluxe in many instances; Bendix 30x5 wheels and

Fig. 272. A Command-Aire 5C3 "Duster"; a team of similar craft sprayed 200,000 acres of cotton in 1929.

Fig. 273.   The Command-Aire 5C3 was an entry in the 1929 Guggenheim safe airplane contest.

brakes, an Eclipse engine starter, wiring for navigation lights, and a metal propeller were standard equipment. The next development in the Command-Aire biplane was the "Hisso" powered model 5C3-A, a rare combination that will be discussed here in the chapter for A.T.C. # 185 in this volume.

Listed below are the model 5C3 entries that were gleaned from various records; this list may not be quite complete, but it does show the bulk of this model that were built.

NC-609 ; 5C3 (# W-   ) Challenger.
NR-900E; 5C3 (# W-74) Challenger.
NR-920E; 5C3 (# W-76) Challenger.
NR-914E; 5C3 (# W-78) Challenger.
NC-922E; 5C3 (# W-82) Challenger.

C-928E; 5C3 (# W-86) Challenger.
NR-925E; 5C3 (# W-88) Challenger.
C-939E; 5C3 (# W-93) Challenger.
NC-946E; 5C3 (# W-95) Challenger.
NC-947E; 5C3 (# W-96) Challenger.
NR-953E; 5C3 (# W-98) Challenger.
NR-949E; 5C3 (# W-99) Challenger.
C-954E; 5C3 (# W-100) Challenger.
C-955E; 5C3 (# W-101) Challenger.
C-956E; 5C3 (# W-102) Challenger.
NR-957E; 5C3 (# W-103) Challenger.
NR-959E; 5C3 (# W-105) Challenger.
NR-961E; 5C3 (# W-107) Challenger.
NC-983E; 5C3 (# W-122) Challenger.
NC-984E; 5C3 (# W-123) Challenger.
NC-985E; 5C3 (# W-124) Challenger.
C-986E; 5C3 (# W-125) Challenger.
NC-987E; 5C3 (# W-126) Challenger.
NR-988E; 5C3 (# W-127) Challenger.
NR-989E; 5C3 (# W-128) Challenger.
C-990E; 5C3 (# W-129) Challenger.
NC-992E; 5C3 (# W-131) Challenger.
NR-993E; 5C3 (# W-132) Challenger.
NR-994E; 5C3 (# W-133) Challenger.
NR-995E; 5C3 (# W-134) Challenger.
NR-996E; 5C3 (# W-135) Challenger.
NR-997E; 5C3 (# W-136) Challenger.
NR-998E; 5C3 (# W-137) Challenger.

The above listed entries with NR prefix letters were 5C3 "dusters" with the Curtiss Flying Service stationed in Houston, Texas.

Fig. 274.   J. Carroll Cone flew a 5C3 to 7th place in the 1929 National Air Tour.

Fig. 275. The Command-Aire 5C3-A with the 150 h.p. Hispano-Suiza engine had exceptional performance.

The Command-Aire biplane had already been developed into quite a number of different models, brought about mainly by the installation of several different engines in the same basic airframe; the version as shown here above, was yet another. This good-looking craft that poses here with such capable air was the model 5C3-A, a three place open cockpit general-purpose biplane of the normal Command-Aire configuration that was powered with the water-cooled 8 cylinder vee-type "Hisso" (Hispano-Suiza) engine, the model A series of 150 h.p. The venerable "Hisso" engine, despite the years already behind it, and the fact that it was "old fashioned", was still quite a good powerplant. Though over-shadowed by the new air-cooled radial engines that were finally making an appearance in number, the Hisso was still the choice of many operators and many private-owners; proof of this was reflected in the extended popularity of various other general-purpose airplane "types" that were being powered with the Hisso engine when the stock-piles of OX-5 engines were dwindling away with alarming rapidity. The Hisso powered Command-Aire was a combination of good looks and good performance, and it shared

all the other inherent attributes of sound construction and good handling characteristics that made the Command-Aire biplane such a favorite.

Command-Aire, Inc., who were justifiably proud of their craft, stressed it's excellent stability loud and long, both in word and in demonstration; even to go so far as to have Wright Vermilya their fearless test-pilot, ride a-straddle the turtle-back of the fuselage, while the airplane was flying along by itself! Another such demonstration, though tame in comparison with that as related above, was the flight made by another Command-Aire biplane which was flown from San Diego to Los Angeles (some 100 miles) without once touching the control stick; just using a little rudder now and then to stay on a heading. Fired up with ambition to spare and a complete confidence in their airplane's ability, Command-Aire entered a standard model of the "Challenger" powered 5C3 in the Guggenheim Safe Airplane Contest of 1929; the performance rendered by this craft in the preliminary tests was quite remarkable for a conventional type airplane that was lacking the so-called "safety aids" such as wing slots, wing flaps, and the like. It can be said without question that the Command-Aire biplane was a safe and gentle airplane, one that could be flown in a relaxed and happy manner while enjoying the beautiful scenery that was unfolding far down below.

The type certificate number for the model 5C3-A as powered with the Hisso A engine of 150 h.p., was issued in July of 1929 and as far as records show, there were only two examples of this model manufactured by Command-Aire, Inc. at Little Rock, Arkansas. Wright Vermilya Jr., Command-Aire chief test-pilot who hailed from Indiana, flew in France during World War 1 and became a test-pilot for the Army about a year or so after the war. In 1926 he was a flight instructor for the Arkansas National Gaurd, and performed many observation and rescue-directing flights during the tragic flood of 1927. Impressed by the possibilities then apparent in commercial aviation, Vermilya joined Command-Aire in 1928 as their test-pilot where he earned a fair living and much nation-wide publicity by flying a Command-Aire biplane while a-straddle the fuselage, and steering it on course by leaning over to one side or the other. The most unusual part of this demonstration was the fact that Vermilya didn't consider this a dangerous stunt!

Listed below are specifications and performance data for the Command-Aire biplane model 5C3-A as powered with the 150 h.p. Hisso engine; length overall 24′6″; hite overall 8′6″; wing span upper and lower 31′6″; wing chord both 60″; wing area upper 169 sq.ft.; wing area lower 134 sq.ft.; total wing area 303 sq.ft.; airfoil "Aeromarine 2A"; wt. empty 1610; useful load 913; payload with 43 gal. fuel was 450 lbs.; gross wt. 2523 lbs.; max. speed 115; cruising speed 98; landing speed 42; climb 790 ft. first min. at sea level; service ceiling 14,000 ft.; gas cap. 43 gal.; oil cap. 5 gal.; cruising range 400 miles; price at the factory field was about $4000.

The fuselage framework was built up of welded chrome-moly steel tubing, lightly faired to shape with fairing strips and fabric covered. The engine coolant radiator which was mounted as the nose-section type on the OX-5 powered 3C3, was mounted below the engine section, just ahead of the landing gear, on the Hisso powered 5C3-A; the nose-section was faired out with a propeller "spinner" and the radiator was shuttered for temperature control. The wing framework was built up of solid spruce spar beams with spruce and plywood built-up wing ribs; the completed framework was fabric covered. Both examples of the model 5C3-A, as shown, had a normally mounted fuel tank in the fuselage, that was just ahead of the front cockpit, with provision for extra fuel capacity in two wing tanks that were mounted in the root ends of each upper wing panel; the total fuel capacity in this arrangement would be close to 70 gallon. The extra strut that was incorporated into the center-section cabane of all the 3C3 models, to eliminate the cross-wires in this bay and provide uncluttered vision straight ahead, was eliminated in the 5C3 models, and bracing was accomplished by the normal cross-wires; apparently the slight improve-

Fig. 276. The "Hisso"-powered Command-Aire shown was originally a 3C3 with an OX-5 engine; both the "Hisso" and the OX-5 were famous engines of the World War I period.

ment in vision forward by use of the extra strut was not actually worth the extra expense involved. The fabric covered tail-group was built up of welded chrome-moly steel tubing; the fin was ground adjustable and the horizontal stabilizer was adjustable in flight. As mentioned times previous, the ailerons on the Command-Aire were of the "slotted hinge" type which were more effective than the "piano hinge" type at lower air-speeds, and were aerodynamically balanced also for lighter "stick loads". The elevators were of the plain unbalanced type, but the rudder had a projecting "balance horn" for aerodynamic balance, which lightened the air-load on the rudder pedals. The split-axle landing gear was of the normal 3-member type and used two spools of rubber shock-cord as shock absorbers; the tail-skid was of the popular spring-leaf type. Special equipment such as wheel brakes, a metal propeller, and navigation lights were available. The Command-Aire registered C-6690 was serial # 515 and originally was powered with an OX-5 engine; the Command-Aire registered NC-607 was serial # W-54. The next development in the Command-Aire biplane was the model 3C3-BT, a two place trainer that was powered with the 113 h.p. Siemens-Halske engine, this model will be discussed under A.T.C. # 209.

Many times it is rather hard to report the facts as they actually were, because of the discrepancy that sometimes arose in the initial recording. The continued repetition of a published discrepancy then later caused it to become as fact, and no one gave pause to question it. One instance of such a discrepancy is noted here in the Command-Aire specification and nothing short of actual measuring and calculation, could really prove the point. All factory released specifications, and specification tables that are listed in books and periodals, list the Command-Aire biplane as follows; wing span upper and lower 31'6"; wing chord both 60"; and total wing area as 303 sq.ft.; breaking this down further to 169 sq.ft. for upper wings and 134 sq.ft. for lower wings. It would seem that 31.5 ft. span times 5 ft. wing chord would come to 157.5 sq.ft., without discounting the lost area for rounded wing tips, which would amount to at least 5.5 sq.ft. and thus leaving 152 sq.ft. for upper wing area. The lower wing being also of 31.5 ft. span times the 5 ft. chord, would also come to 157.5 sq.ft. minus the lost area for rounded wing tips and minus about 15 sq.ft. for area loss at the fuselage; this would leave about 137 sq.ft. for the lower wings. Adding 152 sq.ft. to 137 sq.ft. gives us a total of 289 sq.ft. of wing area, which is 14 sq.ft. shy of all the published figures. This is about the amount displaced by the fuselage at the intersection of the lower wings; perhaps they calculated the fuselage as lifting area. Perhaps this may not seem like much to fuss about, but it should impress one with pity and understanding for the poor historian who wants to do his level best to report actual fact.

Fig. 277.   The Swallow TP-K trainer with a 100 h.p. Kinner K-5 engine, an improved model of the popular TP.

The jolly natured Swallow TP was selling very well and building up quite a popular following in the short time since it's introduction; nearly one hundred had already been built and they could be seen in active service around all parts of the country. Designed especially for economical pilot training, with a basic simplicity of structure and arrangement that allowed for ease of maintenance and repair, the TP was paying off well with more hours on the flight line, and doing it's job admirably without any hint of complaint. Doing it's utmost to tranform a fairly nerve-wracking activity, into one of profit and pleasure. As was the case with all OX-5 powered airplanes at this particular time, there always hung the realization that sooner or later there would be no more engines left in the hoarded stocks, and some steps would have to be taken to prepare a suitable substitute combination. Designed with this in mind, the Kinner powered TP-K was developed and groomed as a companion model in the Swallow TP (Training Plane) series. A promising test-version of the TP was powered with the Siemens-Halske engine, but it was soon evident that the Kinner engine surely would be the better choice in the long run.

The Swallow model TP-K, as shown here in very good likeness, was also a two place open cockpit biplane seating two in tandem, and was powered with the 5 cylinder Kinner K5 engine of 90-100 h.p. Fairly typical to the earlier TP in most respects, the model TP-K was not quite so slab-sided and did present a somewhat neater appearance, which was brought about by the fact that the TP-K was also planned to be offered as an economical sport-type craft for the enjoyment of the private owner. In combination with the Kinner engine, the TP-K was trimmer and a good bit lighter; a combination which translated itself into a more sprightly performance, while still sharing the many amiable flight characteristics that were inherent in the basic TP design. The familiar stiff-legged straight-axle landing gear of the TP was retained on this new model in the interests of economy, but several pleas had been entered for a gear of the split-axle type, which surely would have been the better choice. The type certificate number for the Swallow TP-K as powered with the Kinner K5 engine, was issued in July of 1929 and some 13 or more examples of this model were manufactured by the Swallow Airplane Company at Wichita, Kansas. The Swallow TP was certificated for any powerplant to 110 h.p., and such engine combinations as the Le-Blond 90 and the Wright-Gipsy were proposed, but probably not manufactured; a Warner "Scarab" powered version was already being groomed as the model TP-W.

Listed below are specifications and performance data for the Swallow model TP-K as powered with the 90-100 h.p. Kinner K5 engine; length overall 23'10"; hite overall 8'10"; wing span upper 30'11"; wing span lower 30'3"; wing chord both 60"; wing area upper 150 sq.ft.; wing area lower 146 sq.ft.; total wing area 296 sq.ft.; airfoil USA 27 modified; wt. empty 1170; useful load 530; payload with 28 gal. fuel was 170 lbs.; gross wt. 1700 lbs.; max. speed 100; cruising speed 85; landing speed 35; climb 760 ft. first min. at sea level; service ceiling 16,000 ft.; gas cap. 28 gal.; oil cap. 4 gal.; cruising range 340 miles; price at the factory field was $4123.

The fuselage framework was built up of welded steel tubing in a rigid truss form, faired to shape with wood fairing strips and fabric covered. The cockpit coamings were well padded and the instruments were arranged so as to be seen from either cockpit; a quick-release was provided for the stick controls to forestall dire catastrophe in case of student freeze-up on the controls during a moment of panic. The fuel tank was mounted high in the fuselage, just ahead of the front cockpit and a direct-reading fuel gauge projected through the cowling. The wing framework was built up of solid spruce spar beams with spruce and plywood wing ribs; the completed framework was fabric covered. The upper wing was in two panels that were joined together at the center-line and fastened to the center-section cabane; there were four ailerons that were connected together in pairs by a stranded steel cable. All interplane bracing wires were of stranded steel aircraft cable, and all movable controls were cable operated. The fabric covered tail-group was built up of welded steel tubing; the fin was ground adjustable and the horizontal stabilizer was probably adjustable in flight. The landing gear was built up of welded chrome-moly steel tubing in a rigid frame and was of the straight-axle type, with two spools of rubber shock-cord to absorb the bumps; wheels were 26x4 and there was no provision for wheel brakes. The next development in the Swallow TP trainer series biplane was the Warner powered model TP-W, this model will be discussed in the chapter for A.T.C. # 253.

Listed below are Swallow TP-K entries that were gleaned from various records; this list may not be quite complete, but it does show the bulk of this model that were built.

C-8736; TP-K (# 136) Kinner K5.
C-9892; TP-K (# 192) Kinner K5.
C-9893; TP-K (# 193) Kinner K5.
C-9894; TP-K (# 194) Kinner K5.
NC-9895; TP-K (# 195) Kinner K5.
C-9896; TP- K(# 196) Kinner K5.
C-9897; TP-K (# 197) Kinner K5.
C-9898; TP-K (# 198) Kinner K5.
NC-9899; TP-K (# 199) Kinner K5.
C-9900; TP-K (# 200) Kinner K5.
NC-686H; TP-K (# 201) Kinner K5.
NC-687H; TP-K (# 206) Kinner K5.
NC-689H; TP-K (# 207) Kinner K5.

A.T.C. #187
(7-29)
STEARMAN LT-1

Fig. 278. The Stearman LT-1 with the 525 h.p. "Hornet" engine carried four passengers and extra cargo.

The graceful Stearman model LT-1 (Light Transport) was a combination mail and passenger carrying aerial transport that was developed for double duty from the basic M-2 "Speedmail" design; developed basically in similiar configuration, but it was somewhat bigger and somewhat heavier. The forward fuselage which was fitted with cargo compartments on the M-2, was arranged into a cabin section seating 4 passengers in chummy comfort on the model LT-1; the pilot still operated from an open cockpit well aft, in true "airmail fashion". An ability to carry a payload of some 1200 pounds, allowed for a load of 4 passengers with their baggage, and yet an additional 400 or 500 pounds of airmail and air-cargo; this was indeed the versatility that many smaller air-lines would be seeking in a craft for their use at this particular time. Endowed with inherent Stearman dependability and sure-footed performance, the model LT-1 transport posed well to answer the needs of such air-lines in all parts of the country; that it did not sell in any great number, is only a quirk of fate that often

seems to decide these things.

Introduced early in 1929, the model LT-1 was initially developed for Interstate Air Lines which operated a route from Atlanta to Chicago, and to St. Louis, through Evansville, Indiana; a 1576 mile route (C.A.M. # 30) that had been served daily since November of 1928. The Interstate air-line had already been operating 3 Stearman C3MB on their mail run, so it is evident that satisfaction as guaranteed promotes repeat orders. The first example of the model LT-1 was powered with a Wright "Cyclone" engine, but it may have been the case that the customer showed a preference for Pratt & Whitney engines, so the installation was changed to the 525 h.p. "Hornet" engines. The combination arrangement for mail and passengers too, was utility well suited to the requirements of Interstate and they operated three of these craft; incidently, these were the only examples of this model that were built. Of dependable character and of good sound construction, the LT-1 served well into the "thirties" when the line was absorbed and taken over by American Airways, which was the forerunner to American Airlines of today. The type certificate number for the Stearman model LT-1 as powered with the "Hornet" engine, was issued in July of 1929 and three examples of this model were manufactured by the Stearman Aircraft Company at Wichita, Kansas.

Listed below are specifications and performance data for the Stearman model LT-1 as powered with the 525 h.p. Pratt & Whitney "Hornet" engine; length overall 32'6"; hite overall 12'6"; wing span upper 49'0"; wing span lower 34'6"; wing chord upper 88"; wing chord lower 57"; wing area upper 347 sq.ft.; wing area lower 143 sq.ft.; total wing area 490 sq.ft.; airfoil "Goettingen 398"; wt. empty 3890; useful load 2360; payload with 145 gal. fuel was 1190 lbs.; gross wt. 6250 lbs.; max. speed 138; cruising speed 115; landing speed 58; climb 900 ft. first min. at sea level; service ceiling 13,000 ft.; gas cap. 145 gal.; oil cap. 15 gal.; cruising range at 23 gal. per hour was 690 miles; price at the factory field was $25,800—$25,000.

The fuselage framework was built up of welded chrome-moly steel tubing, and faired to shape with wood fairing strips; the portion of the fuselage from the rear of the pilot's cockpit, clear on up to the engine, was covered in removable aluminum alloy panels, and the balance of the fuselage aft was covered in fabric. A 40 cubic foot compartment ahead of the cabin section was available for mail, cargo, and baggage. The wing framework was built up of spruce and plywood box-type spar beams, with spruce and plywood truss-type wing ribs; the completed framework was fabric covered. Ailerons of the self balancing offset-hinge type were in the upper wing panels, and were operated by tubes and bellcranks through a streamlined push-pull strut; the fuel supply was carried in a gravity-feed tank that was mounted in the center-section panel of the upper wing. The fabric covered tail-group was built up of welded chrome-moly steel tubing, and both rudder and elevators were aerodynamically balanced; the fin was ground adjustable and the horizontal stabilizer was adjustable in flight. The split-axle landing gear of 96 inch tread was of the typical Stearman outrigger type and employed "Aerol" (air-oil) shock absorber struts; Bendix wheels (36 x 8) and brakes were standard equipment. Instead of the typical tail-skid, the LT-1 used a steerable tail-wheel which made ground manuvering much easier for an airplane of this weight and size. Two long engine exhaust tail-pipes were run under the fuselage; this directed exhaust fumes well below the level of the cabin and pilot's cockpit, and also provided manifold-heat for all occupants. Wheel brakes, a metal propeller, and inertia-type engine starter were standard equipment; the LT-1 was also fit-

Fig. 279. A Stearman LT-1 loading mail and passengers at the Wichita Terminal.

Fig. 280. The Stearman LT-1 design was developed from the "Speedmail"; its sure-footed performance was typical.

ted with a full complement of night-flying equipment. The next development in the Stearman biplane was the Wright J6-7 powered model C3R, that will be discussed in the chapter for A.T.C. # 251.

Listed below are model LT-1 entries that were gleaned from various records; this list represents the total number of these aircraft.

NC-8829; LT-1 (# 2001) Hornet.
NC-8832; LT-1 (# 2002) Hornet.
NC-8833; LT-1 (# 2003) Hornet.
Stearman LT-1 serial # 2001 originally fitted with 525 h.p. Wright "Cyclone" engine; all three listed craft operated by Interstate Air Lines.

Fig. 281. *The Travel Air Model E-4000 had a 5 cylinder Wright J6 engine of 165 h.p.; this is the prototype.*

Of all the various "radial engined" combinations that were offered in the standard open cockpit Travel Air biplane for 1929, the model E-4000 was probably the most popular and most numerous of them all. This may have been influenced a great deal by the universal popularity of the Wright engine amongst private owners and flying service operators. Wright engines have long had an excellent service record of performance and dependability, and the new J6 series engines that now came in 5 cylinder (165 h.p.), 7 cylinder (225 h.p.), and 9 cylinder (300 h.p.) versions, posed well to carry on this world-wide tradition. It is also interesting to note that the roster of E-4000 owners runs the gamut of varied uses; flying-salesmen making their rounds, charter service operators flying to out-of-way places on short notice, flying schools teaching advanced piloting techniques, air-taxi for business executives, flying signboards that reminded people of certain products, the nimrod that would head for a little secluded mountain lake to hunt game, and just average private owners who flew for the fun of it, all were engaged in their own pursuits for pleasure and profit with the help of this versatile aircraft. Of the various old Travel

Air biplanes that have been restored to fly again in recent years, the model E-4000 is evidently one of the most popular in this respect.

The model E-4000 as pictured here in various views, was a three place open cockpit biplane of the general-purpose type, that was powered with the 5 cylinder Wright J6 engine of 165 h.p. The production model E-4000 was fitted with the new series wing panels that were of slightly heavier construction and had the rounded wing tips; there were a few examples of the model E-4000 fitted with the old-style wings that had the "elephant ear" ailerons, but these were conversions from craft that were powered previously with different engines. One case as an example, was serial # 861 that was first a model 3000 with "Hisso" engine, and later converted to an E-4000 with the installation of a Wright 5 engine. Planning to ride the wave of the airplane-buying spree that reached a peak in 1929, Travel Air offered their cabin monoplanes in 6 different versions, and the standard open cockpit biplanes in 12 versions, plus 3 different custom-built versions, plus the continuous production of the standard OX-5 powered model 2000. From the

Fig. 282. The roster of E-4000 owners ran the gamut of varied uses; it was very popular with flying salesman and business houses.

stand-point of engineering and development, we can appreciate the work that was involved and can easily imagine that much "midnight oil" was burned in an effort to get this varied selection out before the buying, flying, public. With 3 shifts working around the clock, and all departments anxious to maintain their production quotas, it was not long before the flight-test line was receiving 30 completed airplanes per week. In his mind's eye, this is what Walter Beech had planned for with tireless energy and fierce determination; with his ambition surely now realized, we can imagine he was quite proud of what was accomplished. The type certificate number for the model E-4000 as powered with the Wright 5 engine, was issued in July of 1929 for both a landplane and seaplane version and some 85 or more examples of this model were manufactured by the Travel Air Company at Wichita, Kansas.

Listed below are specifications and performance data for the Travel Air biplane model E-4000 as powered with the 165 h.p. Wright J6 engine; length overall 24'1"; hite overall 8'11"; wing span upper 33'0"; wing span lower 28'10"; wing chord upper 66"; wing chord lower 56"; wing area upper 171 sq.ft.; wing area lower 118 sq.ft.; total wing area 289 sq. ft.; airfoil T.A. # 1; wt. empty (landplane) 1695; useful load 1007; payload with 67 gal. fuel was 392 lbs.; gross wt. 2702 lbs.; max. speed 120; cruising speed 103; landing speed 48; climb 720 ft. first min. at sea level; service ceiling 13,000 ft.; gas cap. 67 gal.; oil cap. 6 gal.; cruising range at 9.5 gal. per hour was 650 miles; price at the factory field was $6425 and later reduced to $5850. The following figures are for the model E-4000 as mounted on Edo model "M" twin-float seaplane gear; wt. empty 1885; useful load 835; payload with 67 gal. fuel was 220 lbs.; payload with 42 gal. fuel was 370 lbs.; gross wt. 2720 lbs.; max. speed 110; cruising speed 95; landing speed 50; climb 700 ft. first min. at sea level; service ceiling 12,500 ft.; cruising range at 9.5 gal. per hour was approx. 600 miles; all other dimensions and data remained the same.

The fuselage framework was built up of welded chrome-moly steel tubing, faired to shape with wood fairing strips and fabric covered. The cockpits were comfortable and well upholstered; there was a baggage compartment of 9 cubic foot capacity located behind the rear cockpit, that was accessible through a door on the left side. The wing framework was built up of solid spruce spars that were routed for lightness, with spruce and plywood built-up wing ribs; the leading edges were covered back to the front spar to preserve the airfoil form, and the completed framework was fabric covered. The ailerons in the upper wing panels were operated differentially by push-pull tubes and bellcranks; each aileron was operated independently, so if one became inoperative, the other could still control the airplane. The main fuel tank was mounted high in the fuselage just ahead of the front cockpit; extra fuel was carried in a tank

Fig. 283. An "elephant ears" Travel Air converted to an E-4000; it originally had OX-5 or "Hisso" engines.

that was mounted in the center-section panel of the upper wing. The fabric covered tail-group was built up of welded chrome-moly steel tubing and sheet steel formers; the vertical fin was ground adjustable and the horizontal stabilizer was adjustable in flight. The landing gear of 78 inch tread was of the normal split-axle type and used two spools of rubber shock-cord in tension to absorb the shocks; Bendix wheels (30x5) and brakes were standard equipment. A metal propeller, navigation lights, and inertia-type engine starter were also available. The next development in the Travel Air biplane was the 6 cylinder Curtiss "Challenger" powered model BC-4000; for discussion of this craft see the chapter for A.T.C. # 189 in this volume.

Listed below are model E-4000 entries that were gleaned from various records; the complete listing would be quite extensive, so we will list only the first twenty or so.

C-9079; E-4000 (# 861) Wright 5.
C-9874; E-4000 (# 1060) Wright 5.
NC-9918; E-4000 (# 1104) Wright 5.
NC-9952; E-4000 (# 1150) Wright 5.
NC-9955; E-4000 (# 1153) Wright 5.
NC-9957; E-4000 (# 1155) Wright 5.
NC-9954; E-4000 (# 1157) Wright 5.
C-9959; E-4000 (# 1158) Wright 5.
NC-647H; E-4000 (# 1223) Wright 5.
NC-675H; E-4000 (# 1232) Wright 5.
NC-676H; E-4000 (# 1233) Wright 5.
NC-678H; E-4000 (# 1235) Wright 5.
NC-697H; E-4000 (# 1238) Wright 5.
CF-AIC; E-4000 (# 1240) Wright 5.
NC-699H; E-4000 (# 1241) Wright 5.
NC-602K; E-4000 (# 1246) Wright 5.
NC-622K; E-4000 (# 1247) Wright 5.
NC-623K; E-4000 (# 1248) Wright 5.
NC-616K; E-4000 (# 1249) Wright 5.
NC-618K; E-4000 (# 1251) Wright 5.
CF-AME; E-4000 (# 1382) Wright 5.
Model E-4000 serial # 861 was converted from a "Hisso" powered model 3000; model E-4000 serial # 1060 was production version prototype; model E-4000 serial # 1240 and serial # 1382 were registered in Canada; model E-4000 serial # 1382 was one of the last examples in this series.

Fig. 284. The new wings of the E-4000 lacked the familiar "elephant ear" ailerons and were of heavier construction. The E-4000 was the most popular of the 4000 series.

Fig. 285.   The Travel Air BC-4000 with the Curtiss   "Challenger" engine was a rare type, shown here in prototype form.

We have already covered a "Challenger" powered Travel Air biplane in our previous discussion of the model C-4000. The model BC-4000 that is in discussion here might be described as somewhat typical, but there were several changes that would cause it to stand well apart from the other "Challenger" powered version. The major outward change, that was certainly quite noticeable, was the landing gear that was of the robust outrigger type; an undercarriage that was trussed in connection with oleo-spring shock absorber struts, and was better able to handle the higher gross loads under which this craft would operate. This new craft was also fitted with the new wing panels that were somewhat stronger and heavier; the well-rounded wing tips did away with the familiar "balance horns" of the ailerons. Other deluxe features of appointments and equipment would definately single out this airplane for the sportsman-pilot, but timing and circumstances joined forces to keep this model from developing any further. Only one example of this model was built (C-9821, serial # 1041) and it was later mounted on Edo twin-float seaplane gear as the model SBC-4000; approval for this seaplane version was issued in November of 1929 on Group 2-154. The type certificate number for the model BC-4000 as powered with the 170 h.p. Curtiss "Challenger" engine, was issued in July of 1929 and it was manufactured by the Travel Air Co. at Wichita, Kan.

Let us just imagine for a moment or two that we were in a vast show-room that was displaying the various "Travel Air" models that were available through 1929; if they were all exhibited on the show-room floor, this is what we'd see. On one side would be arranged, all the various biplanes and on the other side would be all the various monoplanes; if you were a sport at heart like most of us are, you would naturally be attracted to the line-up of biplanes first. The first in line you recognize immediately, it is the standard model 2000 that is powered with the familiar OX-5 engine; this model has been in production for nearly 5 years now and is still available, though somewhat improved now over the first offering of some 1000 airplanes back. We notice a placard saying that this airplane is also available with the water-cooled vee-type "Hisso" engine as the model 3000, and there is a version powered with the Curtiss C-6 engine as the model SC-2000.

Anxious to see the rest we move on, and come to the W-4000 which is powered with the Warner "Scarab" engine; a tidy little craft that would be nice to own, but we are distracted by the B-4000 that is standing right next to it. The 4000 has always been a favorite, and this new model with the robust-looking outrigger landing gear and the rounded wing tips, sure has a business-like air about it, but we sort of miss the old "elephant ears" which to us meant "Travel Air" for many years back. One last turn around this craft and we see that it is also available as a one-seater for carrying mail; they call this Wright J5 powered version the BM-4000. What's this one right here, it has a 7 cylinder Axelson engine and is called the model A-4000; nice looking airplane with trim lines and a reasonable price-tag. Having heard much about the Curtiss "Challenger" engine, we recognized it on the model C-4000; noticing that it is a "staggered radial" like two 3 cylinder engines put together. Right next to it is the model E-4000 that is powered with the new 5 cylinder Wright J6 series engine; Wright engines have a terrific reputation and we just know this model will be very popular.

What's this, another "Challenger" powered Travel Air biplane, but this one seems different just by the way it stands. It has the outrigger landing gear, the rounded wing tips, and is labeled the BC-4000; real nice airplane, and it is available with "floats" too as the model SBC-4000. Well what do you know, here is one powered with the 5 cylinder Kinner K5 engine and is called the model K-4000; sort of impressed, we couldn't help to think that this one would be fun to own. What's this one here, it looks like another B-4000, but it has the new 7 cylinder Wright J6 engine; plenty of class apparent in this one and it's called the model 4D. Moving on we see an engine that looks rather familiar, but the name-plate says "Jacobs" and something about A.C.E. LA-1; looks quite nice, this is the model 4P. We were just thinking that some "efficiency expert" probably decided the three zeros were superfluous in the making up of model designations, when we noticed we had come to the end of the line-up of biplanes. In the aisle we notice a big poster of photos; these are the more or less custombuilt craft that are also available. Look at that "Speedwing", that is the D-4000 with the J5 engine and it sure looks good rounding that pylon. Here's a craft that has it's nose stuck up in the air at what seems to be a 30 degree angle, and the tail-wheel has barely left the ground; no wonder, this is the fabulous B9-4000 and it is powered with the 300 h.p. Wright J6-9 engine. We can just imagine that this one is a real "tiger".

Hurrying along because there is yet much to see, we swing over to the stately monoplanes and look for the familiar 6000 with the Wright J5 engine, but notice that the engine has been changed to the 300 h.p. Wright J6-9; it is available with floats too and they now call it the model S-6000-B. One turn around this classic beauty, and we notice that the next one in line has been fitted with the 450 h.p. "Wasp" engine; a combination with plenty of muscle that is called the A-6000-A, and is also available with twin-float seaplane gear as the SA-6000-A. There are seats for 6 or 7 and a peek inside reveals limousine comforts. This next one is a little different and is a good bit smaller; it is the model 10D and carries 4 with the power of a 7 cylinder Wright J6 engine. A nice-looking airplane that would be suitable for the private-owner who likes to take his family along on flights to here and there. Well, we have seen them all and are preparing to leave, when we are shunted through a side door into a separate room. There she was in all her splendor, the famous Travel Air "Mystery Ship". Certainly everyone has heard or read about the "Mystery R", and as we gaze upon it's sleek lines, that must have been born of divine inspiration, we just know that it is bound to influence aircraft design for many years to come. As we leave with a last look, we harbor the pleasing thought that this was indeed time well spent. Of course this was only imagining, but if you were there, this is what you would have seen.

Listed below are specifications and performance data for the Travel Air biplane model BC-4000 as powered with the 170 h.p. "Challenger" engine; length overall 24'6"; hite overall 9'1"; wing span upper 33'0"; wing span lower 28'10"; wing chord upper 66"; wing chord lower 56"; wing area upper 171 sq.ft.; wing area lower 118 sq.ft.; total wing area 289 sq.ft.; airfoil T.A.#2 upper & T.A.#1 lower; wt. empty 1793; useful load 1007; pay-

Fig. 286. The Travel Air BC-400 had new wings and improved landing gear; it was also available on floats.

load with 67 gal. fuel was 392 lbs.; gross wt. 2800 lbs.; max. speed 122; cruising speed 104; landing speed 50; climb 720 ft. first min. at sea level; service ceiling 12,500 ft. gas cap. 67 gal.; oil cap. 6. gal;.; cruising range 600 miles; price at the factory field was $6500.

The fuselage framework was built up of welded chrome-moly steel tubing, faired to shape with wood fairing strips and fabric covered. The cockpits were comfortable and well upholstered; there was a good-sized baggage compartment in the lower fuselage just behind the rear cockpit that was accessible through a door on the left side. The wing framework was built up of solid spruce spars that were routed to an "I beam" section, with spruce and plywood built-up wing ribs; the completed framework was fabric covered. The new style wings as mounted on this model had well-rounded wing tips and the ailerons were of the Freise inset-hinge type for aerodynamic balance; there were three fuel tanks in this model and they were all mounted in the upper wing. One fuel tank was mounted in the center-section panel, and the other two fuel tanks were mounted in the root end of each upper wing half. The fabric covered tail-group was built up of welded chrome-moly steel tubing and sheet steel formers; the fin was ground adjustable and the horizontal stabilizer was adjustable in flight. The landing gear of 78 inch tread had oleo-spring shock absorbing struts; wheels were 30x5 and brakes were standard equipment. A Curtiss-Reed forged aluminum alloy propeller, navigation lights, and inertia-type engine starter were also standard equipment. The next development in the Travel Air biplane was the model K-4000 as powered with the 100 h.p. Kinner K5 engine; this delightful combination will be discussed in the chapter for A.T.C. # 205.

Fig. 287. An Alexander "Eaglerock" Model A-15 with a 100 h.p. Kinner K-5 engine; the A-15 had a pleasant personality.

The Alexander "Eaglerock" A-series biplane was a versatile general-purpose airplane that had been fitted in combination with many different engines; with the structure calculated and stressed for a variety of power, any engine from 90 to 260 horsepower could be installed. Already approved with power combinations such as the Curtiss OX-5, Wright J5, the Hispano-Suiza A and E, Comet 130-165, the Curtiss "Challenger", and now the 5 cylinder Kinner K5; the "Eaglerock" was also approved with the 5 cylinder Wright J6-165, and then there were many interesting special versions. Among these were combinations with the Ryan-Siemens 125, the Menasco-Salmson 230-260, the Hallett 130, the Curtiss C-6A, the Wright J4-200, the Anzani 120, Floco (Axelson) 150, Jacobs & Fisher 120, the Warner "Scarab" 110, the Dayton "Bear" 120, the MacClatchie "Panther" 150, and the Brownback 90; if there were any other combinations they are not known, but chances are good that there were some. One interesting thing about the average 3 place open cockpit biplane of the general-purpose type, as developed during this period, was the wide variety of engine combinations that were possible; this brought about many assorted airplanes of

great interest. Memory still recalls the time we pedalled well over two miles in anxious haste to see a J4 powered Travel Air that had landed at the local airport, the rewarding sight was well worth the trouble.

The Alexander "Eaglerock" model A-15 was typical to the standard A-series biplane as being produced at this time, except for it's engine installation which in this case was the 5 cylinder Kinner K5 of 100 h.p. The Kinner being a rather light engine in comparison to the average powerplant that was being installed in the "Eaglerock", made it necessary to mount the engine quite a distance forward to gain the proper balance about the c.g.; therefore, the A-15 wound up with a rather long nose that was almost of comical proportion, and nearly always evoked a smile and sometimes a laugh or two. Despite it's somewhat comical appearance, the Kinner powered A-15 was a great little craft with a pleasant personality; beside the enjoyable flight characteristics and economy of operation, it shared all the inherent good qualities that made the "Eaglerock" such a great favorite. The type certificate number for the "Eaglerock" model A-15 as powered with the 100 h.p. Kinner engine, was issued in July of 1929 and some

15 or more examples of this model were manufactured by the Alexander Aircraft Co. at Colorado Springs, Colorado; a division of Alexander Industries. The benevolent J. Don Alexander was the president; D. M. Alexander was V.P. in charge of production; J. A. McInaney was V.P. in charge of sales; and Ludwig Muther was chief of engineering.

Listed below are specifications and performance data for the Alexander "Eaglerock" model A-15 as powered with the 100 h.p. Kinner engine; length overall 25'11"; hite overall 9'10"; wing span upper 36'8"; wing span lower 32'8"; wing chord both 60"; wing area upper 183 sq.ft.; wing area lower 153 sq.ft.; total wing area 336 sq.ft.; airfoil "Clark Y"; wt. empty 1423; useful load 838; payload with 39 gal. fuel was 393 lbs.; gross wt. 2261 lbs.; max. speed 100; cruising speed 85; landing speed 35; climb 550 ft. first min. at sea level; climb in 10 min. was 4800 ft.; service ceiling 10,500 ft.; gas cap. 39 gal.; oil cap. 5 gal.; cruising range at 6 gal. per hour was 500 miles; price at the factory field was $4157, lowered to $ 3907; could also be purchased on time payment plan with 40% down, and balance in 20 semi-monthly payments at 10% finance charges; also available for $2250 less engine and propeller.

The fuselage framework was built up of welded chrome-moly steel tubing, lightly faired to shape with wood fairing strips and fabric covered. The cockpits were deep and well upholstered, and there was a baggage compartment of 5 cubic foot capacity. The wing framework was built up of laminated spruce spar beams that were routed out to an "I beam" section, with spruce and plywood built-up wing ribs; the completed framework

was fabric covered. There were four ailerons that were connected together in pairs by a streamlined push-pull strut; with only slight modification, upper and lower wing panels were interchangeable. The gravity-feed fuel tank was mounted high in the fuselage just ahead of the front cockpit; the front cockpit had an entrance door and dual controls were available. The fabric covered tail-group was built up of welded chrome-moly steel tubing; the vertical fin was ground adjustable and the horizontal stabilizer was adjustable in flight. The split-axle landing gear of 72 inch tread was an extremely rugged chrome-moly steel framework and used two spools of rubber shock-cord to absorb the bumps; wheels were 26x4 and brakes were available. Wiring for navigation lights, first-aid kit, and cockpit covers were standard equipment. A metal propeller, wheel brakes, and Eclipse or Heywood engine starter were available as extra equipment. The next Alexander development was the C-7 "Bullet" which will be discussed in the chapter for A.T.C. # 318.

Listed below are "Eaglerock" A-15 entries that were gleaned from various records; this list may not be quite complete, but it does show the bulk of this model that were built.

C-8294; A-15 (# 901) Kinner K5.
NC-718H; A-15 (# 903) Kinner K5.
NC-711H; A-15 (# 904) Kinner K5.
NC-727H; A-15 (# 906) Kinner K5.
NC-719H; A-15 (# 907) Kinner K5.
NC-742H; A-15 (# 908) Kinner K5.
NC-743H; A-15 (# 909) Kinner K5.
NC-745H; A-15 (# 910) Kinner K5.
NC-749H; A-15 (# 911) Kinner K5.
NC-762H; A-15 (# 912) Kinner K5.

Fig. 288. The A-15 shared the typical features that made the "Eaglerock" a long-time favorite; the abundance of wing area and the sturdy undercarriage should be noted.

Fig. 289. An "Eaglerock" A-15 modified from a Model A-2 with an OX-5 engine.

NC-517V; A-15 (# 913) Kinner K5.
NC-726W; A-15 (# 966) Kinner K5.
NC-638W; A-15 (# 968) Kinner K5.
NC-6232; A-15 (# 658) Kinner K5.

Serial # 901 was factory prototype for model A-15; serial # 658 was originally an A-2 with OX-5 engine that was converted to A-15 with the Kinner K5 engine.

Fig. 290. A Curtiss "Fledgling" with a "Challenger" engine, a commercial version of the trainer built for the Navy.

The lovable and almost comical-looking Curtiss "Fledgling" was an amiable airplane of contagious personality that had trained thousands of student pilots at the famous Curtiss Flying Service schools; busy schools that always buzzed with activity and were scattered just about all over the country. At first meeting, the "Fledgling" posed as a rather ungainly looking craft of overly large dimensions, with it's acres of wing area and a literal forest of wing struts and wires, but this all transformed into the surprising phenomonon of a graceful and gentle airplane when off the ground and in the air. Many of the pilots who learned to fly in the friendly "Fledgling" will proclaim readily and loudly to this day, that a safer, more dependable, or easier to fly airplane was never built, before or since. Just about perfect in every respect as a training craft, it was inherently rugged in all the right places to be able to soak up the hard knocks and unintentional mistreatment, to which it was generally exposed in the normal course of pilot training. Well over a hundred of the "Fledgling" trainers had been built and all but 3 or 4 were in operation

with the Curtiss Flying Service schools that were located in just about every state in the union; the Curtiss training centers were part of a system that was the largest "flying service" in the world. Of the 3 or 4 "Fledglings" that were not in service with Curtiss, one was privately owned as a sport-plane and the others were used in pilot training with other privately-owned flying schools.

Ruth Nichols, a gentle and very capable aviatrix of this period, flew a "Challenger" powered "Fledgling" biplane on a 46 state promotion tour in the early summer of 1929; this round-robin tour was to be a shake-down flight for the new "Fledgling" and a promotion of additional flying-school sites that were planned by the Curtiss Flying Service. Accompanied by Robb Oertel in a "Challenger" powered Curtiss "Robin", Ruth Nichols enjoyed the tour immensely and proclaimed the "Fledgling" as an unbeatable platform from which to view the wonderful scenery far down below, especially if you were in no particular hurry. Her one annoying problem was to keep the "Robin" in sight, which had a tendency to pull away from her with it's higher cruising speed and leave her all to herself.

The "Fledgling" biplane was originally designed by Curtiss engineers to participate in a competition set forth by the Navy to select a training craft that would be suitable to serve as both a primary trainer, and an advanced trainer; performing it's duties both as a landplane and as a float-mounted seaplane. Of the 14 designs entered in the competition, the Curtiss "Fledgling" was adjudged the winner and awarded a contract for a good number of these versatile machines. The Navy-type "Fledgling" (N2C-1) was powered with the Wright J5 engine, and beside serving as a primary trainer, was also used in the advanced phases of training that included bomb-dropping, gunnery training for both pilot and rear-gunner, and radio techniques in spotting duties; all training phases could be performed as a landplane or as a seaplane.

The commercial version of the "Fledgling" trainer was powered with the 6 cylinder Curtiss "Challenger" engine of 170 h.p., and all provisions for military equipment were naturally omitted; but the "Fledgling" was easily convertible to the military version should the need arise. Among the many useful features plainly visible in the "Fledgling" design, good vision is evident by the excessive stagger between the upper and the lower wing panels, with a trailing edge cut-out in the center-section of the upper wing for vision upward. It is also plainly evident by the large number of quickly removable inspection plates and panels, that everything was designed for quick inspection and easy maintenance, in order to keep the "hangar time" down to the barest minimum.

The Curtiss "Fledgling" trainer was a 2 place open cockpit biplane seating two in tandem, with an abundance of wing area that was done up in the old fashioned double-bay layout to support the unusually long span of the wings; there used to be a story that

Fig. 291. U. S. Navy "Fledglings" on a training flight; the amiable characteristics were popular with pilots and the ease of servicing was popular with maintenance crews.

many will remember, about the old "Jenny" not being rigged properly if a bird could escape through the wires and wing struts, the Curtiss "Fledgling" was almost as bad. Certainly not one designed for good looks, nor rigged for high speed, the "Fledgling" was first a trainer that was easy to fly and it's amiable flight characteristics were like a friendly hand that guided many thousands through their primary flight training; a craft with a personality that was almost infectious, it will surely remain in memory as an all-time favorite. The type certificate number for the Curtiss "Fledgling" trainer as powered with the 170 h.p. Curtiss "Challenger" engine, was issued in August of 1929 and well over 100 of this model were manufactured by the Curtiss Aeroplane & Motor Co. of Garden City, Long Island, N. Y.; the "Fledgling" was manufactured in a plant of the Buffalo, New York division. After 10 years time, there were still 30 or more "Fledglings" in active service, which is sufficient testimonial for the dependability and popularity of this airplane.

Listed below are specifications and performance data for the Curtiss "Fledgling" as powered with the Curtiss "Challenger" engine at a 175 h.p. rating; length overall 27′8″; hite overall 10′4″; wing span upper 39′1″; wing span lower 39′5″; wing chord both 60″; wing area upper 188 sq.ft.; wing area lower 177 sq.ft.; total wing area 365 sq.ft.; airfoil "Curtiss C-72"; wt. empty 1990; useful load 696; payload with 40 gal. fuel was 236 lbs.; gross wt. 2686 lbs.; max. speed 104; cruising speed 88; landing speed 45; climb 670 ft. first min. at sea level; climb to 5,000 ft. was 9 min.; climb to 10,000 ft. was 22.7 min.; service ceiling 14,100 ft.; gas cap. 40 gal.; oil cap. 3.5 gal.; cruising range at 10 gal. per hour was 346 miles; cruise was based on 85% of max. power; the following figures are for the "Fledgling" (Gaurdsman) that was convertible to the military version; wt. empty 2005; useful load 696; crew wt. 380; gross wt. 2701 lbs.; max. speed 103; cruising speed 87; landing speed 46; climb 645 ft. first min. at sea level; service ceiling 13,400 ft.; all other figures and dimensions remained more or less the same; aircraft with modifications to allow conversion to military-type craft were covered under a Group 2 approval numbered 2-59 issued 7-9-29; the load factor of safety at high incidence was 7.4 for both versions.

The fuselage framework was built up of welded chrome-moly steel tubing in truss form, faired to shape and fabric covered. Inspection plates and removable panels were placed in numerous positions for quick inspection and easy maintenance to all vital components; all movable parts and points of wear were provided with lubrication fittings. Steps were placed for easy entrance to the cockpits, and the seats were adjustable; the seats were provided with wells for a parachute pack. The wing framework was built up of solid spruce spar beams with spruce and birch plywood built-up wing ribs; the leading

Fig. 292. A "Challenger"-powered "Fledgling" on skis.

Fig. 293. The friendly nature of the "Fledgling" trainer endeared it to thousands of student-pilots who remember this awkward-looking two-bay biplane with fondness.

edge of the upper wing was covered with aluminum alloy sheet and the leading edge of the lower wing was covered with plywood. The reason for plywood on the lower leading edge is because metal sheet had a tendency to dent; the completed framework of the wings was fabric covered. There were four ailerons that were connected together in pairs by a streamlined push-pull strut; the ailerons were operated independently by push-pull metal tubes, and the rudder and the elevators were operated by stranded cable. The fuel tank was mounted in the fuselage just ahead of the front cockpit, and a small tool locker was in the turtle-back just behind the rear cockpit. The fabric covered tail-group was built up of welded chrome-moly steel tubing; the fin was ground adjustable and the horizontal stabilizer was adjustable in flight. Both the rudder and the elevators were aerodynamically balanced by large projecting "balance horns". The wide-track landing gear of 87 inch tread was of the split-axle type with two long telescopic legs using a combination of oleo and rubber in compression for the shock absorbers; wheels were 28x4 and the tail-skid was steerable. A Curtiss-Reed metal propeller was provided and a hand crank inertia-type engine starter; the crank handle for winding up the starter was fastened by clips to the engine cowling and a fire-extinguisher bottle was fastened to the cockpit cowling. Everything in the make-up of the "Fledgling" trainer spoke of efficiency with simplicity. The next Curtiss development was the twin-engined "Condor" transport; see the chapter for

A.T.C. # 193 in this volume.

Listed below are "Fledgling" entries that were gleaned from various records; a complete listing of all the "Challenger" powered "Fledgling" would be quite extensive, so we submit a partial listing of 20 or so.

NC-540; Fledgling (# B- 1) Challenger.
NC-8660; Fledgling (# B- 2) Challenger.
NC-8661; Fledgling (# B- 3) Challenger.
NC-8662; Fledgling (# B- 4) Challenger.
NC-8663; Fledgling (# B- 5) Challenger.
NC-8664; Fledgling (# B- 6) Challenger.
NC-8665; Fledgling (# B- 7) Challenger.
NC-8666; Fledgling (# B- 8) Challenger.
NC-8667; Fledgling (# B- 9) Challenger.
NC-8668; Fledgling (# B-10) Challenger.
NC-8669; Fledgling (# B-11) Challenger.
NC-8671; Fledgling (# B-12) Challenger.
NC-8672; Fledgling (# B-13) Challenger.
NC-8673; Fledgling (# B-14) Challenger.
NC-8674; Fledgling (# B-15) Challenger.
NC-8675; Fledgling (# B-16) Challenger.
NC-8676; Fledgling (# B-17) Challenger.
NC-8677; Fledgling (# B-18) Challenger.
NC-8678; Fledgling (# B-19) Challenger.
NC-8679; Fledgling (# B-20) Challenger.
Serial # B-5 and # B-18 were later converted to model J-1 with Wright J6-5-165 engine; NC-483K (B-87) was registered to W. P. Draper; NC-486K (B-90) to C. R. Van Etten Jr.; NC-653M (B-100) to Sidles Airways of Lincoln, Nebraska; all other "Fledglings" registered to the Curtiss Flying Service; the earliest version registered was C-7992 (# 3) which was powered with "Challenger" engine # 6.

Fig. 294. An all-metal "Flamingo" G-2-W with a Pratt and Whitney "Wasp" engine; the fuselage was of steel tubing.

The all-metal airplane with it's lasting qualities, had been the fond dream of aeronautical engineers for over ten years now; many varied examples in shape and structure of this type were developed and built in this length of time, but none of the engineers could quite agree, yet, as to what form the all-metal airplane would finally take. At this stage of development, certain methods of structure and certain configurations were beginning to be accepted as more or less the ideal, but engineers still felt the need for more experimentation in order to decide. Some all-metal aircraft were structures completely built up in aluminum alloy of riveted open section stock and sheet, and some of these craft were built up in a combination of welded steel tube framework and aluminum alloy covering. The all-metal "Flamingo" was such a composite structure, and was one of the finest examples of the light transport type airplane; an airplane with a performance and utility well comparable to the best, with the added bonus feature of a sound metal structure that would stand up for years of service, in all climates, with a negligible amount of deterioration. The "Flamingo" aircraft were first sold with a 90 day warranty on the structure, but actual daily service on a number of

air-lines proved this to be just a token gesture, so they extended the warranty to six years; they at the plant felt reasonably sure that the "Flamingo" was good for ten years at least without any replacements in the structure or outer skin being necessary, regardless of the climates, the housing, or the manner of use. Nothing quite like this had ever been offered before. Three air-lines were soon using the "Flamingo" light transport on daily schedules, and were showing a reasonable profit on their investment; these carriers were Embry-Riddle, which served Cincinnati-Indianapolis-Chicago; the Mason & Dixon Air Line, which served Cincinnati to Detroit; and U. S. Airways which served Kansas City to Denver. With a rosy future looming ahead, things looked very promising for the Metal Aircraft Corp. but developments in the next year or so almost bankrupted the nation, and with this economic sag fell the further manufacture and development of the "Flamingo" airplanes.

The prototype "Flamingo" G-MT-6, as shown here, was a Thomas E. Halpin development that was designed and engineered by Ralph R. Graichen, and was introduced in the early part of 1928. The success of this design, proven by excellent characteristics and performance, was forecast almost immediately and the Metal Aircraft Corp. was formed as a successor to the Halpin Development Co. to manufacture these craft in quantity. Redesigned slightly over the G-MT-6 (Graichen-Metal Transport-6 place), the first production model G-1 was rolled off the line towards the latter part of the year, and others were soon to follow. Early development of the "Flamingo" series kept the seating down to 5 or 6, and it was tried in both the Pratt & Whitney "Wasp" and "Hornet" powered versions; these early examples were awarded Group 2 approvals and are listed at the end of this chapter. The model G-2-W as later approved, was an 8 place high wing cabin monoplane and was powered with the 9 cylinder P & W "Wasp" engine of 450 h.p. The type certificate number for the model G-2-W was issued in August of 1929 and some 21 at least were known to be built in the various versions by the Metal Aircraft Corp. at Lunken Airport, in Cincinnati, Ohio. Henry C. Yeiser Jr. was the president; Julius Fleischmann was 1st V.P.; Powell Crosley Jr. was 2cd V.P.; Thomas E. Halpin was 3rd V.P. and general manager; Ralph R. Graichen was 4th V.P. and the chief engineer; and John B. Hollister was the secretary.

Listed below are specifications and performance data for the Metal "Flamingo" model G-2-W as powered with the 450 h.p. P & W "Wasp" engine; length overall 32'8"; hite overall 9'6"; wing span 50'0"; wing chord 96"; total wing area 365 sq.ft.; airfoil M-12;

wt. empty 3370; useful load 2430; payload with 150 gal. fuel was 1265; gross wt. 5800 lbs.; max. speed 135; cruising speed 115; landing speed 60; climb 850 ft. first min. at sea level; service ceiling 14,000 ft.; gas cap. 150 gal.; oil cap. 11-14 gal.; cruising range at 22 gal. per hour was 745 miles; price at the factory field was $21,000. The following figures are for the 8 place "Hornet" powered model G-2-H; wt. empty 3570; useful load 2430; payload with 150 gal. fuel was 1265; gross wt. 6000 lbs.; max. speed 145; cruising speed 122; landing speed 62; climb 1200 ft. first min. at sea level; service ceiling 16,000 ft.; all other dimensions and data were typical; price at the factory field was $23,800. The fuselage framework was built up of welded chrome-moly steel tubing and was faired to shape with "dural" channel strips to which the corrugated aluminum alloy skin was riveted. The cabin was fitted with 8 seats and there was a lavatory, and a 40 cubic foot baggage compartment to the rear of the cabin section; six of the seats were quickly removable for hauling bulky freight loads. The passenger door was on the right side and to the rear of the main cabin, and the pilot had an entry door up front on the left side. The wing framework of semi-cantilever design, was built up of girder-type spar beams of "dural" (duralumin) members riveted together in an "I beam" section, with wing ribs of flanged duralumin stamped-out sheet, and the completed framework was covered with corrugated "dural" skin. The skin of both

Fig. 295. The "Flamingo" G-MT-6 prototype was a Halpin development designed by Ralph Graichen.

Fig. 296. The "Flamingo" carried eight with good performance; its rugged dependability proved worthy in the bush country.

the fuselage and the wing was corrugated on 3 inch centers for added stiffness and strength. The tail-group was built up of riveted channel sections of duralumin stock, and was also covered with corrugated metal skin; the fin was ground adjustable and the horizontal stabilizer was adjustable in flight. The landing gear was of the outrigger type and used "Aerol" shock absorber struts; the wheel tread was 120 inches and the heavy duty wheels were 32x6. Two gravity-feed wing tanks of 75 gallon capacity each, were mounted in the root end of each wing, flanking the fuselage. Dual control was provided in a swing-over control wheel, with two sets of rudder pedals and separate pedals for the wheel brakes. A ground-adjustable metal propeller, Bendix wheels and brakes, and inertia-type engine starter were standard equipment; complete night-flying equipment was available.

When the "Flamingo" was finally retired from active air-line service in the mid-thirties, they were soon after given a new lease on life by enterprising "bush pilots" who recognized the rugged and dependable airplane as one that was well suited to their type of work. One of these pilots was the fearless Jimmy Angel, a roving "bush pilot" who was flying in So. America on various assignments. While looking for a legendary river-of-gold, he flew daringly into a deep narrow canyon and came upon a waterfall that was plummeting down thousands of feet from the clouds, to the jungle floor far down below. The column of water dropped over 2600 feet straight down to a ledge below, and then another 500

feet or so to the floor of the jungle. Here was a giant waterfall of some 3200 feet that was twenty times higher than Niagara Falls, and at least twice as high as the Empire State building! Thrilled at the sight and bent on further exploration of the falls, Jimmy Angel landed his "Flamingo" on the jungle floor near the base of the falls, and became mired and stuck fast in the soggy ground that was hidden by a deceptive cover of lush green grass. Stuck fast and nearly swallowed up by the jungle ooze, the trio aboard had no choice but to leave it in it's jungle grave. After a harrowing two-week journey through the jungle on foot, they reached their camp safely, and the fallen "Flamingo" is probably still there; a stout-hearted symbol of the "Flamingo's " rugged nature.

Listed below are "Flamingo" entries that were gleaned from various records; this list is not complete, but it does show the bulk of the aircraft that were built.

X-4487; G-MT-6 (# 1) Wasp 420, 5 pl., proto, Halpin Development Co.

C-7690; G-1 (# 1) Wasp 450, 5 pl., production proto, Metal A/C Corp. Group 2-19 (9-22-28).

NC-588; G-2 (# 2) Wasp 450, 6 pl., Group 2-63 (5-14-29).

NC-9304; G-2-H (# 3) Hornet 525, 6 pl., Group 2-67 (5-13-29).

NC-656E; G-2-W (# 4) Wasp 450, 8 pl., Group 2-62 (7-31-29).

        ; G-2-W (# 5) Wasp 450, 8 pl., Group 2-62 (7-31-29).

        ;G-2-W (# 6) Wasp 450, 8 pl., Group 2-62 (7-31-29).

Fig. 297. This late version of the G-2-W operated on the U.S. Airways route from Kansas City to Denver.

C-662E; G-2-W (# 7) Wasp 450, 8 pl., Group 2-62 (7-31-29).

NC-655E; G-2-W (# 8) Wasp 450, 8 pl., Group 2-62 (7-31-29).

NC-316H; G-2-H (# 9) Hornet 525, 8 pl., Group 2-75 (6-12-29).

C-317H; G-2-W (#10) Wasp 450, 8 pl., Group 2-62 (7-31-29).

NC-9487; G-2-W (# 11) Wasp 450, 8 pl., Group 2-62 (7-31-29).

NC-9488; G-2-W (# 12) Wasp 450, 8 pl., A.T.C. # 192.

NC-9489; G-2-W (# 13) Wasp 450, 8 pl., A.T.C. # 192.

NC-9490; G-2-W (# 14) Wasp 450, 8 pl., A.T.C. # 192.

NC-9491; G-2-W (# 15) Wasp 450, 8 pl., A.T.C. # 192, unverified.

NC-265K; G-2-W (# 16) Wasp 450, 8 pl., A.T.C. # 192.

NC-266K; G-2-W (# 17) Wasp 450, 8 pl., A.T.C. # 192.

NC-267K; G-2-W (# 18) Wasp 450, 8 pl., A.T.C. # 192.

C-268K; G-2-W (# 19) Wasp 450, 8 pl., A.T.C. # 192.

C-269K; G-2-W (# 20) Wasp 450, 8 pl., A.T.C. # 192.

NC-279K; G-2-H (# 21) Hornet 525, 8 pl., A.T.C. # 192.

A.T.C. #193
(8-29)
CURTISS "CONDOR", CO

Fig. 298. The huge Curtiss "Condor" (CO) transport with two 12 cylinder Curtiss "Conqueror" engines; the twin-engined "Condor" biplane was an unusual craft in many ways.

The huge Curtiss "Condor" model CO transport was a commercial version of the U. S. Army "Condor" bomber; following a re-design of the basic military concept into a passenger-carrying transport, the first commercial "Condor" was flight-tested during the summer of 1929. Flight tests and numerous shake-downs, which also included an active participation in the National Air Tour for 1929 (flown by W. J. Crosswell to 4th place), pointed to numerous modifications that were needed, and consequently, these changes and improvements were incorporated into the production models soon coming off the line. The first few saw service in 1930 with the newly organized Transcontinental Air Transport (T.A.T.), and before long one saw service with Eastern Air Transport (E.A.T.); carrying 18 passengers in complete comfort and luxury, at a high enough cruising speed to offer fast schedules between distant points. The monstrous "Condor" could be considered as quite conventional except for it's huge proportions; truly a large craft that was over 16 feet high, nearly 58 feet long, and had a wing span of close to 100 feet. Aerodynamically it was also unusual because of the fact that it was a biplane; all but one other of the large transports available at this time were monoplanes. Also unusual was the fact that it was the only large transport to use liquid-cooled engines, and only used two engines, whereas most other transports used air-cooled engines and were tri-motored craft.

Excepting for interior arrangement, the "Condor" commercial transport bears a marked resemblance to it's military sistership; engines were the 12 cylinder geared (2 : 1) Curtiss "Conqueror" (GV-1570) of 600 h.p. each and were mounted in long streamlined nacelles that were built into the top side of the lower wing. Cabin interior was large, comfortable, and quite plush; divided into 3 sections, each section seated 6 passengers, and lavatory facilities were available in the center portion of the cabin section. A complete ventilation and heating system was provided for passenger comfort; the pilot's compartment, being in the extreme front end of the fuselage, offered the maximum in visibility. Altogether, this craft poses as an interesting milestone in the progress of air transportation; credit for the design and development of the "Commercial Condor" was mainly due to one T. P. Wright who was chief of the Curtiss design and engineering section, and to George Page, and Alexander Noble, project engineers. The type certificate number for the "Conqueror" powered "Condor" model CO was issued in August of 1929 and amended and re-issued in October of 1929 for added equipment and a higher permissable gross load. The "Condor" was manufactured by the Curtiss Aeroplane & Motor Co. at Garden City, Long Island, New York; a division of the Curtiss-Wright Corp.

Listed below are specifications and performance data for the Curtiss "Condor" trans-

Fig. 299. The "Condor" carried 18 passengers in comfort; performance was excellent for a craft so large.

port, model CO; length overall 57′6″; hite overall 16′3″; span upper and lower 91′8″; chord both 108″; total wing area 1510 sq.ft.; airfoil "Curtiss C-72"; gap between panels 126″; dihedral lower only 3 deg.; wt. empty 11,352 (11, 574); useful load 6326; payload 2876; gross wt. 17,678 (17,900) lbs.; max. speed 139; cruising speed 120; landing speed 50; climb 925 ft. first min. at sea level; service ceiling 17,000 ft.; gas cap. 446 gal.; oil

Fig. 300. The Curtiss "Condor" prototype on an early flight; commercial version stemmed from a bomber.

Fig. 301.   Clarence Chamberlin, famous transatlantic flyer, "barnstorming" in the Midwest .with a "Condor".

cap. 38 gal.; range 550 miles. Weights in brackets were as per certificate issued 8-29. The following weights, wt. empty 11,574; useful load 6326; gross wt. 17,900 were for serial # G-1, G-2, and G-3. The following weights, wt. empty 11,818; useful load 6082; gross wt. 17,900 were for serial # G-4 and up. The fuselage framework of the huge "Condor" conforms to standard "Curtiss" practice, embodying a combination of welded chrome-moly steel tubing and plate at highly stressed points, with aluminum alloy tubing in the lower stressed areas; tubes that were riveted to special wrap-around clamps to make the joint. The fuselage structure used K type bracing in the cabin section to avoid having a tube pass through the window area, and the aft section was in Warren truss form. The fuselage frame was faired to shape with formers and fairing strips and then fabric covered; that is, except for the arched cabin roof which was of ribbed aluminum alloy. Cabin walls were sound-proofed and insulated with a 3 inch blanket of Dry-Zero. The main cabin was fully 6 feet wide and had a head-room of 6 foot and 8 inches; displacing at least 865 cubic feet. Double seats were on the star-board side and single seats were on the port side, with an aisle running the full length. Two main entrance doors were provided, with additional exits in case of emergency. Engine nacelles were of a structure similiar to that of the fuselage, but were covered completely with aluminum alloy pa-

nels; the engines were mounted in the forward section of the nacelles and in the rear section of each nacelle housed a fuel tank of 192 gallon capacity, and a baggage compartment. A gravity-feed fuel tank of 62 gallon capacity was mounted in the upper wing, and fuel from the nacelle tanks could be transferred to the wing tank by engine-driven fuel pumps. Each engine swung a large slow-turning three-bladed propeller that was 13 feet in diameter, and the engine coolant radiators were mounted in a streamlined housing, just above and to the rear of the engines; adjustable shutters were provided for temperature control. The wing framework which was more or less of the three-bay type, was made up of spars that were built up of welded chrome-moly steel tubing in a girder-type form, with wing ribs of aluminum alloy tubes that were riveted to the spar beams. The leading edge of each wing was covered to preserve airfoil form with aluminum alloy sheet and the completed framework was then fabric covered. The ailerons were constructed in a manner somewhat similiar to the wing structure, and also fabric covered. There was a plywood covered walk-way between the fuselage and the engine nacelle on each side, for servicing the engines. The "Condor" had a biplane tail-group with double rudders, one in the slipstream of each engine; construction of the tail-group was in a combination of steel and aluminum alloy tubing that was welded and riveted together, and then fabric covered.

The horizontal stabilizers were adjustable in flight, and the rudders had a "compensator" of rubber shock-cord which was adjustable to relieve the strain of "holding rudder" due to high engine torque or when operating on one engine. The split-axle landing gear was of a 3 member unit built of chrome-moly steel tubing on each side, that employed large "oleo-spring" struts which provided a 14 inch deflection on extra hard landings; Bendix wheels (54x12) and brakes were provided, and the tread was over 20 feet. Other standard equipment included a steam-generator for the heating system, a ventilation system to provide fresh air, two Curtiss-Reed ground adjustable metal propellers, radio equipment for sending and receiving, carbon-dioxide fire extinguishers in the main cabin and in the nacelles, (those in the nacelles were operated from the cockpit), adjustable cabin seats, shatter-proof glass throughout, a complete lighting system for the interior and the exterior, engine-driven generators, two 12 volt batteries, engine-driven fuel pumps, and Eclipse electric engine starters. Truly indeed, the big, beautiful "Condor" was the ultimate in luxurious air-travel at this time. In the latter "thirties", Clarence Chamberlin, who flew the Atlantic Ocean from New York to Germany in 1927, was barnstorming the mid-west in one of these old "Condors"; a tired looking and well-worn craft that had surely seen it's hey-dey, but was still impressive enough to command ones attention. Rides were being sold to "see the city from the air for $3.00", and a good many people were parting with hard-earned money to enjoy the experience. After exchanging a hello with the famous Chamberlin, we decided to take a ride to be able to say we had flown with the intrepid trans-Atlantic flyer. Seated in the forward section, it looked like the "props" weren't turning fast enough to get us off even though we knew the engines were "revving" wide; after many rumbles and jolts we did get off, climbing at a good clip, and we were on our way. Though certainly not new to flying, the flight was enjoyed immensely because of the majestic size of the "Condor" and besides, we had one of the world's greatest pilots at the wheel.

The next Curtiss development after the "Condor", was the cargo-carrying "Conqueror-Falcun", which will be discussed in the chapter for A.T.C. # 213. Listed below are Curtiss "Condor" CO (2 Conqueror) entries that were gleaned from the Aeronautical Chamber of Commerce aircraft register of 1930:

NC-185H; Condor CO (# G-1) 2 Conqueror
NC-725K; Condor CO (# G-2) 2 Conqueror
NC-984H; Condor CO (# G-3) 2 Conqueror
NC-726K; Condor CO (# G-4) 2 Conqueror
NC-727K; Condor CO (# G-5) 2 Conqueror
NC-728K; Condor CO (# G-6) 2 Conqueror
Recorded entries had not been found for any additional aircraft of this model, but it is a good possibility that more were built.

A.T.C. #194
(8-29)
STINSON "JUNIOR", SM-2AC

Fig. 302. The Stinson "Junior" Model SM-2AC had a 225 h.p. Wright J6 engine and carried four.

The Stinson "Junior" model SM-2AC was not very much different from previous SM-2 versions, except that constant development of this series brought about increases in horsepower to better the performance, and various other changes and improvements to the structure and to the airplane as a whole, to make it more attractive as a personal-type airplane for the private-owner or the business-man. Improvements both in appearance and comfort, and also in utility and performance. The new J6 series of the Wright "Whirlwind" engine were offered in the 5 cyl., 7 cyl., and 9 cyl. versions which were rated at 165, 225, and 300 h.p. The model SM-2AC was powered with the "Whirlwind" J6-7 which developed 225 h.p., and though performance of this new combination was almost identical to that of the J5 powered SM-2AB, the new series engine had much to offer in the way of performance and reliability over the old J5 "Whirlwind" which was soon to be out of production. It was now definately established that the "Junior" was of a size and power combination most suitable to the purposes for which it was generally used, so further development of this series for the next 5 years or so, paralleled along the same line without any major changes. Stinson Aircraft was now comfortably located in their new plant at Wayne, Mich. which was a modern lay-out of some

90,000 square feet, and the adjoining factory airfield was most always a bee-hive of activity due to the delivery of new airplanes which were coming off the line in rapid succession, and also busy with experimentation in new models that would grace the airports of the country in the coming year.

The type certificate number for the model SM-2AC as powered with the new 7 cylinder Wright J6 engine of 225 h.p., was issued in August of 1929 and this approval was amended in November of 1929 to include the model SM-2ACS which was typical except that it was mounted on Fairchild twin-float seaplane gear. The SM-2AC was manufactured by the Stinson Aircraft Corp. at Wayne, Michigan and some 20 or more of this model were built. An overall average of 3 airplanes per week were rolling off the production line, and Stinson had now become the undisputed leader in the manufacture of large commercial cabin monoplanes. The genial "Eddie" Stinson often engaged in good-natured boasting about his airplanes, and one could certainly agree that he had much to be happy about at this particular time.

Listed below are specifications and performance date for the Stinson "Junior" model SM-2AC as powered with the 7 cyl. Wright J6 engine of 225 h.p.; length overall 28'6"; hite overall 8'4"; wing span 41'6";

wing chord 75"; total wing area 238 sq.ft.; airfoil "Clark Y"; weight empty 2091 lbs.; useful load 1126; payload with 70 gal. fuel was 510 lbs.; gross wt. 3217 lbs.; max. speed 132; cruising speed 113; landing speed 49; climb 850 ft. first min. at sea level; service ceiling 16,000 ft.; gas cap. 70 gal.; oil cap. 6 gal.; cruising range 600 miles; price at the factory field was $11,000. The following figures are for the float-equipped SM-2ACS seaplane; wt. empty 2396; useful load 1126; payload with 70 gal. fuel was 510 lbs.; gross wt. 3522 lbs.; max. speed 120; cruising speed 100; landing speed 55; climb 700 ft. first min. at sea level; service ceiling 12,000 ft.; price at the factory was about $13,500. with "Fairchild" floats.

The fuselage framework on the SM-2AC was quite typical to previous models and was gusseted with chrome-moly steel plate and sheet stock at every joint to make a structure of exceptional strength and rigidity. The cabin interior was upholstered in tasteful combinations and all windows were of shatterproof glass that could be rolled up and down; there was a sky-light in the roof of the cabin for vision overhead. The engine was muffled by a large-volume collector-ring that was mounted on the front of the engine; this muffler cut down engine noise a great deal and allowed normal conversation amongst occupants in the cabin at all times. The wing framework was also typical, and we could point out a few details that were missed in previous discussions of the "Junior". The ailerons were of the Freise balanced-hinge type and were built up in a welded steel tube framework that was fabric covered. The thick-sectioned wing bracing struts were of heavy gauge chrome-moly steel tubes that were encased in a balsa wood fairing; these fairings were shaped to an "Eiffel 380" airfoil section for added lift and stability. The heavy

wing spars were of solid spruce and were routed to an "I beam" section; the leading edges of the wing were covered with duralumin sheet back to the front spar to preserve the airfoil form. Wheel brakes were standard equipment and the tail-wheel was steerable for easy ground manuvering. The cabin doors were large and there was a convenient step on each side for exit and entry. The next development in the Stinson monoplane line was the model SM-1FS which was the familiar 6 place "Detroiter" on twin floats, this craft will be discussed in the chapter for A.T.C. # 212.

Listed below are SM-2AC entries that were gleaned from various records; this list may not be complete, but it does show the bulk of this model that were built.

X-8432; SM-2AC (# 1055) Wright J6-7-225.
C-8475; SM-2AC (# 1070) Wright J6-7-225.
NC-462H; SM-2AC (# 1076) Wright J6-7-225.
NC-452H; SM-2AC (# 1093) Wright J6-7-225. Grp. 2-143 on floats.
NC-454H; SM-2AC (# 1095) Wright J6-7-225.
NC-461H; SM-2AC (# 1099) Wright J6-7-225.
NC-467H; SM-2AC (# 1103) Wright J6-7-225.
NC-470H; SM-2AC (# 1105) Wright J6-7-225.
NC-471H; SM-2AC (# 1106) Wright J6-7-225.
NC-474H; SM-2AC (# 1109) Wright J6-7-225.
NC-478H; SM-2AC (# 1113) Wright J6-7-225.
NC-456H; SM-2AC (# 1117) Wright J6-7-225.
NC-481H; SM-2AC (# 1118) Wright J6-7-225.
NC-485H; SM-2AC (# 1119) Wright J6-7-225.
NC-482H; SM-2AC (# 1120) Wright J6-7-225.
NC-488H; SM-2AC (# 1121) Wright J6-7-225.
NC-489H; SM-2AC (# 1124) Wright J6-7-225.
NC-402M; SM-2AC (# 1127) Wright J6-7-225.
NC-403M; SM-2AC (# 1128) Wright J6-7-225.
NC-405M; SM-2AC (# 1129) Wright J6-7-225.
There were 14 unlisted aircraft in the SM-2 series, some of these may have been SM-2AC; others were SM-2AA and SM-2AB.

Fig. 303. The clamor for higher performance caused power increases which pulled the Stinson "Junior" out of the lightplane class; the SM-2AC had performance to please the most demanding.

Fig. 304. The Spartan C3-165 with Wright J6 engine; the craft was noted for its good behavior and manners.

The Spartan C3-165 as shown here, was the latest development in the popular C3 series as built for nearly the past two years. Though still quite typical in general form, the new C3 was faired-out better to a more buxom shape, had a new stout undercarriage and was powered with the 5 cylinder Wright J6 engine of 165 h.p.; altogether impressing one as a youngish matron that still looked rather pretty, but was beginning to put on weight. The power increase of some 40 h.p. over the previous models, had not actually improved the performance to any degree because of the extra bulk taken on and the added weight of airframe and useful load. The configuration and aerodynamic proportion was still pretty much as originally conceived, and though the general appearance had been changed somewhat, the new C3 was still a very neat and well-mannered airplane. Easy to fly and inherently stable, the C3 had always been a favorite with the flying-salesmen and the business-men who appreciated the chance to fly relaxed; concerned only to get to their appointed destination with the minimum in effort or piloting technique. Willis Brown, himself a flying-salesman for many years, fashioned the Spartan biplane with this very thing in mind, and the busy salesman or preoccupied executive, and the sight-seeing private owner were very grateful for it.

Willis C. Brown, designer of the Spartan biplane, had been in Europe negotiating contracts for the "Walter" engine because suitable American engines in the medium power range had not yet been available; upon his return to the U. S. he discovered to his dismay that the Spartan biplane had been redesigned somewhat, and a program of testing several newly developed American engines had already been initiated. This unwarranted behavior naturally had it's effects and caused enmity because of the principle involved, causing a breech in relations; tactics such as these convinced Brown to resign, sell out his interests and pull out of the company. When Brown severed his connections with Spartan Aircraft, he remained in Tulsa as distributing agent for the Walter engines here in the U. S., but Walter engines were not selling so Brown went on to Warner Aircraft (engines) as sales manager. Actually, the unwarranted redesign of the Spartan biplane was detrimental only in matter of principal; as far as the airplane itself was concerned, it was a step in the right direction. Retaining the same basic design and the same aerodynamic proportion, the new Spartan was still a stable and well-mannered airplane in the air, and had now acquired some modifications that would only be natural in the process of continuous improvement for better performance and increased utility. A very good example of the average general-purpose airplane, as developed to this date, the Spartan was rugged and dependable and was used nation-wide to perform a wide variety of services. A Spartan C3-165 was flown to 9th place by J. W. Welborn in the National Air Tour for 1929 amongst a field of determined contenders; not normally of a competitive nature, the Spartan could be spurred into a good showing in contest. A good stable breed of airplane with no hidden tricks, the Spartan biplane usually struck one at first meeting by it's air of quiet elegance, further enhanced by it's good behavior and good manners in the air.

First approved as the Spartan C3-5 on a Group 2 approval numbered 2-79 issued 6-14-29, the C3-5 as powered with the Wright J6-5-165 was later approved for a type certificate number issued 8-9-29. In 10-30-29 the type certificate was amended for the C3-165 to allow two different fuel loads and varying weights, as later explained. Some 40 or more examples of this model were manufactured by the Spartan Aircraft Co. at Tulsa, Okla. W. G. Skelly, well-known Oklahoma oil-man was chairman of the board and acting president; Rex B. Beisel was V.P. in charge of engineering; George Hammond was chief engineer; and L. R. Dooley was in charge of sales. Of the 120 or so Spartan biplanes that were built in 7 versions, at least 40 were still in active service some 10 years later.

Listed below are specifications and performance data for the Spartan model C3-165 as powered with the 165 h.p. Wright J6 engine; length overall 23'10"; hite overall 8'10"; wing span upper & lower 32'0"; wing chord both 60"; wing area upper 151 sq.ft.; wing area lower 140 sq.ft.; total wing area 291 sq.ft.; airfoil "Clark Y"; wt. empty 1650 (1633); useful load 968 (878); payload 360 (363) lbs.; gross wt. 2618 (2511) lbs.; figures in brackets are weights for 49 gal. fuel without fuselage tank; max. speed 118; crusing speed 100; landing speed 47; climb 800 ft. first min. at sea level; climb in 10 min. was 6900 ft.; service ceiling 12,000 ft.; gas cap. 49-65 gal.; 49 gal. without fuselage tank, and 65 gal. with center-section tank and fuselage tank; oil cap. 6.5 gal.; cruising range at 10 gal. per hour was 475-600 miles; price at the factory field was $6750, later reduced to $5975.

The fuselage framework was built up of welded chrome-moly steel tubing, heavily faired to shape with wood fairing strips and fabric covered. The cockpits were deep and well upholstered, and a baggage compartment of 6.8 cubic foot capacity was in the turtleback section of the fuselage, just in back of the rear cockpit. The wing framework was

Fig. 305. A Spartan with a 165 h.p. J6 engine, designated C3-5; several were built on Group 2 certificate.

built up of solid spruce spar beams that were routed-out to an "I beam" section, with spruce and plywood built-up wing ribs; the leading edges were covered with "dural" sheet and the completed framework was fabric covered. There were four ailerons that were connected together in pairs by a streamlined push-pull strut; interplane struts were of streamlined chrome-moly steel tubing, and interplane bracing was of streamlined steel wire. The main fuel tank of 49 gal. capacity was mounted in the center-section panel of the upper wing, an extra fuel tank of 16 gal. capacity was mounted high in the fuselage just ahead of the front cockpit. The fabric covered tail-group was built up of welded chrome-moly steel tubing; the fin was ground adjustable and the horizontal stabilizer was adjustable in flight. The split-axle landing gear of 83 inch tread was of two long telescopic legs with oleo-spring shock absorbers; wheels were 30x5 and a steerable tail-wheel was standard equipment. A metal propeller, Bendix brakes, booster magneto, navigation lights, and dual controls were also standard equipment. The next development in the Spartan biplane series was the C3-225 which will be discussed in the chapter for A.T.C. # 286.

Listed below is a partial listing of Spartan C3-165 entries that were gleaned from various records; a complete listing would be rather extensive so we submit about 20 or so.

C-8075; C3-165 (# 101) J6-5-165.
NC-72M; C3-165 (# 110) J6-5-165.
NC-283M; C3-165 (# 111) J6-5-165.
C-8070; C3-165 (# 112) J6-5-165.
NC-8082; C3-165 (# 113) J6-5-165.
NC-8071; C3-165 (# 114) J6-5-165.
NC-8072; C3-165 (# 115) J6-5-165.
NC-864H; C3-165 (# 116) J6-5-165.
NC-865H; C3-165 (# 117) J6-5-165.
NC-866H; C3-165 (# 118) J6-5-165.
NC-257K; C3-165 (# 119) J6-5-165.
NC-285M; C3-165 (# 120) J6-5-165.
NC-286M; C3-165 (# 121) J6-5-165.

NC-287M; C3-165 (# 122) J6-5-165.
NC-289M; C3-165 (# 124) J6-5-165.
NC-571M; C3-165 (# 125) J6-5-165.
NC-572M; C3-165 (# 126) J6-5-165.
NC-573M; C3-165 (# 127) J6-5-165.
NC-575M; C3-165 (# 129) J6-5-165.
NC-576M; C3-165 (# 130) J6-5-165.

The registration number for serial # 123 and # 128 is not known; serial # 101 first had Curtiss "Challenger" engine as X-8075; serial # 102 thru' # 109 had Axelson and Curtiss "Challenger" engines; the last Spartan C3-165 was recorded as serial # 155.

The following will try to explain how error can sometimes crop into even the best reporting when relying on human memory instead of cold facts and figures. In our account of the Spartan C3-1 (ATC 71) in U.S. CIVIL AIRCRAFT, Vol. 1 we reported that nearly 100 of the Siemens-Halske powered C3-1 had been built; as this figure was received from former company sources we assumed they were correct, but this figure we learned was taken from memory and not from records. As it turned out, it just wasn't so. Actually, records show that no more than 15 examples of the C3-1 were built, and about 35 of the Walter powered C3-2 were built. The numbering of Spartan aircraft started with # 55, so adding the 50 airplanes that were built in this series to the # 55 would be 105, and that is probably how the impression was formed that nearly 100 airplanes were built. From records that are now available, it would seem that not over 120 Spartan biplanes were built in total, in 7 different versions. When records are not available on a certain airplane, the historian often turns to people once closely associated with this particular craft, in search of information, but memory alone cannot always span 30 years or so in any great clarity, then it is logical that such presumed fact and figures tend to be approximate and an occasional error will crop up.

PITCAIRN "SUPER MAILWING", PA-7

Fig. 306. A Pitcairn "Super Mailwing" PA-7S with a 225 h.p. J6 engine; the "Super Sport" carried three.

The jaunty Pitcairn "Mailwing" had an aura of romance around it that was brought about by it's almost flawless service record on various mail-routes about the country; an aura built up by the almost legendary stories that came from "Mailwing" pilots in the two years since it's introduction to carrying the mails. Pilots were known to often exaggerate a point here and there for emphasis in the telling of a story, the degree of exaggeration varying with the man's personality, but they all wouldn't elaborate on the same points, unless they were true. Therefore, we would feel that the "Mailwing" was well deserving of the praise and the love that was heaped upon it. Without question, the "Mailwing" series were a cocky and handsome breed of airplane endowed with plenty of heart, and a performance eager enough and lively enough to satisfy even the most critical; it's success as a mail-carrier was a shining example for all to strive for.

The Pitcairn "Mailwing" biplanes were trim and beautiful airplanes that were designed with sure-footed and flashing performance, and the new PA-7 "Super Mailwing" series was certainly no exception to this set rule.

The new series were every bit as beautiful, if not more so, and many improvements were incorporated into it's basic make-up that was developed from priceless experience gained with the models PA-5 and PA-6. First developed as the PA-6B with the Wright J5 engine, this new "Mailwing" was fitted with the newly developed N.A.C.A. "low drag" engine cowling, and the fuselage was heavily faired to conform to the new lines of flow. For some reasons, use of this "cowling" was discontinued in service and the J5 engine was later replaced with the new 7 cylinder Wright J6 of 225 h.p. This then became the model PA-7M in the mail-carrying version, and the model PA-7S in the 3 place open cockpit sport version.

The Pitcairn "Super Mailwing" model PA-7 series, was a 1 or 3 place open cockpit biplane with a configuration and dimension that were basically typical to the two earlier versions, with numerous slight improvements in detail. Though operating with a slightly heavier gross load and somewhat less wing area, nearly all of the snap and sure-footed manuverability was retained in this new design. The model PA-7M, or mail-carrying ver-

Fig. 307. An earlier "Mailwing" development was the J5-powered PA-6B; note N.A.C.A. low-drag cowling.

sion, had a large cargo compartment of some 42 cubic feet that was fire-proof and weather-proof, with a capacity for some 550 lbs. of paying load. The model PA-7S, or "Super Sport Mailwing" version, was a 3 place open cockpit type primarily leveled at the sportsman-pilot; the front cockpit could be closed off with a metal panel when not in use, and a slight gain in speed could be very well noticed. The PA-7S "Super Sport Mailwing" was now also one of the select few aircraft that had performed the "outside loop"; taking it's place along-side others on the mantle of fame. The type certificate number for the Pitcairn model PA-6B was issued in August

of 1929 and re-issued in November of 1929 for the models PA-7M and PA-7S. Some 35 or more examples of these models were manufactured by Pitcairn Aircraft, Inc. at Bryn Athyn, Penna.; early in 1930, operations were moved to Pitcairn Field in Willow Grove, Pa., with general offices in Philadelphia. Harold F. Pitcairn was the president; Geo. S. Childs was V.P.; the brilliant Agnew E. Larsen was chief of design and engineering; and veteran James Ray was the chief pilot in charge of test and development.

Listed below are specifications and performance data for the Pitcairn "Super Mailwing" series model PA-7 as powered with the

Fig. 308. The mail-carrying version of the "Super Mailwing" was the PA-7M; payload was over 500 pounds.

225 h.p. Wright J6 engine; length overall 23'9"; hite overall 9'6"; wing span upper 33'0"; wing span lower 30'3"; chord upper 54"; chord lower 48"; wing area upper 142 sq.ft.; wing area lower 101 sq.ft.; total wing area 243.5 sq.ft.; airfoil "Pitcairn # 3"; wt. empty 1821; useful load 1129 (1229); payload with 70 gal. fuel was 459 (559) lbs.; gross wt. 2950 (3050) lbs.; weights in brackets are for PA-7M; max. speed 135; cruising speed 115; landing speed 57; climb 900 ft. first min. at sea level; service ceiling 16,000 ft.; gas cap. 70 gal.; oil cap. 10 gal.; cruising range at 14 gal. per hour was 520 miles; price at the factory field was $8500; performance with the N.A.C.A. low-drag cowling was about 10 m.p.h. better than figures given here.

The fuselage framework was built up of welded chrome-moly steel tubing, with the longeron tubes of square section to make a better joint without having to round-out and bevel the ends of the connecting tubes; the framework was heavily faired to shape with formers and fairing strips, then fabric-covered. The cargo compartment of the mail-version was lined with metal to make it fire-proof, and the metal hatch-cover was de-signed to make it weather-proof; the sport-version was comfortably upholstered and the cockpit in front could be closed off with a metal panel when not in use. The fuel tank was mounted high in the fuselage just ahead of the front cockpit, and the oil tank was mounted on the engine firewall. The wing fra-mework was built up of heavy-sectioned so-lid spruce spar beams that were routed for lightness, with spruce and plywood built-up wing ribs; the leading edges were covered with aluminum alloy sheet and the completed framework was fabric covered. Ailerons were on the lower panels only, and were operated by bellcranks and metal push-pull tubes. The fabric covered tail-group was built up of welded chrome-moly steel tubing; the fin was ground adjustable and the horizontal stabili-zer was adjustable in flight. The split-axle landing gear of 78 inch tread was of the robust outrigger type and used oleo-spring shock absorbers; wheels were 30x5 and Ben-dix brakes were standard equipment. A metal propeller and a hand crank inertia-type engine starter were also standard equipment; the mail carrying version could be fitted with a full

Fig. 309. The PA-6B "Super Mailwing" preparing for flight on Southern mail route; Pitcairn biplanes were ideal for "Pony Express" hops on short-haul feeder routes.

complement of night-flying equipment. The next development in the Pitcairn "Mailwing" series was the 300 h.p. model PA-8, which will be discussed in the chapter for A.T.C. # 364.

Listed below are PA-6B, PA-7M, and PA-7S entries that were gleaned from various records; this list is not complete, but is does show the bulk of these models that were built.

NR-213M; PA-7S (# 50) Wright J6.
NC-68M; PA-7S (# 51) Wright J6.
NC-69M; PA-7S (# 52) Wright J6.
NC-70M; PA-7S (# 53) Wright J6.
NC-71M; PA-7S (# 54) Wright J6.
     ;   (# 55) Wright J6.
NC-877M; PA-7M (# 56) Wright J6.
NC-876M; PA-7M (# 57) Wright J6.
NC-824N; PA-7S (# 58) Wright J6.
NC-87M; PA-7M (# 59) Wright J6.
NC-825N; PA-7M (# 60) Wright J6.
NC-826N; PA-7M (# 61) Wright J6.
NC-351V; PA-7M (# 140) Wright J6.
NC-377V; PA-6B (# 141) Wright J5.
NC-378V; PA-7M (# 142) Wright J6.
NS-53W; PA-7S (# 143) Wright J6.
NC-54W; PA-7S (# 144) Wright J6.
     ;   (# 145)
NC-94W; PA-7M (# 146) Wright J6.
NC-95W; PA-7S (# 147) Wright J6.
NC-96W; PA-7S (# 148) Wright J6.
NC-97W; PA-7S (# 149) Wright J6.

Serial # 40 through # 49 were of the Wright J5 powered PA-6B type; serial # 50 (NR-213M) was prototype for PA-7 series; we cannot account for the serial number pattern change from # 61 to # 140; serial # 152 was a PA-7B; serial # 150 was first of the PA-8 series.

Fig. 310. A DeHavilland "Moth" 60-GM with an 85 h.p. DeHavilland "Gipsy" engine, an American version of the popular British sport trainer; wing slots in upper wing made it spin-proof.

The DeHaviland "Moth" undoubtedly had a more interesting background of world-wide popularity and achievements than any light airplane in the world; built in great number, it was more than likely used by someone in nearly every country in the world at one time or another. Designed and developed in 1925 as the D.H. 60 for training purposes, it was used by private owners, flying-clubs, and the R.A.F., throughout the whole British empire. Many notable flights have been accomplished with the D.H. "Moth" and over 10,000,000 miles of flying in all parts of the world, under any and every condition, had helped to develop a craft of unbeatable dependablility with inherent safety. Though quite frisky and manuverable with delightful flight charact-eristics, the "Moth" was also very stable and the automatic "wing slots" made it practi-cally stall-proof and spin-proof. Capt. De-Haviland in a demonstration, deliberately "stalled" a D.H. "Moth" in flight at some 200 feet up, and let it "mush" to the ground while still holding the stick all the way back. The airplane hit the ground with quite a jar, there's no doubt about that, but DeHaviland was not hurt and the craft was damaged only slightly. With the average craft built at this time, this sort of caper would have been next

to sheer suicide. Standing as a record for many years, not one person had ever been in-jured in a "slotted" Moth, so the safety that was built into the "Moth" trainer can hardly be disputed. Fitted with folding-wings the "Moth" biplane could be easily stored in the space of a one-car garage; with all these de-sirable features that fairly bristled out of this little craft, the "Moth" was certainly desti-ned to become a very popular airplane for sport or pilot-training.

The Moth Aircraft Corp. of Lowell, Mass. was incorporated in 1928 to manufacture the D.H. "Moth" under license agreement with the DeHaviland Aircraft Co., Ltd. of Great Britain. Several airplanes were brought over from England for test and demonstration, and production of the American version of the D.H. "Moth" was commenced in the early summer of 1929; by November of that year, over 120 had already been built. The D.H. "Moth" model 60-GM was a perky 2 place open cockpit biplane seating two in tandem, and was powered with the famous 4 cylinder aircooled in-line D.H. "Gipsy" engine of 85 h.p. The configuration of this airplane, though quite average, was definately "British" in appearance and harbored several interesting innovations. Quite novel was the use of wooden interplane bracing struts, and also the automatic "wing slots", which were like auxiliary "air-foils" that drooped-out forward at low air-speeds and high angles of attack to bend the air-flow around the wing. Changing the flow to restore the lift around the wing and prolong the interval before a complete "stall" of the wing section. Actually, some lift was retained even at extremely high angles of attack, and the "stall" of the wing could be better described as a "mush", with the airplane settling in a nose-high condition. There was no tendency to "spin" off, and control could be immediately regained by dropping the nose of the airplane. The "wing slots" certainly were not fool-proof, but were a great advancement in the search for inherent safety in an airplane.

The "Moth" was very easy to fly, with a friendly and forgiving nature; one flying-school operator with the strictest confidence in him-self and his equipment, offered to teach anyone to fly in the space of one day. Starting bright and early in the morning, he guaranteed to have you up for your first "solo" by sun-down, with the help of a patient and understanding "Moth", of course. Al Krapish, veteran test-pilot for Moth Aircraft, flew a 60-GM to 20th place in the hotly-contested National Air Tour for 1929. During the National Air Races of 1929 which were held at Cleveland, Frank Courtney flew a "Moth" to 4th place in the Miami to Cleveland Air Derby; the first 4 places in the

Toronto to Cleveland Derby were won by "Moth", and the 5 "Moth" that were in the Cleveland to Buffalo Efficiency Race made a handsome showing. "Moth" took first place twice in the Australian Pursuit Race for men; and took first place in one event and 2cd and 3rd in another event of the Australian Pursuit Race for women. During the National Air Races for 1930 held at Chicago, Laura Ingalls flew a "Moth" to 3rd place in the Women's Dixie Derby from Washington, D. C. to Chicago; a "Moth" was 3rd in a 25 mile speed race against aircraft of much higher top speeds, and a "Moth" came in 2nd in one dead-stick landing event. The "Moth" was an eager competitor by nature, and one did not have to be a thousand-hour pilot to make a good showing in any event. To top this off, we must make mention that the capable Laura Ingalls "looped" a D.H. "Moth" 344 times to establish a new record for women.

Sales and distribution of the popular D.H. "Moth" was handled by the huge Curtiss Flying Service organization, and in early 1930 the Moth Aircraft Corp. was absorbed as a division of the expanded Curtiss-Wright Corp.; manufacturing operations were then moved to Robertson, Mo. in the sprawling plant site of the Curtiss-Robertson Airplane Div. The Wright Aero. Corp., also an affiliated company in the Curtiss-Wright complex, was now manufacturing the D.H. "Gipsy" engine under a license agreement, in a 90 h.p.

Fig. 311. The "Moth" had world popularity and impressive accomplishments; nearly 200 were built in the U.S.

version called the Wright-Gipsy; an engine which powered the later version of the "Moth" called the 60-GMW. The sprawling Curtiss Flying Service system, which had flying-schools and service-stations all over the country, was now called the Curtiss-Wright Flying Service and often used the "Moth" trainer in their primary pilot-training courses. Nearly 200 of the "Moth" trainers were built and it became a familiar addition to the American aviation scene; even some 10 years later, there were at least 55 "Moth" in active flying service, and NC-916M (serial # 117) built at Lowell, Mass. is still flying to this day in it's original configuration. The type certificate number for the D.H. "Moth" model 60-GM and the model 60-GMW as powered with the 85 h.p. DeHaviland "Gipsy" engine and the 90 h.p. Wright-Gipsy engine, was issued in August of 1929 for both a landplane and a float-mounted seaplane version. Some 168 or more examples of the "Moth" were manufactured by the Moth Aircraft Corp. at Lowell, Mass., and the Moth Aircraft Div. at Robertson, Mo. Officers of the Moth Aircraft Corp. were Minton M. Warren as president; J. Edwin Morrow as general manager; the well-known Kenneth Unger was manager of sales; and veteran Al Krapish was chief pilot in charge of test and development. Officers of the Moth Aircraft Div. of the Curtiss-Wright Corp. were Minton M. Warren as president; with Ralph S. Damon as V.P. and general manager.

Listed below are specifications and performance data for the D.H. "Moth" as powered with the 85 h.p. D.H. "Gipsy" engine; length overall 23'11"; hite overall 8'9"; wing span upper & lower 30'0"; wing chord both 52.4"; wing area upper 125 sq.ft.; wing area lower 118 sq.ft.; total wing area 243 sq.ft.; airfoil "R.A.F. 15 Modified"; span with wings folded 9'10"; wt. empty 1027; useful load 623; payload with 23 gal. fuel was 295 lbs.; gross wt. 1650 lbs.; max. speed 102; cruising speed 85; landing speed 40; stall speed 40; climb 700 ft. first min. at sea level; climb to 5,000 ft. was 9 min.; climb to 10,000 ft. was 21 min.; service ceiling 16,000 ft.; gas cap. 23 gal.; oil cap. 2.4 gal.; cruising range was 360 miles or about 20 miles per gallon; price at the factory field was $4500. The following figures are for the 60-GMW as powered with the Wright-Gipsy; wt. empty 1045 (1143); useful load 605 (527); payload with 23 gal. fuel was 272 (200) lbs.; gross wt. 1650 (1670) lbs.;

max. speed 105 (98); cruising speed 88 (80); landing speed 40; climb 700 ft. first min. at sea level; the figures in brackets are for seaplane version; all other dimensions and data remained the same; price at the factory field for the 60-GMW was $3960; Edo float installation was extra; automatic "wing slots" were $240 extra.

The fuselage framework was built up of welded chrome-moly steel tubing, lightly faired to shape and fabric covered. There was an entrance door to the front cockpit, and a 2 cu. ft. baggage locker was behind the rear cockpit. The wing framework was built up of solid spruce spars that were routed to an "I beam" section, with girder-type wing ribs that were built up of spruce and plywood; the completed framework was fabric covered. Wing bracing struts were spruce members that were shaped to a streamlined section, vaguely reminiscent of the old DH-4; ailerons were on the lower wing panels only and were differentially operated. The gravity-feed fuel tank was mounted in the center-section panel of the upper wing and was actually shaped like a deep airfoil section; the wings rotated on the rear spar hinges for folding, and a telescopic jury-strut was provided to keep the wings "in rig" while in the folded position. The fabric covered tail-group was built up of welded chrome-moly steel tubing; the fin was ground adjustable and the horizontal stabilizer was adjustable in flight. The split-axle landing gear of 66 inch tread used rubber donut rings in compression for shock absorbers; wheels were 24x4 and no brakes were provided. The D.H. "Moth" was built by Curtiss-Wright into 1931 and further production of the series was suspended because of the lack of sales brought on by the crippling "depression years"; though none have been manufactured for at least 30 years, the perky little "Moth" is still well remembered.

Listed below is a partial list of "Moth" 60-GM entries that were gleaned from various records; a complete listing would be too extensive, so we submit a listing of about 20 or so.

NC-810E; 60-GM (# 1-C) Gipsy 85.
NC-811E; 60-GM (# 1-D) Gipsy 85.
NC-830E; 60-GM (# 1-E) Gipsy 85.
NC-831E; 60-GM (# 1-F) Gipsy 85.
NC-832E; 60-GM (# 1-G) Gipsy 85.
NC-318H; 60-GM (# 1-H) Gipsy 85.
NC-319H; 60-GM (# 1-I) Gipsy 85.
NC-320H; 60-GM (# 1-J) Gipsy 85.

NC-372H; 60-GM (# 1-K) Gipsy 85.
NC-373H; 60-GM (# 1-L) Gipsy 85.
  ; 60-GM (# 1-M) Gipsy 85.
NC-825H; 60-GM (# 1-N) Gipsy 85.
NC-826H; 60-GM (# 1-O) Gipsy 85.
NC-827H; 60-GM (# 1-P) Gipsy 85.
NC-828H; 60-GM (# 1-Q) Gipsy 85.
NC-829H; 60-GM (# 1-R) Gipsy 85.
NC-229K thru NC-237K for # 41 thru # 49; NC-556K thru NC-566K for # 50 thru # 60; NC-961H thru NC-970H for # 61 thru # 70; NC-55M thru NC-64M for # 71 thru # 80; NC-580M thru NC-589M for # 81 thru # 90; NC-131M thru NC-140M for # 91 thru # 100; NC-900M thru NC-919M for # 101 thru # 120; NC-922M thru NC-941M for # 121 thru # 140; NC-215V thru NC-224V for # 141 thru # 150; NC-713M thru NC-717M for # 151 thru # 155; NC-572N thru NC-578N for # 156 thru # 162; NC-583N is # 163; NC-966K is # 164; NC-590N thru NC-595N for # 165 thru # 170; serial # 1-C and up, serial # 41 thru # 150 were 60-GM; serial # 151 thru # 170 were 60-GMW.

Fig. 312. A New Standard Model D-29 with a "Cirrus" Mark III engine; the craft was an excellent trainer.

The New Standard model D-29 was an "ugly duckling" that was a much better airplane than one would think, but human nature being what it is, we tend to shy away from what we think is ugliness and attach our affections to beauty, whether it is possessed of merit or not. It has many times been said of airplanes that "the uglier they are the better they are apt to fly", and many ugly-looking airplanes were veritable "princesses of beauty" with their manner in the air. Many times, lines of beauty in an airplane can only be achieved by some sacrifice in ideal proportion, or by some sacrifice to amiable flight characteristics; no such problems of compromise were involved in the design of the D-29 trainer series. The model D-29 was planned to be an ideal training craft first, and let the rest be damned.

Charlie Day, brilliant engineer and designer of the New Standard series, liked to fly obedient airplanes of pleasant nature, and naturally, when he designed a new craft, he used all his vast knowledge of aerodynamics and practical engineering to make it come out just so; the D-29 design was certainly a good example of obedience and pleasant nature. An airplane especially designed for pilot training in the primary phases that led to first "solo", must first of all be easy to fly with

an inherent forgiving nature to offset dire results from occasional error, it must be stable enough to eliminate constant correction on the part of the perspiring fledgling pilot, and it must have a good "feel" that will eventually "get through" to the aspiring pilot and give him the confidence needed in his quest to master the flying-machine. The D-29 design was endowed with all these requisites right down to the letter, and even more; when spurred on by a good firm hand, it was manuverable enough to perform all sorts of "acrobatics" and the structure was rugged enough throughout to even stand the abnormal strain of "outside loops". Simplicity in manufacture, and ease of maintenance and repair were also taken into consideration; these requisites also governed the configuration and had much to do with the final appearance of this craft. To sum it up in short, the requirements of the manufacturer, the operator, and the student-pilot, had to be blended together to bear an advantage to all.

Introduced about mid-1929, the New Standard model D-29 trainer was a 2 place open cockpit biplane seating two in tandem, in one elongated cockpit that separated the occupants only by a small panel and a windshield; this sort of cockpit arrangement was used on several other trainers, and it's use was probably governed by the simplicity of it's manufacture and the ease of inspection afforded to the cockpit area. A large amount of interplane stagger and a trailing edge cut-out in both the upper and lower wings afforded good visibility in all directions, and the rugged framework was strong enough to soak up plenty of abuse that was dished out by the average trainee. Powered with the 4 cylinder Cirrus Mk. 3 engine of 85 h.p., the D-29 certainly had sufficient performance for this type of work, but it was looked upon as underpowered, and perhaps this was to the detriment of this little craft and kept it from being fully developed. From the records available, it is doubtful if this model went beyond the prototype stage; the Kinner powered version was more readily accepted. The type certificate number for the model D-29 as powered with the 85 h.p. Cirrus Mk. 3, was issued in August of 1929 and at least one example of this model was manufactured by the New Standard Aircraft Corp. at Paterson, New Yersey, Charles L. Auger Jr. was the president; Charles Healy Day was V.P. and chief engineer; and Louis G. Randall was general manager.

Listed below are specifications and performance data for the New Standard trainer model D-29 as powered with the 85 h.p. Cirrus Mk. 3; length overall 24'11"; hite overall 9'0"; wing span upper & lower 30'0"; wing chord both 54"; wing area upper 126 sq.ft.; wing area lower 122 sq.ft.; total wing area 248 sq.ft.; airfoil "Clark Y"; wt. empty 1097; useful load 535; payload with 24 gal. fuel was 195 lbs.; gross wt. 1632 lbs.; max. speed 88; cruising speed 75; landing speed 37; climb 620 ft. first min. at sea level; climb to 5,000 ft. was 12.5 min.; take-off 8 sec. & 245 ft.; service ceiling 10,000 ft.; gas cap. 24 gal.; oil cap. 3.5 gal.; cruising range was 300 miles;

price at the factory field was tentatively set at $4250.

The fuselage framework was built up of open section duralumin members that were riveted and bolted together into a simple structure that could be easily repaired with ordinary hand tools; the framework was faired to shape with wood fairing strips and fabric covered. The cockpit was one elongated opening, but was divided by separate windshields; there was a baggage locker of 2.5 cu. ft. capacity. The wing framework was built up of laminated spruce spar beams, with basswood and plywood built-up wing ribs; the leading edges were covered with plywood veneer and the completed framework was fabric covered. There were four ailerons that were connected together in pairs by a streamlined push-pull strut; the ailerons were a riveted framework of open section duralumin members and fabric covered. The gravity-feed fuel tank was mounted in the center-section panel of the upper wing and was actually shaped like a deep airfoil section; the upper and lower wing panels were identical and could be interchanged. All interplane struts were of streamlined duralumin tubing, and interplane bracing was of streamlined steel wire tie-rods. The fabric covered tail-group was built up of riveted duralumin members of open section; the vertical fin was ground adjustable and the horizontal stabilizer was adjustable in flight. The split-axle landing gear of 78 inch tread was of chrome-moly steel tubing in the rugged outrigger type and used oil-draulic shock absorber struts; wheels were 24x4 and no brakes were provided. The next development in the New Standard trainer series was the Kinner powered model D-29A which will be discussed in the chapter for A.T.C. # 216.

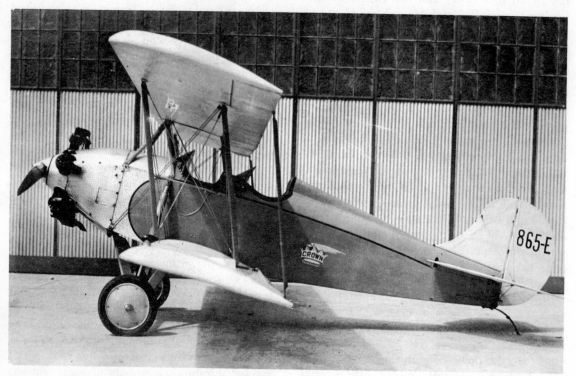

Fig. 313. The Crown B-3 "Custom-Bilt" with a 5 cylinder Kinner K-5 engine was designed as a sport trainer.

The Crown model B-3 was a two place open cockpit biplane seating two in tandem; a ruggedly built biplane of the sport-trainer type that was designed exclusively for use of the 5 cylinder air-cooled Kinner K5 engine of 90-100 h.p. Modified slightly from the basic design developed by W. B. "Bert" Kinner a year or so earlier as the Kinner "Airster", the present design was improved to some extent, and now known as the Crown "Custombilt". The model B-3 was a modification conceived and developed by M. M. Brockway, and engineered by Earl Marsh. The rather plain and quite normal configuration of the new B-3 was slightly beyond the average in at least one respect; the fuselage was of all-wood semi-monococque construction, but the balance of the airplane was more or less conventional. The original Kinner "Airster" was designed to carry three, but the Crown modification was especially designed as a sport-trainer and seating was limited to two. The Crown B-3 was first awarded a Group 2 approval numbered 2-61 which was issued in April of 1929 for serial # 37 (C-865E) as a

two place airplane of 1800 lb. gross; this approval was superseded by A.T.C. # 199 which was issued in August of 1929 for this and all subsequent aircraft. The amount of the model B-3 that were built has been hard to determine; there is no evidence of more than 2 or 3 of this craft.

The Crown Aircraft Corp., manufacturers of the "Custombilt" B-3, was incorporated as a subsidiary division of the Crown Motor Carriage Co. of Los Angeles, Calif. Z. S. Freeman was the president; M. M. Brockway was V.P. in charge of production and sales; Earl Marsh was the chief engineer, and Lee Brusse was the chief pilot. Manufacture of the Crown B-3 was carried out in a building added to the main plant building of the Crown Motor Carriage Co.; many employees of the "motor carriage" plant also worked in the aircraft division when certain jobs called for their skills. Purchasers and prospective buyers of the Crown sport-trainer were most always offered a dealership for their area; this type of deal was not too unusual and was pretty much normal practice

Fig. 314. This 1929 Crown B-3 was still in service many years later; note the various modifications.

with many aircraft manufacturers during this stage of aviation development.

Listed below are specifications and performance data for the Crown "Custombilt" B-3 as powered with the Kinner K5 engine; length overall 21'6"; hite overall 7'6"; wing span upper and lower 28'; chord both 54"; wing area upper 126 sq.ft.; wing area lower 110 sq.ft.; total wing area 236 sq.ft.; airfoil "U.S.A. 27 Mod." & "Clark Y"; wt. empty 1243; useful load 513; payload 173; gross wt. 1756 lbs.; max. speed 112; cruising speed 96; landing speed 35; climb 850 ft. first min. at sea level; service ceiling 14,000 ft.; gas cap. 25 gal.; oil cap. 4 gal.; range 400 miles; price at the factory was $5250. The fuselage framework was semi-monocoque in design and was built in a composite of wood longitudinals and diagonals, with laminated wood bulkheads and formers, that were covered with plywood veneer for stiffness and then fabric covered. The front cockpit had a large door for easy entrance and exit, which otherwise would have been quite difficult. The fuel tank was high up in the top half of the fuselage, just ahead of the front cockpit. The wing framework for all four panels was a composite of built-up spruce and metal spar beams, and built-up wood and metal wing ribs; the completed framework was fabric covered. The upper wing was in two panels that were joined

together at the center-line and then fastened to a center-section cabane of inverted vee struts; there was no separate center-section panel as would normally be the case. There were four ailerons that were connected in pairs by a push-pull strut. The fabric covered tail-group was built up of welded steel tubing and was braced externally with streamlined steel tie-rods; The fin was ground adjustable and the horizontal stabilizer was adjustable in flight. The robust landing gear of six foot tread, was of the long-leg "oleo" type using Gruss air-oil shock-absorber struts; wheels were 26x4, and the tail-skid was of the leaf-spring type with a removable hardened shoe. Wheel brakes, metal propeller, and engine starter were offered as optional equipment.

Fig. 315. An early Crown B-3 designed by Bert Kinner; note early Kinner engine with uncovered valve action.

Fig. 316. A Parks P-2 with Axelson engine, designed as a sportplane, also offered the 130 h.p. "Comet" engine.

As we had mentioned in the discussion previous, "Parks" airplanes had actually started out as student-built projects in the classroom, that were patterned after the designs of one aircraft manufacturer or another; most likely leaning more towards the design and configuration of the popular KR "Challenger" biplane. This school program eventually led to the development of the OX-5 powered Parks P-1, of which a good number were built. Intended for use in the primary phases of pilot training at Parks Air College, the model P-1 was quite satisfactory, but they also needed a higher powered craft that would be more suitable for the secondary phases of pilot training, which required an airplane with more performance and better manuverability in order to perform the more intricate manuvers and other advanced practices in piloting techniques.

The need for a craft to fit the more demanding requirements led directly to the development of the Parks P-2. The Parks model P-2 was also a three place open cockpit biplane of the general-purpose type, and was quite typical to the P-1 in most respects, except for the engine installation which in this case was the 7 cylinder air-cooled Axelson engine of 115-150 h.p. Though not exactly a raring "tiger" with this engine combination, there was ample power reserve for good performance and a fairly responsive manuverability that was adequate enough to handle the necessary requirements in the secondary phases of pilot training. Though primarily designed to be used at Parks Air College for service on the flight line, the model P-2 was also offered to the small operator for general purpose uses, or as a sport-craft for the average private-owner. Beside the installation of an aircooled powerplant, there was one other major change that was incorporated into the design of the model P-2, and that was the switch of the landing gear configuration from the straight-axle type to one of the split-axle type, which was certainly an improvement and would be more popular. The type certificate number for the Parks model P-2 as powered with the Axelson engine, was issued in August of 1929 and some 6 or more examples of this model were manufactured by the Parks Aircraft Corp. at E. St.Louis, Illinois; a subsidiary of Parks Air Lines, Inc. which owned the Parks Airport and the Parks Air College. This whole enterprise later became a division of the Detroit Aircraft Corp. The model P-2 was first issued an approval on Group 2-99 (7-24-29) for serial # 2951 and up, but this approval was superseded by A.T.C. # 200. The model P-2 was also offered in a version powered with the 130 h.p. "Comet" engine, but it is doubtful that any were built.

Listed below are specifications and performance data for the Parks biplane model P-2 as powered with the 115-150 h.p. Axelson engine; length overall 23′6″; hite overall 9′3″; wing span upper 30′0″; wing span lower 28′6″; wing chord both 63″; wing area upper 154 sq.ft.; wing area lower 131 sq.ft.; total wing area 285 sq.ft.; airfoil "Aeromarine 2A modified"; wt. empty 1458; useful load 860; payload with 50 gal. fuel was 350 lbs.; gross wt. 2318 lbs.; max. speed 115; cruising speed 98; landing speed 42; climb 750 ft. first min. at sea level; service ceiling 14,000 ft.; gas cap. 50 gal.; oil cap. 5.5 gal.; cruising range 490 miles; price at the factory field was $6000. The 1930 version of the Parks P-2 was called the Ryan "Speedster", and weight changes were as follows; wt. empty 1520; useful load 920; payload with 50 gal. fuel was 410 lbs.; gross wt. 2440 lbs.; all other dimensions and data remained the same, and there was no appreciable change in the performance figures for the latter version.

The fuselage framework was built up of welded chrome-moly steel tubing, deeply faired to shape with wood fairing strips and fabric covered. The fuel tank was mounted high in the fuselage, just ahead of the front cockpit; the front cockpit had a door on the left side for easy entry, and there was a small baggage compartment just behind the rear cockpit. The wing framework was built up of solid spruce spar beams with spruce and plywood built-up wing ribs; the completed framework was fabric covered. There were only two ailerons on the model P-2, and they were connected to the lower wing panels. The fabric covered tail-group was built up of welded chrome-moly steel tubing; the fin was ground adjustable and the horizontal stabilizer was adjustable in flight. The landing gear of 65 inch tread was of the normal split-axle type and used rubber shock-cord in tension for absorbing the bumps; the tail-skid was a shock-cord sprung steel tube with a removable hardened shoe. Wheel brakes and a metal propeller were standard equipment. Navigation lights, extra instruments, and inertia-type engine starter were available as optional equipment. The next development in the Parks biplane was the model P-2A as powered with the 5 cylinder Wright J6 series engine; this model will be discussed under A.T.C. # 276.

Listed below are Parks P-2 entries that were gleaned from various records; this list may not be complete, but it does show the bulk of this model that were built.

X-8386; Parks P-2 (# 2931) Axelson.
NC-9295; Parks P-2 (# 2951) Axeslon.
NC-332K; Parks P-2 (# 2971) Axelson.
NC-902K; Parks P-2 (# 2972) Axelson.
NC-965K; Parks P-2 (# 2981) Axelson.
NC-8488; Parks P-2 (# 104) Axelson.
The last entry has company serial number as used by the Detroit Aircraft Corp. Model P-2 serial # 2972 was later modified to P-2A with installation of Wright J6-5-165 engine.

Fig. 317. The 3-place Parks P-2 was ideal for carefree sport flying back in the days of the rumble-seat roadster.

APPENDICES

BIBLIOGRAPHY

BOOKS:

Aircraft Year Book for 1929—1930—1931
Conquering The Air by Archibald Williams
Air Power For Peace by Eugene Wilson
Modern Aircraft by Maj. Victor Page
A Chronology of Michigan Aviation by Robert S. Ball
Wings For Life by Ruth Nichols
Under My Wings by Basil Rowe
Jane's All The World's Aircraft for 1929 by C. G. Grey
Flight by Year
Around The World In 8 Days by Post & Gatty
The Ford Story by Wm. T. Larkins

PERIODICALS:
Flying
The Pilot
Aviation
Aero Digest
American Airman
Air Progress
Lightplane Review

Western Flying
Popular Aviation
American Modeller
Model Airplane News
Air Transportation
Antique Airplane News
Journal of A.A.H.S.

SPECIAL MATERIAL:

Licensed Aircraft Register by Aeronautical Chamber of Commerce of America, Inc.
Characteristics Sheets by Curtiss Aeroplane & Motor Co.

CORRESPONDENCE WITH THE FOLLOWING FIRMS & INDIVIDUALS:

Chas. W. Meyers
George A. Page Jr.
Dave Jameson
Peter M. Bowers
Pan American World Airways
Sikorsky Aircraft Div.
Northwest Orient Airlines
Gordon S. Williams

PHOTO CREDIT

Aeromarine-Klemm Corp. — Figs. 61, 62, 63.

Alexander Film Co. — Figs. 135, 137, 141, 142, 143, 287, 288.

American Air Lines — Figs. 190, 278, 309.

Anderson Photo — Fig. 120.

Gerald Balzer — Figs. 24, 59, 96, 97, 99, 107, 232, 236, 238, 261; 223, 224 (Advance Aircraft); 121 (Anderson); 36 (M. R. Bailey); 124 (Ball Studio); 117 (Bourdon Aircraft); 2 (Brunner-Winkle); 313 (Crown Aircraft); 247 (Cunningham-Hall); 93, 94 (Denkelberg Photo); 316 (Detroit Aircraft); 37, 39 (Hughes); 171 (Ireland Aircraft); 227 (Kreutzer); 168 (Laird Airplane Co.); 228 (Le Blond Aircraft); 296 (Metal Aircraft); 32 (Mono Aircraft); 254 (Parks Air College); 233, 234 (B. F. Parrish); 306, 308 (Pitcairn Aircraft) 13, 277 (Edgar B. Smith); 95 (Thompson Photo); 241, 284 (Travel Air Co.); 185 (Western Air Express).

Beech Aircraft Corp. — Figs. 44, 159.

Bellanca Aircraft — Fig. 98.

Boeing Airplane Co. — Figs. 15, 16, 113, 114, 183, 187, 269, 270.

Peter M. Bowers — Figs. 28, 31, 45, 67, 68, 105, 128, 148, 163, 176, 178, 198, 219, 229, 244, 245, 282, 283, 292, 314; 151 (Curtiss-Robertson); 179 (J. Mathiesen); 106 (B. C. Reed); 38, 52, 248, 255, 256 (A. V. Schmidt); 158 (U. S. Post Office); 79, 104, 184, 242, 246 (Gordon S. Williams); 80, 140, 186 (Williams/Schmidt).

Butler Aircraft Corp. — Fig. 122.

Chadbourne Aircraft Sales — Fig. 307 (Pitcairn Aircraft).

Consolidated-Fleet — Fig. 66.

Continental Oil Co. — Fig. 147.

Cresswell Photo — Fig. 56.

Curtiss Airplane & Motor Co. — Figs. 298, 299.

Fairchild Aircraft Div. — Figs. 201, 202, 203, 204.

Ford Motor Co. — Fig. 111.

Fokker Aircraft Co. — Fig. 209.

General Airplanes — Fig. 47.

Gulf Oil Co. — Fig. 149.

Marion Havelaar — Figs. 3, 157, 216, 217, 274.

Stephen J. Hudek — Figs. 7, 65, 92, 131, 136, 150, 210, 214, 297, 304; 22 (H. W. Arnold); 119 (Butler Aircraft Corp.); 264 (Curtiss A & M Co.); 55, 110, 180, 181, 182, 211, 212, 213, 215, 237, 252 (Ford Motor Co.); 49 (General Airplanes); 239 (Nicholas-Beazley); 235 (Royal Canadian Air Force); 29, 30, 285 (Travel Air Co.); 174, 175 (Winstead Photo).

Institute of Aeronautical Sciences — Fig. 253.

Ireland Aircraft — Figs. 172, 173.

Joseph P. Juptner — Fig. 301.

Kug-Art Photo — Figs. 257, 258, 315.

Lockheed Aircraft Corp. — Figs. 6, 8.

Lodder Photo Service — Fig. 249 (Cunningham-Hall Aircraft Corp.).

Macdonald Photo — Figs. 40, 41, 42, 260.

Charles W. Meyers — Figs. 70, 72, 73, 218, 221, 231; 71 (Barton); 220, 222 (Great Lakes Aircraft).

Ken M. Molson — Fig. 108.

Moth Aircraft — Fig. 311.

National Aviation Museum, Ottawa, Canada — Figs. 48, 101, 161, 189, 191.

Nicholas-Beazley — Fig. 240.

Ralph Nortell — Figs. 43, 112.

North American Aviation — Fig. 208.

Northwest Air Lines — Figs. 46, 85, 169, 170.

Roy F. Oberg — Figs. 103, 134, 200, 294.

Ortho Print, N. Y. — Fig. 312.

Will D. Parker — Figs. 132, 133.
PILOT — Fig. 286.
Pratt & Whitney Aircraft — Figs. 5, 17, 115, 116, 280, 295.
Earl C. Reed — Fig. 69.
Reid Studio — Fig. 25. —
Le Roy Robbins — Figs. 12, 144, 146.
Sikorsky Aircraft Div. — Figs. 81, 82, 83, 84, 188.
Edgar B. Smith — Figs. 14, 87, 88.
Smithsonian Institution, National Air Museum — Figs. 1, 4, 9, 10, 11, 19, 20, 21, 23, 26, 50, 53, 54, 57, 58, 64, 74, 75, 76, 100, 102, 109, 123, 125, 138, 152, 153, 154, 160, 162, 164, 167, 192, 193, 194, 195, 196, 197, 199, 205, 225, 230, 243, 251, 259, 262, 263, 265, 272, 273, 275, 279, 290, 291, 293, 300, 302, 303, 310, 317.
Spartan Aircraft Co. — Fig. 305.
Stearman Aircraft Co. — Figs. 89, 127, 130.
Stevens Photo — Fig. 118.
Travel Air Co. — Fig. 281.
United Air Lines — Figs. 18, 266, 267, 268.
Truman C. Weaver — Figs. 145; 33, 34, 35 (Clayton Folkerts).
Gordon S. Williams — Figs. 27, 60, 77, 78, 90, 91, 126, 139, 155, 156, 165, 166, 177, 206, 207, 226, 250, 289; 86, 129 (Stearman Aircraft Co.).
R. D. Wolff — Figs. 51, 271, 276.

INDEX